PHYSICS RESEARCH AND TECHNOLOGY

FERROELECTRICS AND SUPERCONDUCTORS

PROPERTIES AND APPLICATIONS

PHYSICS RESEARCH AND TECHNOLOGY

Additional books in this series can be found on Nova's website
under the Series tab.

Additional E-books in this series can be found on Nova's website
under the E-books tab.

PHYSICS RESEARCH AND TECHNOLOGY

FERROELECTRICS AND SUPERCONDUCTORS

PROPERTIES AND APPLICATIONS

IVAN A. PARINOV
EDITOR

Nova Science Publishers, Inc.
New York

Copyright © 2012 by Nova Science Publishers, Inc.

All rights reserved. No part of this book may be reproduced, stored in a retrieval system or transmitted in any form or by any means: electronic, electrostatic, magnetic, tape, mechanical photocopying, recording or otherwise without the written permission of the Publisher.

For permission to use material from this book please contact us:
Telephone 631-231-7269; Fax 631-231-8175
Web Site: http://www.novapublishers.com

NOTICE TO THE READER

The Publisher has taken reasonable care in the preparation of this book, but makes no expressed or implied warranty of any kind and assumes no responsibility for any errors or omissions. No liability is assumed for incidental or consequential damages in connection with or arising out of information contained in this book. The Publisher shall not be liable for any special, consequential, or exemplary damages resulting, in whole or in part, from the readers' use of, or reliance upon, this material. Any parts of this book based on government reports are so indicated and copyright is claimed for those parts to the extent applicable to compilations of such works.

Independent verification should be sought for any data, advice or recommendations contained in this book. In addition, no responsibility is assumed by the publisher for any injury and/or damage to persons or property arising from any methods, products, instructions, ideas or otherwise contained in this publication.

This publication is designed to provide accurate and authoritative information with regard to the subject matter covered herein. It is sold with the clear understanding that the Publisher is not engaged in rendering legal or any other professional services. If legal or any other expert assistance is required, the services of a competent person should be sought. FROM A DECLARATION OF PARTICIPANTS JOINTLY ADOPTED BY A COMMITTEE OF THE AMERICAN BAR ASSOCIATION AND A COMMITTEE OF PUBLISHERS.

Additional color graphics may be available in the e-book version of this book.

Library of Congress Cataloging-in-Publication Data

Ferroelectrics and superconductors : properties and applications / editor, Ivan A. Parinov.
p. cm.
Includes bibliographical references and index.
ISBN 978-1-61324-518-7 (hardcover)
1. Ferroelectric devices. 2. Superconductors. I. Parinov, Ivan A.
TK7872.F44F463 2011
537'.2448--dc23
2011014216

Published by Nova Science Publishers, Inc. † New York

CONTENTS

Preface   vii

Chapter 1 Effect of Deformation Temperature by Torsion under
Pressure on the Microstructure, Texture and Flux Pinning
of Bi2212-Base Materials   1
M. F. Imayev , R. R. Damino, D. B.Kabirova, M. Reissner,
W. Steiner, M. V. Makarova and P. E. Kazin

Chapter 2 Orientation Effects in Novel Composites Based
on Relaxor-Ferroelectric Single Crystals   45
V. Yu. Topolov, A. V. Krivoruchko, P. Bisegna
and S. V. Glushanin

Chapter 3 Pyroelectric Effect at Thermally Activated Phase Transitions
in Ferroelectrics, Ferroelectrics-Relaxors and Antiferroelectrics   81
Yu. N. Zakharov, A. G. Lutokhin and I. P. Raevski

Chapter 4 Designing of Multiferroic Materials Based on Perovskite
and Spinel-Like Compounds: Reactivity and Regions
of Structure Stability; Phase Formation and Stepwise Optimization
of Technology; Relaxation Dynamics, UHF Absorption
and Secondary Periodicity of Ferromagnetic Properties   109
L. A. Reznichenko, O. N. Razumovskaya, L. A. Shilkina,
I. A. Verbenko, K. P. Andryushin, A. A. Pavelko,
A. V. Pavlenko, V. A. Alyoshi, S. P. Kubri, A. I. Miller,
S. I. Dudkina, P. Teslenko, G. Konstantinov, M. V. Talanov,
A. A. Amirov, A. B. Batdalov, V.M. Talanov, N.P. Shabelskaya
and V.V. Ivanov

Chapter 5 Investigations of Conductive and Mechanical Properties
of Superconductive Composites Based on Asymptotical
Methods and Phases with Negative Stiffness   145
Ivan A. Parinov and Shun Hsyung Chang

Chapter 6	Investigation of the Structural Features of High Temperature Superconducting Crystals by the Channeling Method and Resonant Nuclear Reactions *V. S. Malyshevsky*	**207**
Chapter 7	Improved Finite Element Approaches for Modeling of Porous Piezocomposite Materials with Different Connectivity *A. V. Nasedkin and M. S. Shevtsova*	**231**
Chapter 8	Computer Modeling of Resonance Characteristics of the Piezoelectric Cylindrical and Bimorpf Transducers *T. G. Lupeiko, A. N. Soloviev, B. S.Medvedev, A. S.Pahomov,* *M. P. Petrov, V. S. Chernov and D. V. Nasonova*	**255**
Chapter 9	Superconducting Magnetic Energy Storage Coil Integration to Autonomous Rail Locomotive Electric Power Transmission *V. N. Noskov and M. Yu. Pustovetov*	**273**
Index		**281**

PREFACE

The modern science, techniques and technologies use widely ferroelectrics and superconductors. The novel structures and composites based on these materials allow processing numerous devices and goods demonstrating unique properties and characteristics.

The developed preparation techniques, complex compositions and the practice requirements, demanding permanent improvement of the material properties and creating devices with given characteristics, push forward further modernization of theoretical, experimental and model investigations above materials, composites and their applications. The book covers a set of prospective directions in theoretical, model and experimental studies of ferroelectric and superconductive materials and composites including some their applications. In particular there are present the next themes: (i) consideration of the effect of deformation temperature by torsion under pressure on the microstructure and superconducting properties of Bi2212 ceramics and composites on their base, (ii) study of the properties of single crystals of perovskite-type relaxor-ferroelectric solid solutions with analysis of interconnections between the electromechanical properties of the single-crystal component and composite on its base for the single crystal/polymer and single crystal/porous polymer composites, (iii) quasistatic and dynamic pyroelectric measurements used for studying the phase transitions of different types in perovskite ferroelectrics, ferroelectrics–relaxors and antiferroelectrics, (iv) study of behavior of the multiferroics combining the ferroelectric and magnetic properties based on optimization of processes of synthesis and recrystallization sintering, (v) study of conductive and mechanical properties of superconductive composites by taking into account the effects of inclusion covering (for fibers and grains) on the composite conductivity, effects of non-ideal contacts of inclusions and matrix, structural non-regularity and cluster conductivity, and also discussion of anomalous changes of the composite properties when inclusions have negative elastic moduli, (vi) investigation of $YBa_2Cu_3O_7$ crystals by the channeling method and calculation of the basic parameters of channeling ions for the $YBa_2Cu_3O_7$ crystal structure by using the approach to description of the channeling in multicomponent crystals based on numerical solution of the diffusion equations, (vii) development of the effective moduli method on the base of finite-element technique for inhomogeneous piezoelectric media with obtainment of respective constitutive equations and calculation of the full set of effective moduli for porous and polycrystalline composite piezoceramics with various connectivity demonstrating wide porosity range, (viii) presentation of the results of computer modeling based on the finite-element software ACELAN of resonance-frequency characteristics for emitters of cylindrical and bimorphic types of piezo-transducers, (ix) development of scheme for superconducting magnetic energy storage (SMES) coil integration to autonomous rail locomotive electric power transmission and discussion of the sufficient energy capacitance for SMES on board of diesel-electric shunt locomotive.

This collection continues the ideas before published books Piezoceramic Materials and Devices, I. A. Parinov (Ed.). New York: Nova Science Publishers. 2010. ISBN 978-1-60876-459-4, and Piezoelectric Materials and Devices, I. A. Parinov (Ed.). New York: Nova Science Publishers. 2011. ISBN 978-1-61122-817-5 and also touch upon theoretical and experimental studies of superconductive materials and their applications. The chapters have been prepared by internationally recognized teams from Russia, Austria, Italy and Taiwan in the field of theoretical, model and experimental methods and investigations of different ferroelectric and superconductive materials, composites and novel devices designed on their base.

The book addresses to students, post-graduate students, scientists and engineers, taking part in the development and research of ferroelectric, superconductive and related materials.

Ivan A. Parinov (Ed.)
Rostov-on-Don, February 2011

In: Ferroelectrics and Superconductors
Editor: Ivan A. Parinov

ISBN: 978-1-61324-518-7
©2012 Nova Science Publishers, Inc.

Chapter 1

EFFECT OF DEFORMATION TEMPERATURE BY TORSION UNDER PRESSURE ON THE MICROSTRUCTURE, TEXTURE AND FLUX PINNING OF Bi2212-BASE MATERIALS

M. F. Imayev[*,1], *R. R. Daminov*[1], *D. B.Kabirova*[1], *M. Reissner*[2], *W. Steiner*[2], *M. V. Makarova*[3] *and P. E. Kazin*[3]

[1]Institute for Metals Superplasticity Problems of Russian Academy of Sciences, Ufa, Russia
[2]Institute of Solid State Physics, Vienna University of Technology, Wien, Austria
[3]Chemistry Department, Moscow State University, Moscow, Russia

ABSTRACT

A systematic investigation of the effect of high-temperature deformation by torsion under pressure on the microstructure and superconducting properties of Bi2212 ceramics as well as composites on its base was carried out. It has been established that degree of basal plane texture increases with deformation temperature, reaching its maximum near to incongruent melting temperature. Quantitative investigation of sizes of colonies of grains of matrix phase and particles of non superconducting phases was curried out. The analysis of mechanism of basal plane texture formation was curried out too.

Deformed materials demonstrate similar and strongly non-uniform dependence of superconducting properties (J_c, B_{irr}, $\langle E \rangle$) on deformation temperature. In composites the introduced inclusions exert noticeable contribution to the flux pinning only after deformation at low temperatures. With increasing temperature they grow and lose their efficiency. In undoped Bi2212 and composites the best superconducting properties have samples deformed near incongruent melting temperature. The non-uniform dependence of the superconducting properties on deformation temperature are explained on the basis of conception that up to four types of flux pinning centers can operate in the material: (i) introduced inclusions, (ii) intracolonial lattice defects (point defects, dislocations,

---

*E-mail: marcel@imsp.da.ru, 1Institute for Metals Superplasticity Problems of Russian Academy of Sciences, 39, Khalturina Str., 450001, Ufa, Russia.

stacking faults), (iii) low-angle colony boundaries, (iv) particles of secondary phases occurring at decomposition of the Bi2212 phase near melting temperature. The number of each pinning centers depends on deformation temperature. Local maxima of superconducting properties form when at least two different types of pinning centers operate. The maximum properties are formed near melting temperature due to large extent of low-angle boundaries and increased number of secondary particles occurring at peritectic melting of Bi2212 phase.

1. INTRODUCTION

Superconducting $Bi_2Sr_2CaCu_2O_{8+x}$ (Bi2212) ceramics is an advanced material concerning its practical application. It is characterized by a rather high current carrying capacity in strong magnetic fields at 4.2 K (J_c >10^5 A/cm^2, 10 T) [1]. However, in spite of its rather high transition temperature (T_c ~ 90 K) the critical current density of Bi2212 rapidly decreases at temperatures above 20 K [2]. There are mainly two facts responsible for this decrease: the large anisotropy of the material [3-6] and the low density of defects [7].

One of the methods for improving the density of defects in Bi2212 is hot plastic deformation. It provides stronger texture and increased density of dislocations which, as known, may act as strong centers for pinning of magnetic flux [8-11]. Usually such deformation methods as hot isostatic pressing (HIP) [11], HIP plus sinter-forging [12], or only sinter-forging [13, 14] are used with a deformation temperature range of 815-850°C. However, since simple procedures of deformation based on uniaxial compression cannot provide high strain values, it is rather difficult to form strong texture together with a high density of dislocations.

The use of complex loading procedures of deformation is more advanced for increasing strain values. One of such procedures is torsion under pressure [15] where quasi-hydrostatic pressure prevents destruction of the material and torsion leads to deformation up to very high strains [16, 17]. Recently it has been revealed that applying low quasi-hydrostatic pressure (slightly above 0.6 MPa) increases the onset temperature for melting of Bi2212 by almost 60°C, which essentially expands the useable temperature range for hot deformation [18]. An open question is the influence of the deformation performed near the melting temperature of Bi2212 phase on structure and superconducting properties.

Creation of composites is one more efficient method for increasing pinning energy of the magnetic flux and, consequently, a value of J_c, in the Bi-based superconducting materials. The compatibility of a number of chemical elements and disperse non-superconducting phases with Bi2212 and Bi2223 has been studied rather thoroughly by now. The studies [19-21] have allowed revealing some phases which are not only chemically compatible with Bi2212 and Bi2223 but also increase superconducting properties of composites at high temperatures and fields. These particles are formed either in the process of melt-texturing growth or specially added to the starting precursor [22-28].

It is assumed that if the size of these particles is much higher than the coherence length of the superconductor (actually it takes place almost always if the material was exposed to melt-texturing or hot deformation) they are not efficient for pinning of magnetic vortices. In this case the pinning is mostly conditioned by the structure features of the particle-matrix interface [29] or such lattice defects as dislocations and stacking faults [23, 30]. As a rule, the lattice defects appear at cooling since the particle and the matrix have different coefficients of thermal expansion. In this connection the efficiency of particles as vortex pinning centers can be improved if we try to increase the density of defects connected with them.

One of the methods for increasing the density of defects in the vicinity of the particles is plastic deformation. This method is well known in literature and is based on the fact that during deformation of two-phase materials the particles always promote trapping of dislocation tangles around themselves. These tangles are stable till very high temperatures [31]. The plastic deformation of high-T_c superconducting composites Y123/Y211 [32-37] and Bi2212/MgO [38] has been studied. At first the composites have been melt-textured and then hot compressed by a small strain value ($\varepsilon < 10\%$). After deformation the density of dislocations and stacking faults in the vicinity of particles is indeed high. The increase of J_c as a result of the deformation was observed in References [33, 34, 37, 38]. However a degradation of the superconducting properties was observed in [36]. This degradation has been explained by the influence of the initial microstructure, in particular regarding the size, concentration and distribution of the Y211 particles, which affect the deformation behaviour and subsequent oxygenation process of the sample.

It is known that during deformation the density of dislocations trapped by the particles depends on the particle size but much more essentially it depends on the interparticle distance [31]. The melt-texturing initiates the considerable growth of particles and increases the distance between them. Moreover, during melting there can occur agglomeration of particles at inner interfaces [39]. So, it is desirable to exclude melting from processing of composite. For preserving a finer particle size and increasing the density of dislocations connected with the particles it is advisable to process textured composite by means of plastic deformation.

The particle size and their distribution mostly depend on deformation temperature [31]. Moreover, the deformation temperature is one of the most important parameters effecting texture and dislocation density. That is why for more precise regulation of the size of particles and their distribution, as well as the density of dislocations and the texture sharpness the temperature range of the plastic deformation to be carried out should be as wide as possible. The Bi2212-base composites are most interesting for such studies since phase Bi2212 is characterized by the evident lack of natural centers of flux pinning.

Unfortunately, the brittleness of the Bi2212 limits the temperature range of its deformation when simple uniaxial compression is used. In order to expand the deformation temperature range (both to lower and higher temperatures) one should apply complex loading schemes – torsion under quasi-hydrostatic pressure [40]. On one hand, the quasi-hydrostatic pressure prevents initiation and growth of cracks and provides an opportunity to decrease the deformation temperature. On the other hand, it increases the melting temperature of the Bi2212 [18] that provide increasing the deformation temperature essentially. Due to such a double effect from the quasi-hydrostatic pressure the deformation temperature range of the Bi2212 was almost 150°C (from 795 to 940°C) [40]. That is why this deformation scheme is most suitable for studying the total influence of particles and lattice dislocations on superconducting properties of the composite.

The aim of the present paper was to study the influence of temperature of deformation by torsion under pressure on microstructure and superconducting properties of Bi2212-base materials. Four materials were studied: 1) Bi2212 ceramics prepared from commercial powder; 2) composite Bi2212/Mg$_{1-x}$Cu$_x$O prepared from commercial powders; 3, 4) undoped Bi2212 ceramics and Bi2212/Sr$_2$CaWO$_6$ composite prepared from sol-gel powders.

2. EXPERIMENTAL

To obtain Bi2212 ceramics (material No 1) commercial $Bi_{1.98}Sr_{1.88}Ca_{1.03}Cu_{2.00}O_{8+x}$ powder (Hoechst) was used. The starting precursor for composite $Bi2212/Mg_{1-x}Cu_xO$ (material No 2) was a mixture of commercial $Bi_{1.98}Sr_{1.88}Ca_{1.03}Cu_{2.00}O_{8+x}$ Hoechst powder with admixture of 3.45 wt% (6.16 vol.%) MgO particles (with mean diameter approximately equal to 20 nm). Powders for undoped Bi2212 (material No 3) and composite $Bi2212/Sr_2CaWO_6$ (material No 4) were prepared as follows. The starting precursor was $Bi_{1.98}Sr_{1.88}Ca_{1.03}Cu_{2.00}O_{8+x}$ (Bi2212) Hoechst powder. To obtain fine and homogeneous powder it was treated by sol-gel method. For this purpose initial Bi2212 powder was dissolved in 5 M solution of HNO_3 with subsequent adding of 2.5 mole citric acid per 1 mole of cations. To obtain composite $Bi2212/0.25Sr_2CaWO_6$ the solutions of ammonium tungstate $(NH_4)_{10}H_2W_{12}O_{47}$, strontium and calcium nitrates in an amount corresponding to nominal composition $\{Bi_{1.98}Sr_{1.88}Ca_{1.03}Cu_{2.00}O_{8+x} - 0.25Sr_2CaWO_6\}$ were added to the part of the obtained citric acid solution. The solutions were evaporated on sand bath till obtaining vitreous substances and calcinated at 800°C for 6 h.

To obtain pellets of 10 mm diameter and about 2 mm height the powders were pressed at room temperature under 200 MPa and sintered in air at 855°C for 24 h. High-temperature deformation by torsion under pressure was carried out on a home-made complex loading machine U-10/KM-50. The prepared pellets were plastically deformed in the temperature range 795 – 940°C by compression between two anvils superimposed by a torsion which was performed by rotation of one of the two anvils around the compression direction (Inset in Figure 1). To prevent the interaction with the anvils the samples of Bi2212-base materials were placed between two plates of monocrystal (100)MgO with dimension about 15×15 mm^2. Deformation was carried out in the mode of constant load or constant pressure. In the latter case, to maintain a constant pressure value the load was gradually increased taking into account the current area of the sample. The pressure was initiated via applying a load at 795°C (Figure 1). After reaching the desired deformation temperature at a heating rate of 10°C/min the samples were annealed for 30 minutes under load to allow equilibration of temperature in the sample, followed by torsion under pressure. The twist rate was $1.5 \cdot 10^{-3}$, $3 \cdot 10^{-3}$ or $7.7 \cdot 10^{-3}$ rpm. The twist angle (α) varied from 45° to 180°. After the turn of the upper anvil in respect to the lower one by the angle α the torsion procedure was ceased, and the furnace was cooled down to room temperature at a rate of 5°C/min. At 795°C the pressure was released (Figure 1). Afterwards the samples were annealed in air at 850°C for 96 h and cooled by air quenching. The considerable number of samples were deformed under the following conditions (150 kg load, twist rate $1.5 \cdot 10^{-3}$ rpm and $\alpha = 90°$) which we hereafter named as "standard conditions".

X-ray diffraction (XRD) data were collected using a DRON-3.0 diffractometer with filtered Cu-K_α radiation within the 2θ range 20° – 60°. The degree of basal plane texture was determined by analysis of XRD data from the broad faces of the samples using the Lotgering method [41]. The Lotgering factor $F = (P - P_0)/(1 - P_0)$ was calculated from the (1 1 5) and (0 0 10) peaks, with $P = I_{(0\,0\,10)}/[I_{(0\,0\,10)}+I_{(1\,1\,5)}]$ determined from measurements of the textured and $P_0 = I_{0\,(0\,0\,10)}/[I_{0\,(0\,0\,10)}+I_{0\,(1\,1\,5)}]$ from measurements of the powder. For the Bi2212 powder used in this study P_0 is 0.18. The error in determining the F factor didn't exceed 0.01. Moreover the rocking curve of the (0 0 10) peak was measured in a ω–scan mode for a

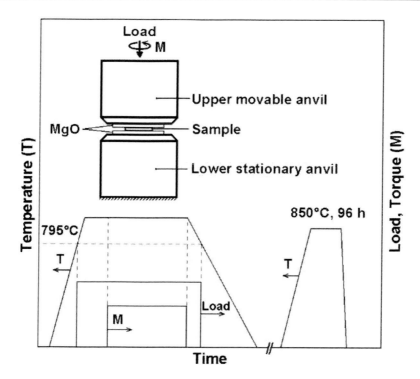

Figure 1. Procedure of the deformation by torsion under pressure and processing thermal cycle as well as diagram of load and torque application.

number of deformed samples. Rocking curves were recorded by step scans with step size 0.5° (θ) and counting time of 5 s.

Microstructure and chemical composition of the phases were studied using two JEOL scanning electron microscopes JSM-840 and JXA-6400. The latter one equipped with a wavelength dispersive spectrometer (WDS). The investigation of the microstructure was performed both on samples polished by diamond paste and on samples electrochemically etched by 5% solution of perchloric acid in butylalcohol at a voltage of 0.1 V.

Quantitative analysis of microstructure was carried out on the transversal sections of the samples after cutting them along a diameter. The mean size of colonies of grains of Bi2212 phase was calculated by measurement of length (L) and thickness (H) of individual colonies (Figure 2). Besides, the mean number of grains per colony was measured. The number of colonies taken into account was approximately 300. Glagolev's point count method [42] was used to determine the total volume fraction of secondary phases (V_{ns}).

A vibrating sample magnetometer (PAR 150A) was used for measuring superconducting properties at 4.2 K and 30 K in fields up to 7 T parallel to the direction of compression. Bars with dimensions $4 \times 1 \times h$ mm^3, with thickness h varying between 1.0 and 0.2 mm according to the different deformation conditions used, were cut from the central portion of the samples, the largest (4 mm) and smallest (h) sides being along the radius and the compression axis, respectively. From hysteresis loops the critical current densities J_c were calculated within Bean's critical state model [43], assuming that the characteristic length of current flow is either the colony or the overall sample size. In addition, the time dependence of the magnetic moment was measured at 1 T and both 4.2 K as well as 30 K for about 30 min. From these

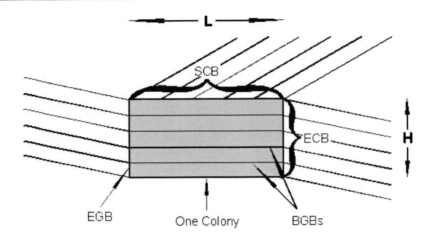

Figure 2. A schematic of two-dimensional diagram of the microstructure of Bi2212 phase. A typical colony with dimensions of $L \times H$ within which basal-plane grain boundaries (BGBs) separate adjacent Bi2212 grains that have essentially perfect c-axis tilt alignment. Colonies are terminated by either edge colony boundaries (ECB) or SCTILT boundaries (SCB). The edge-grain-boundary (EGB) is also shown.

relaxation measurements mean effective activation energies were determined within the Anderson-Kim model of flux creep [44].

AC susceptibility was measured on a home-made susceptometer [45] in the temperature range 15 – 100 K with field amplitudes of 1, 10 and 100 Oe and a frequency of 27 Hz. The magnetic field was oriented perpendicular to the direction of compression. The critical current density at 60 K and 77 K was estimated from imaginary part of the susceptibility ($H_0 = 100$ Oe) using again Bean's model for flux penetration into a slab. The sample thickness was taken as the characteristic size assuming that supercurrent can flow across the whole sample.

3. Bi2212 Ceramics

3.1. Results

3.1.1. Phase Purity and Texture

Tables 1, 2 show deformation conditions, microstructure parameters and texture quality of the samples under study. Into these tables are used the next designations: T_d is the deformation temperature, ω is the twist rate, α is the twist angle, p_o and p_e is the pressure at the initial and final moment of deformation, respectively, h_o and h_e is the sample thickness at the initial and final moment of deformation, respectively, microstructure parameters (L is the length, and H is the height of colonies, V_{ns} is the volume fraction of nonsuperconducting phases) and texture quality (F is the Lotgering factor, $FWHM$ is the full width at half maximum of the rocking curve).

Figure 3 shows XRD patterns of Bi2212 samples after sintering and after deformation at different temperatures T_d. A small amount of secondary phase $(Sr, Ca)_{4-y}Bi_2O_z$ is present in the initial sintered state. The deformed samples almost completely consist of the Bi2212 phase although according to literature [46] Bi2212 should decompose peritectically in air above 870°C. The preservation of the Bi2212 phase until 940°C is due to an increase of the melting temperature of this phase caused by the applied quasi-hydrostatic pressure. As shown by Imayev et al. [18] over the pressure range 5 – 35 MPa the melting onset temperature of Bi2212 varies from 941 to 943°C. Moreover, the deformation results in formation of a strong texture, verified by the increased intensity of the (0 0 1) peaks. The dependence of the factor F on deformation temperature is shown in Figure 4. In the initial sintered state the texture is weak, $F = 0.44$. In the course of deformation the factor F grows from 0.73 at 795°C to 0.97 at 875°C. Further until 930°C the value of F remains at about 0.96 – 0.97. At 940°C F decreases to 0.94.

Table 1. Deformation conditions of samples under study

Sample No.	Deformation conditions					
	T_d (°C)	ω (rpm)	α (deg)	Applied Load (kg)	p_o (MPa)	p_e (MPa)
17	795	$1.5 \cdot 10^{-3}$	90	150	19.1	10.5
16	815	$1.5 \cdot 10^{-3}$	90	150	19.1	9.4
15	835	$1.5 \cdot 10^{-3}$	90	150	19.1	8.6
26	845	$1.5 \cdot 10^{-3}$	90	150	19.1	8.2
1	855	$1.5 \cdot 10^{-3}$	90	150	19.1	7.7
2	865	$1.5 \cdot 10^{-3}$	90	150	19.1	8.1
3	875	$1.5 \cdot 10^{-3}$	90	150	19.1	7.4
5	885	$1.5 \cdot 10^{-3}$	90	150	19.1	7.2
6	895	$1.5 \cdot 10^{-3}$	90	150	19.1	7.0
7	905	$1.5 \cdot 10^{-3}$	90	150	19.1	6.7
8	915	$1.5 \cdot 10^{-3}$	90	150	19.1	6.7
9	930	$1.5 \cdot 10^{-3}$	90	150	19.1	6.7
11	940	$1.5 \cdot 10^{-3}$	90	150	19.1	6.7
79	875	$1.5 \cdot 10^{-3}$	135	150	19.1	8.0
80	875	$1.5 \cdot 10^{-3}$	180	150	19.1	6.9
28	905	$1.5 \cdot 10^{-3}$	45	150	19.1	8.0
29	905	$1.5 \cdot 10^{-3}$	180	150	19.1	6.7
62	895	$1.5 \cdot 10^{-3}$	135	150	19.1	8.0
64	895	$1.5 \cdot 10^{-3}$	180	150	19.1	6.7
65	895	$1.5 \cdot 10^{-3}$	90	50	6.4	2.5
70	895	$1.5 \cdot 10^{-3}$	90	200	25.5	9.1
66	895	$1.5 \cdot 10^{-3}$	90	300	38.2	13.5
112	895	$1.5 \cdot 10^{-3}$	90	450	57.3	20.0
103	895	$3 \cdot 10^{-3}$	90	150	19.1	6.7
30	905	$1.5 \cdot 10^{-3}$	135	150	19.1	8.0
3Z	895	$7.7 \cdot 10^{-3}$	90	150	19.1	6.7
0	Undeformed sample				0	0

Table 2. Microstructure parameters and texture quality of the samples under study

Sample No.	Sample thickness		Microstructure parameters			Texture quality	
	h_o (mm)	h_e (mm)	L (µm)	H (µm)	V_{ns} (%)	F	$FWHM$ (deg)
17	2.00	1.10	2.78	0.84	15	0.73	
16	2.00	0.98	2.89	0.87	12	0.76	9.7
15	2.00	0.90	3.50	0.90	13	0.80	
26	2.00	0.85	4.43	0.98	12	0.84	9.5
1	2.02	0.79	4.53	1.00	13	0.93	
2	1.81	0.75	4.33	0.97	17	0.93	8.7
3	2.01	0.72	4.74	0.96	13	0.97	
5	1.97	0.65	4.60	1.10	16	0.96	8.1
6	1.97	0.62	4.46	0.98	19	0.97	7.3
7	1.94	0.35	4.71	1.06	17	0.97	
8	1.91	0.35	4.24	1.01	16	0.97	
9	1.86	0.33	3.87	0.96	25	0.96	
11	1.98	0.30	3.50	0.90	27	0.94	
79	2.00	0.64	3.93	0.95	17	0.95	
80	2.00	0.59	4.04	0.99	19	0.95	
28	2.00	0.36	4.33	0.92	17	0.86	
29	2.00	0.21	2.80	1.12	30	0.93	
62	2.00	0.53	3.60	0.88	20	0.97	
64	2.00	0.50	4.17	1.01	20	0.97	
65	2.00	0.74	3.49	1.08	17	0.89	
70	2.00	0.56	4.31	1.03	12	0.94	
66	1.80	0.47	3.43	0.97	12	0.97	
112	1.90	0.44	2.95	0.85	12	0.98	
103	1.90	0.37	3.87	1.09	12	0.96	
30	2.00	0.22	-	-	-	0.91	
3Z	1.98	0.37	2.82	1.02	12	0.91	
0	2.00	2.00	3.02	1.11	12-19	0.44	10.0

In Figure 5 the relation between F and the full width at half maximum (FWHM) of the (0 0 10) peak for a number of deformed samples is given. With increasing F FWHM decreases non-uniformly. The FWHM value is changed slightly up to $F \sim 0.8 - 0.9$, but for $F > 0.90$ it decreases more rapidly. It is known that the FWHM value of a rocking curve gives a first estimate of the amount of colonies (at least 50%) being oriented within a certain degree (FWHM). The rapid decrease of FWHM for $F > 0.90$ indicates that in this range even a relatively small increase in F leads to a significant increase in the relative fraction of low-angle colony boundaries. For example, after deformation at $T_d = 865°C$ ($F = 0.93$) and $T_d = 895°C$ ($F = 0.97$) about 50% of the total amount of the Bi2212 colonies are oriented with respect to the compression axis within an angle of 8.7° and 7.3°, respectively.

At $T_d = 875$, 895 and 905°C with increasing twist angle α up to 90° F grows to about 0.97 (Figure 6). For higher α-values F depends strongly on the deformation temperature. For $T_d = 875°C$ with increasing α F slightly decreases to about 0.95. For $T_d = 905°C$ a more clear decrease of texture is present: F drops to $0.91 - 0.93$. On the contrary, at 895°C the texture is more stable and until $\alpha = 180°$ F remains at 0.97.

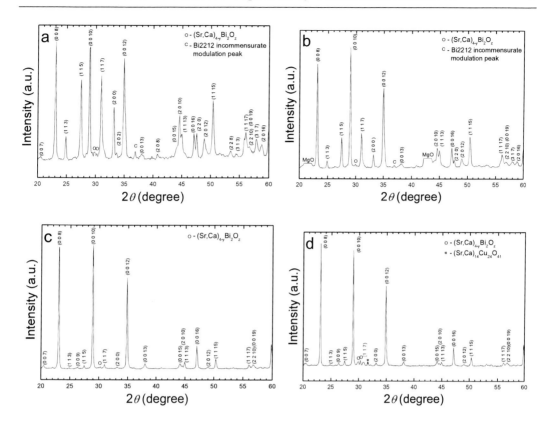

Figure 3. XRD patterns of the Bi2212 samples (a) as-sintered at 855°C and deformed under standard conditions at (b) 795°C, (c) 895°C, and (d) 940°C.

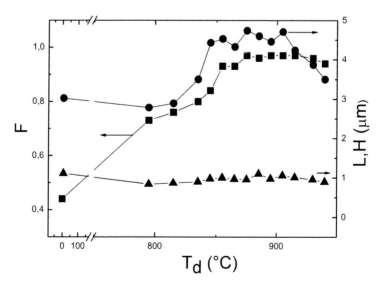

Figure 4. Dependence of Lotgering factor F (square), mean colony length L (circle) and thickness H (triangle) on deformation temperature T_d under standard conditions.

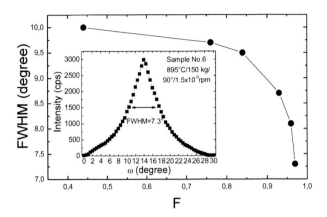

Figure 5. Relation between the FWHM of the (0 0 10) peak and Lotgering factor F for samples deformed under standard conditions. Inset: Rocking curve of the sample 6.

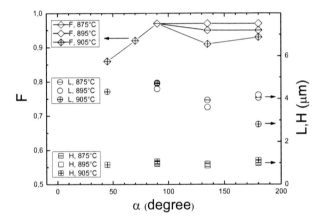

Figure 6. Dependence of Lotgering factor F, mean colony length L and thickness H on twist angle α for different deformation temperatures. Load is 150 kg, twist rate is $1.5 \cdot 10^{-3}$ rpm.

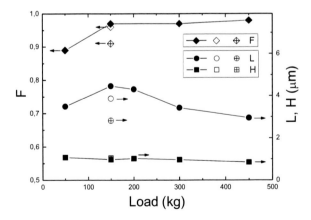

Figure 7. Dependence of Lotgering factor F, mean colony length L and thickness H on load at $T_d =$ 895°C for twist rate $1.5 \cdot 10^{-3}$ rpm. Open symbols and symbols with crosses correspond to the samples deformed with twist rate $3 \cdot 10^{-3}$ rpm and $7.7 \cdot 10^{-3}$ rpm, respectively. The twist angle is 90°.

The influence of the load on F is shown for $T_d = 895°C$ in Figure 7. With increasing load up to 150 kg F increases rapidly up to 0.97. Above 150 kg the growth of F becomes slower leading to 0.98 at 450 kg. An increase of the twist rate from $1.5 \cdot 10^{-3}$ to $7.7 \cdot 10^{-3}$ rpm at a load of 150 kg leads to a decrease in F from 0.97 to 0.91.

3.1.2. Microstructure

The comparative investigation of etched metallographic sections before and after annealing at 850°C for 96 h have shown that annealing doesn't change the size and aspect ratio of the constituent elements of the microstructure. That means that all changes in microstructure visible in a scanning electron microscope are only due to deformation.

Under all deformation conditions a microstructure (typical for Bi2212) which consists of colonies of plate-like grains is present (Figure 8). A schematic two-dimensional diagram of this microstructure is shown in Figure 2 [47]. Formation of fine equiaxial grains was not observed under the deformation conditions being studied.

The deformation leads to formation of a distinct metallographic texture, with the [001] axis of the colonies approximately parallel to the axis of compression (Figure 8). The microstructure is rather uniform along the sample radius: the difference in mean colony size between the center and the periphery of the sample is not more than 5%. The averaged quantitative parameters of the microstructure are given in Table 2.

Figure 4 shows the dependence of the mean length L and the mean thickness H of the colonies on deformation temperature. Whereas H is approximately 1 μm and is independent of T_d, the behavior of L is more complex: it grows in the temperature range 815 – 845°C to approximately 4.5 μm, and stays nearly constant up to 915°C. Above this temperature the growth of the colony length becomes smaller leading to $L \sim 3.5$ μm at 940°C.

The influence of the twist angle α is shown in Figure 6 for samples deformed at 875, 895 and 905°C. Again there is no influence on H but on L. For $T_d = 875$ and 895°C a scatter in L below 1 μm is present. Only for $T_d = 905°C$ and $\alpha = 180°$ the value L is decreased by almost 2 μm (Figure 6).

At 895°C with increasing load up to 150 kg the colony length grows from 3.49 to 4.46 μm. Above 150 kg L decreases to 2.95 μm at 450 kg (Figure 7). For $T_d = 895°C$, 150 kg load and $\alpha = 90°$ the increase of the twist rate by a factor of five (from $1.5 \cdot 10^{-3}$ rpm to $7.7 \cdot 10^{-3}$ rpm) led to a decrease in L from 4.46 to 2.82 μm while the colony thickness did not change and remained at about 1 μm. The mean number of grains per colony does not depend on deformation conditions and is equal to about 3.

In the initial sintered state the total amount of Cu-free and Bi-free phases (V_{ns}) is 12 – 19% (Table 2). In the samples deformed with $\alpha = 90°$, under 150 kg load and twist rate $1.5 \cdot 10^{-3}$ rpm (standard conditions) with increasing temperature up to 915°C V_{ns} is approximately the same as for the sintered-only state. However, above 915°C V_{ns} increases strongly and at 940°C reaches 27%.

At $T_d = 895°C$ and $\alpha = 90°$ a change in load from 50 to 450 kg does not influence V_{ns}. The volume fraction remains approximately constant in the range 12 – 19%. Varying the twist angle between 45 and 180° at 150 kg also exerts only weak influence on V_{ns} up to $T_d = 895°C$ (Table 2). For $T_d = 905°C$ V_{ns} is stable up to $\alpha = 90°$, but a further increase in the twist angle leads to a sharp increase reaching 30% at $\alpha = 180°$.

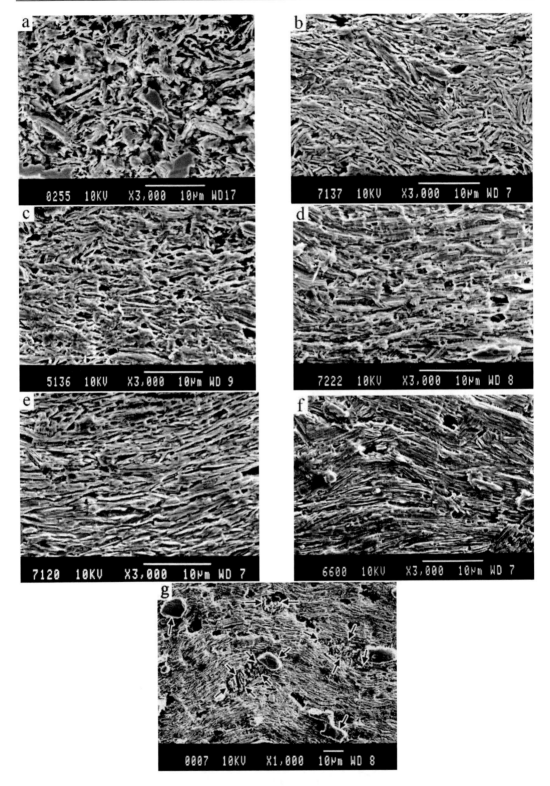

Figure 8. SEM micrographs of the Bi2212 samples (a) as-sintered at 855°C and deformed under standard conditions at T_d = 795°C (b), 815°C (c), 895°C (d), 915°C (e) and 940°C (f). The axis of compression is vertical.

The increase of V_{ns} during deformation under standard conditions at the higher temperatures 930°C and 940°C (samples 9 and 11) and also at 905°C for high torsion angle α = 180° (sample 29) is due to a decomposition of the Bi2212 phase. Since the sample is kept under quasi-hydrostatic compression extra oxygen slowly flows out from the sample resulting in a loss of thermal stability and a decomposition of the Bi2212 phase [18]. The outflow of extra oxygen becomes significant at high temperatures (samples 9 and 11) and for the high-torsion tests at hardly lower temperatures (sample 29).

As a rule, the particles of the secondary phases are of anisotropic shape (Figure 8f). They are elongated perpendicular to the compression axis and are characterized by an essential spread in size. In the undeformed sample and samples deformed below 915°C the average size of the particles varies within 7 – 10 μm along and 10 – 15 μm in the direction perpendicular to the compression axis. Above 915°C the particle size increases and at 940°C the size is 15 and 22 μm along and perpendicular to the compression axis, respectively.

The investigation of micrographs at low magnification revealed an interesting feature of the influence of the nonsuperconducting particles on material flow. Whereas in samples with low V_{ns} an almost uniform laminar material flow is observed in samples with high V_{ns} (samples 9, 11 and 29) in regions where coarse particles accumulate the material flow becomes wave-like leading to the so-called "crimping" of the microstructure. It is known that crimping is caused by a non-uniform deformation in different layers of the material [48]. It is more prominent at 940°C where the wave amplitude is maximal (Figure 8f).

Thus, from the data it is seen that the length of colonies changes only slightly with load, α, and the twist rate. Most sharp texture ($F = 0.97$) over a wide temperature range (875-895°C) is obtained by deformation under standard conditions (150 kg load, twist rate $1.5 \cdot 10^{-3}$ rpm and $\alpha = 90°$).

3.1.3. Superconducting Properties

A comparison of the width of the hysteresis loops for sintered and powdered samples have shown that approximately 50% of the measured magnetization is due to inter- and 50% to intragranular currents present in the bulk samples. Because a precise separation of these two contributions is not possible, the magnetization loops were analysed first by assuming that only intragranular currents are flowing in the sample and second that the signal comes only from intergranular currents. In the first case the mean colony size L was introduced as characteristic length into Bean's formula for cylindrical geometry, whereas in the second case the sample dimensions perpendicular to the field direction (4×1 mm) were used in Bean's equation for rectangular geometry. Figures 9 and 10 show the dependences of the thus determined current densities on deformation temperature for 4.2 and 30 K at 1.5 T. Apart from the fact that the absolute values are quite different, both analyses show more or less the same dependence on deformation temperature (Figures 9 and 10). At 4.2 K J_c increases slowly up to 865°C, followed by a sharp rise. Between 875°C and 895°C J_c again increases slowly, but now followed by a decrease leading to a local minimum around 905°C and 915°C depending on the type of analysis. For still higher temperatures J_c further increases with the highest values found at 940°C. Similar behaviour is also found for the mean effective activation energy which shows a strong enhancement between 865°C and 915°C, with a maximum at 895°C (Inset in Figure 9). At 30 K a reduction in J_c at 905°C is found (Figure 10). Above 915°C J_c stays more or less constant.

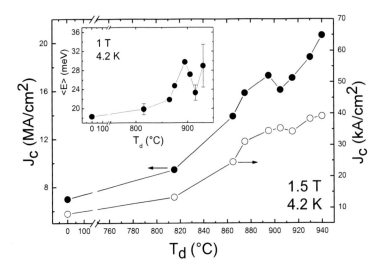

Figure 9. Dependence of J_c at 1.5 T and 4.2 K interpreted as both intra- (full circle) and inter-(open circle) granular current density on deformation temperature T_d under standard conditions. Inset: Dependence of mean effective activation energy at 1 T and 4.2 K on deformation temperature T_d.

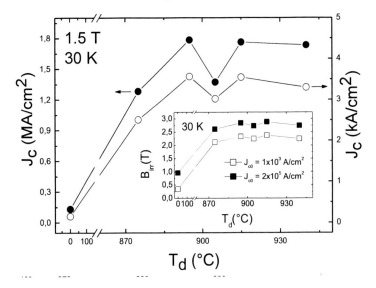

Figure 10. Dependence of J_c at 1.5 T and 30 K on deformation temperature T_d under standard conditions. Inset: Dependence of irreversibility field at 30 K determined for J_c criterion of $1\cdot 10^3$ and $2\cdot 10^5$ A/cm^2 on deformation temperature T_d. Full (open) symbols: J_c interpreted as intra- (inter-) granular current density.

A drastic increase of the irreversibility field is obtained after deformation. The influence on T_d is small (Inset in Figure 10).

The results of the AC susceptibility measurements at 100 Oe for samples deformed under standard conditions at T_d = 865, 905 and 940°C are shown in Figure 11. In advance to annealing at 850°C no pronounced peaks in the $\chi''(T)$ curves for T_d = 905°C and 940°C. This points on weak connectivity between the grains. Annealing at 850°C appreciably improves intergranular contacts. Intergranular peaks appear on the $\chi''(T)$ curves for the samples

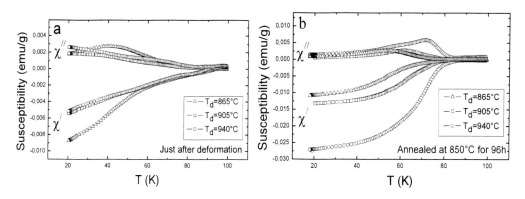

Figure 11. Temperature dependence of AC susceptibility at an AC field of 100 Oe (a) for the samples Bi2212 just after deformation under standard conditions at different temperatures and (b) after deformation and subsequent annealing at 850°C for 96 h.

Table 3. The onset temperature of superconducting transition (T_c) and critical current density (J_c) at T = 60 K and 77 K for samples deformed under standard conditions

T_d (°C)	After deformation			After deformation and annealing at T=850°C for 96 h		
	T_c (K)	J_c at 60 K (A/cm^2)	J_c at 77 K (A/cm^2)	T_c (K)	J_c at 60 K (A/cm^2)	J_c at 77 K (A/cm^2)
815	-	-	-	92	800	40
855	84	-	-	92	530	30
865	95	700	100	88	1300	30
875	95	1300	250	89	3000	270
895	94	1500	300	91	4000	400
905	92	-	-	87	3900	270
915	93	1100	160	91	3000	280
940	92	-	-	88	4300	270

deformed at 905°C and 940°C. For the sample with T_d = 865°C the peak shifts to higher temperatures. The sample deformed at 940°C has the most pronounced peak which shifts to higher temperatures as compared with all other samples. The calculated values of J_c at 60 K and 77 K for samples deformed under standard conditions before and after annealing are given in Table 3. Just after deformation the current-carrying capacity of the samples is low. Annealing improves the current carrying capability in all samples. The dependence of J_c at 60 K on the deformation temperature is similar to ones at 4.2 K and 30 K. There appears a local maximum at 895°C, followed by a decrease and a further increase above 915°C.

The influence of different preparation conditions is shown in Figure 12. Optimal conditions for T_d = 895°C are found for the so-called standard conditions.

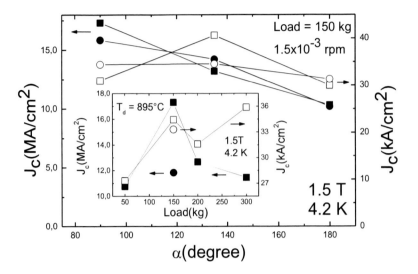

Figure 12. Influence of tilt angle α on critical current density for samples deformed at 875°C (squares) and 895°C (circles). Inset: load and twist rate [(squares) – $1.5 \cdot 10^{-3}$ rpm, (circles) – $3 \cdot 10^{-3}$ rpm] dependence of critical current density. Full (open) symbols: J_c interpreted as intra- (inter-) granular current density.

3.2. Discussion

The studies have shown that under all deformation conditions the colonies are plate-like ones. Whereas the length of colonies changes their thickness remains almost unchanged. This proves that a colony is a stable structural unit which deforms as a whole. Therefore in analyzing the deformation mechanisms one should take into account intercolony sliding (ICS). Consequently the total deformation (e_Σ) of Bi2212 can be represented as a combination of intercolony sliding and intracolonial deformation (e_{ID}) [49]:

$$e_\Sigma = e_{ICS} + e_{ID} \qquad (1)$$

Since Bi2212 does not form twins during deformation, three mechanisms might contribute to e_{ID}, namely intracolonial dislocation slip (IDS), grain boundary sliding and diffusion creep [49]. Grain-boundary sliding can be neglected, because it would cause a disintegration of colonies into grains and, as a result, a decrease in colony thickness should be observed. The contribution of diffusion creep to the intracolonial deformation is hardly essential too. In this case the increase (decrease) in the colony length should be accompanied by an appropriate decrease (increase) in the colony thickness which is not observed (Table 2, Figure 4). Therefore intracolonial deformation can be concluded to be mainly due to IDS. The observed bending of colonies which is connected with the presence of lattice dislocations and more precisely with prevailing of one sign edge dislocations in the bent sections supports this assumption. Thus, the operating deformation mechanisms are ICS and IDS.

Essentially, only two independent systems of dislocation slip operate in Bi2212. These systems are [100](001) and [010](001) [50]. The von Mises criterion [51] requiring the operation of five independent slip systems is not fulfilled and compatibility of deformation of neighbouring colonies is absent. Only ICS can provide the possibility to preserve the compatibility of deformation of colonies and is responsible for the high plasticity of this material. Therefore the main deformation mechanism of Bi2212 is ICS. IDS is only a minor mechanism which accommodates ICS.

Concerning the basal texture, it is known that it can result from three different mechanisms: IDS [52], grain boundary sliding [16, 53, 54] and oriented growth of grains/colonies due to applied stress [54, 55].

IDS by two slip systems cannot provide rotation of colonies as a whole relative to other colonies. It can only initiate the appearance of an additional texture component with an [hk0] -axis oriented along the sample radius. Therefore in spite of its presence in some deformation regimes (for example, at increased twist rates) IDS does not become an important deformation mechanism and cannot be considered as main mechanism for basal texture formation.

Under special conditions the formation of basal texture can also be attributed to a preferred growth of colonies with the [001]-axis oriented parallel to the axis of compression. Consequently, the Lotgering factor F should correlate with L, but this correlation is not always observed. For example, within the temperature range $845 - 915°C$ (150 kg load) the colony length and, correspondingly, the colony aspect ratio L/H essentially exceeds the correspondent value for the undeformed sample and the growth of texture correlates with the growth of L. However, for the sample deformed at $895°C$ (450 kg load), the colony length is similar to that in the undeformed one but the texture is maximal ($F = 0.98$). Therefore the colony growth during deformation seems to be a negligible process in the formation of texture in Bi2212 ceramics. Thus, it can be concluded that the main mechanism of deformation and formation of basal texture is ICS, due to which the plate-like colonies are rotated and oriented in such a way that their [001] direction is parallel to the compression axis.

Concerning the change of size of the colonies during the deformation, the colony thickness H is found to be a rather stiff parameter which does not depend on the special deformation conditions. The constancy of H can be explained by comparison with Y123. It is known that the basal-plane-faced grain boundaries in the Y123 ceramics grow due to initiation and movement of ledges [56, 57]. Most likely the mechanism of plate thickening operating in Y123 is also observed in Bi2212. But, since the unit cell of Bi2212 consists of much more atoms and because the lattice parameter c is about three times larger, the generation of ledges on the broad faces of Bi2212 colonies is much more difficult than in the Y123 ceramics. Therefore the amount of ledges generated in Bi2212 is expected to be small and the rate of colony thickening should be very slow.

The constancy of the colony thickness proves one more fact, namely the absence of dynamic recrystallization in all deformation regimes under study. That means, no new colonies are formed. There occurs only separation of the initially existing colonies by transverse intracolonial subboundaries which, as known, appear due to development of IDS and climb of dislocations [49].

In conclusion, the colony length L depends on three processes, namely: (i) thermal activated growth of colonies, (ii) formation of transverse subboundaries in the colonies, and (iii) the retarding action of other phase particles and pores. Depending on deformation conditions the contributions of these processes vary and, as a result, either a decrease or increase of L is observed.

3.2.1. Effect of Deformation Conditions on Microstructure and Texture

The T_d dependence of L and F is very similar (Figure 4). This indicates that both phenomena are related to each other and are determined by processes occurring at the colony boundaries. Both L and F start to increase at the same temperature, indicating that within the temperature range 820 – 835°C the intercolony diffusion accelerates. As shown by Imayev et al. [57] the necessary condition for the onset of grain growth in the Y123 ceramics during annealing is the presence of a liquid film at the grain boundaries. In Bi2212 ceramics the temperature of the onset of colony growth also correlates with the temperature of the appearance of an eutecticum which is equal to 825°C [58]. Thus, it can be assumed that in Bi2212 ceramics colony growth and acceleration of texture formation are also caused by the appearance of a thin liquid film at the colony boundaries.

For the deformation temperatures from 855 to 915°C the growth of L is almost similar ($\approx$ 4.5 μm), i.e. the expected thermal activation for colony growth is absent. In this temperature range the amount of secondary phases remains about 12 – 19% as in the undeformed sample. Therefore a retarding influence of particles can be excluded as reason for the constancy of L between 855 and 915°C. Because there are no other changes in the microstructure, the most probable reason for a retarding growth of L is the formation of a strong basal plane texture (F = 0.93 – 0.97). The formation of strong texture sharply increases the number of low-angle boundaries which are less mobile than high-angle ones [59, 60]. In this connection the relative constancy of L over the temperature range 855 – 915°C might be explained by the low mobility of low-angle ECB type colony boundaries (see Figure 2).

Above 915°C L decreases. Two reasons might be responsible for this decrease, which are both connected with the increase of the amount of secondary phases. Firstly, secondary phases hinder migration of colony boundaries. Secondly, particles of secondary phases are stress concentrators which promote IDS by reducing ICS (at the same e_y). The growth of lattice dislocations density contributes to the formation of new transverse subboundaries and an increase of their misorientation during deformation causes disintegration of the colonies into smaller fragments. The crimping of the microstructure and a weaker texture above 915°C (and at 905°C for $\alpha = 180°$) associated with this crimping are features of a reduction of ICS at the sites of the precipitates.

Within the temperature range 875 – 905°C the influence of strain on the microstructure (parameter L) is rather weak, because of the constant amount of secondary phases. This is attributed to the fact that at these temperatures within a rather wide range of twist angles the steady flow stage is attained and the contribution of IDS to the total deformation is small. The factor influencing the ratio between the operating deformation mechanisms (the contribution of IDS is increased and the contribution of ICS is decreased) is the growth of the amount of secondary phases. In particular, at 875°C and 895°C (up to $\alpha = 180°$) and at 905°C (up to $\alpha = 90°$) the amount of secondary phases is approximately the same (see Table 2). The situation changes with increasing strain at 905°C, where for $\alpha = 180°$ the amount of secondary phases increases up to 30%. Since particles precipitate mainly at colony boundaries the contribution of IDS to the total deformation is increased and that of ICS is decreased. The action of IDS contributes to the increase of the amount of subboundaries, thus decreasing L. The increased amount of secondary phases also results in a reduced texture.

An increase in load does not change the amount of secondary phases at 895°C. Therefore the change in L with load is caused by the influence of the compression stress on IDS. At 50 kg the compression occurs slowly, leading to a disappearance of the initial porosity and consequently the colonies have no time to grow. Under a 150 kg load the porosity disappears rather quickly, but since the activity of IDS is still low and the tendency to form subboundaries is also small L reaches a maximum. Above 150 kg the density of lattice dislocations increases, leading also to an increase in the amount of transverse subboundaries, and, therefore, to a decrease of L. In spite of the decrease in L the increase in load does not cause a reduction of texture.

Whereas the increase in the twist rate at 895°C is not accompanied by an increase of the amount of secondary phase, a considerable decrease in L is observed. Two reasons might be responsible for that: firstly, the time of deformation, namely the time of temperature processing was five times less. Therefore the colonies simply had no time to grow. Secondly, the increase in strain rate leads to activation of IDS. That is why the decrease in L might also be connected with the formation of a larger amount of transverse subboundaries as compared to the conventional strain rate.

3.2.2. Superconducting Properties

To get information about the influence of changes in the microstructure on the superconducting properties, the results of the magnetic measurements have to be related to these changes in proper way. The problem is that mean critical current densities can only be calculated from the hysteresis loops by using the Bean model where information about the geometry of the relevant super-current flow is necessary. Whereas in case of pure bulk pinning this dimension is the mean grain size, for intergranular currents the sample shape has to be used in the analysis. Both intra- and intergranular currents are determined by different types of pinning centres. Because a separation of both components is extremely complicate [61], the magnetic data are often interpreted either as being due only to intra- or only to intergranular current flow, and the other contribution is assumed to be negligible. In knowing that in our case both current systems contribute to about 50% to the magnetic signal we have analysed our data in both ways. Although neither of the results gives true J_c values, the dependence on the various preparation conditions is very similar. The combination of the obtained critical current densities with the results of the microstructure investigations allow us to propose a model of pinning in the investigated material. Checked is the model by the results of the magnetic relaxation measurements, which give mean effective activation energies, independent of the geometry of the samples.

After deformation the critical current density at 4.2 K strongly enhances (Figures 9, 10). With increasing T_d a steady increase (by more than a factor of two) up to $T_d = 940$°C with slightly enhanced values in the temperature range $T_d = 875 - 895$°C is found. This behaviour can be related to the observed changes in microstructure.

Between 815°C – 895°C a correlation of texture and superconducting properties is present. With increasing F critical current density J_c, mean effective activation energy $\langle E \rangle$ and irreversibility field B_{irr} grow. The slight increase in J_c and B_{irr} can be explained by the decrease in the amount of weak links at intercolony boundaries due to a sharp increase in the relative fraction of low-angle colony boundaries. Indeed at 865°C the value $F = 0.93$ corresponds to FWHM = 8.7°, i.e. the most colony boundaries of the sample deformed at 865°C are of low-angle type. However such an explanation is not sufficient. Within the same temperature range there occurs an essential increase in $\langle E \rangle$, most likely connected with the increase in the amount of low-angle boundaries, too. It is well known that low-angle boundaries can be strong centers of magnetic flux pinning [8, 62, 63]. That is why the increase in J_c and B_{irr} is attributed not only to the decrease in the amount of weak links at intercolony boundaries, but also to increase the mean effective pinning energy. Both factors are caused by the increasing number of low-angle colony boundaries.

Above 895°C the correlation between superconducting properties (J_c, B_{irr} and $\langle E \rangle$) and texture (factor F) is lost. In particular, in the range 905 – 915°C the superconducting properties show a reduction, although F and L are almost the same as at 895°C. Above 915°C the superconducting properties increase whereas F decreases. The value of L also decreases but only slightly.

This complex dependence of superconducting properties on deformation temperature can be explained by the scheme shown in Figure 13, which is based on the two simplified assumptions: (i) the deformation temperature influence the density of defects acting as pinning centers of magnetic flux, instead of the pinning energy on the corresponding defects, and (ii) there exist three main types of pinning centers – intracolonial lattice defects (point defects, dislocations and stacking faults), low-angle colony boundaries and particles of secondary phases. The dependence of the number of pinning centers on deformation temperature for a definite defect type is presented for simplicity by straight lines, though it is evident that in reality it may have a more complex shape. Line N_{ld} corresponds to the number of pinning centers in the system of intracolonial lattice defects. Since there are no phase transformations in the material and the colony size is rather stable, the density of lattice defects usually decreases with rising deformation temperature [49]. Therefore N_{ld} decreases with rising deformation temperature. Above T_3 the decomposition of the Bi2212 phase starts, which is accompanied by precipitation of particles of nonsuperconducting phases (Figures 3d, 8f and Table 2). It results in appearance of pinning centers associated with particles (N_p). As the amount of particles grows with increasing temperature N_p grows, too. Above T_1 the pinning centers associated with low-angle colony boundaries appear. The number of low-angle colony boundaries N_{lacb} starts at T_1, grows with increasing temperature to a definite value and above T_2 it shows a plateau, due to the relatively high and constant degree of texture. The temperature T_1 corresponds to a texture of $F \sim 0.90$ (FWHM $\sim 9°$) and is within the range 845 – 855°C. Similar to line N_p the end of line N_{lacb} is also near the melting temperature (T_m). The sum of these curves gives the total number of pinning centers (dashed line N_{total}). The shape of line N_{total} resembles the temperature dependences of J_c, $\langle E \rangle$ and B_{irr} and allows one to explain the temperature dependence of the superconducting properties consistently. The scheme shows that the features in the T_d dependence of J_c (Figure 9) can only be explained if the action of at least two types of pinning centers is assumed. The peak at 895°C forms when the parameter F becomes rather high and stable ($F = 0.97$). At this temperature the density of lattice defects is still sufficiently high. That is why the maximum at 895°C (Figure 9) is caused by the action of low-angle colony boundaries and intracolonial lattice defects. The minimum at 905 – 915°C is connected with the fact that within this temperature range, due to the recovery inside of the colonies, the density of lattice defects is rather low and mainly only low-angle colony boundaries contribute to pinning. Above 915°C the amount of particles increases sharply. However, the texture is only slightly weakened ($F \geq 0.94$), and the parameter L is minimal. Therefore the increase up to 940°C is due to the action of low-angle colony boundaries and increasing number of nonsuperconducting particles.

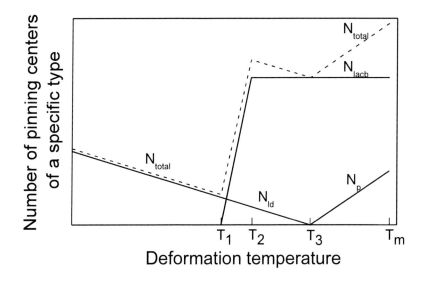

Figure 13. Scheme illustrating the dependence of the number of pinning centers of a specific type on deformation temperature of Bi2212 ceramics. N_{ld} is the number of pinning centers in the system of the intracolonial lattice defects (point defects, dislocations and stacking faults), N_p is the number of pinning centers in the system of the particles of nonsuperconducting phases formed during decomposition of the Bi2212 phase near the melting point, N_{lacb} is the number of pinning centers in the system of the low-angle colony boundaries, N_{total} (dashed line) denotes the total number of pinning centers, T_1 is the temperature at which $F \sim 0.90$ (FWHM $\sim 9°$) is achieved, T_2 is the temperature at which the texture reaches a high and relatively constant value ($F = 0.97$); T_3 and T_m is the temperature of decomposition onset and macroscopic melting of Bi2212 phase, respectively.

The present analysis of lattice defects responsible for pinning is only qualitative. Naturally for a quantitative description of pinning one should know the density and allocation of each defect type. Moreover, it is necessary to establish the position, density, size and morphology of nonsuperconducting phases occurring at decomposition of the Bi2212 phase. For more precise identification of defects and determination of their contribution to the changes in superconducting properties it is planned to perform TEM studies.

Since low-angle boundaries are strong pinning centers, one of the methods for improving the superconducting properties of the Bi2212 ceramics is a decrease of the mean colony size with preservation of a strong texture ($F > 0.94$). We assume that a strongly textured structure with submicrocrystalline and nanocrystalline grain/colony size in combination with high density of nanosized particles of nonsuperconducting phases would have much higher J_c in strong magnetic fields and at high temperatures. Moreover, to increase J_c at 77 K it is also necessary to optimize the post-deformation annealing process.

3.3. Conclusions

[1] The main mechanism of hot plastic deformation and basal texture formation is intercolony sliding. Intracolonial dislocation slip is an accommodative mechanism of deformation. Under the effect of external stresses the plate-like shaped colonies rotate and stack with the [001] axis parallel to the compression axis. The most strong

basal plane texture ($F = 0.97 – 0.98$) is formed during deformation in the temperature range $T_d = 875 – 940°C$.

[2] In all deformation regimes under study the colony thickness does not change, only their length varies. This proves the absence of dynamic recrystallization. Only transversal intracolonial subboundaries form, which divide the colonies into parts.

[3] The dependence of the superconducting properties (J_c, $\langle E \rangle$, B_{irr}) on deformation temperature for the deformed Bi2212 ceramics can be explained by the common action of at least two different types of pinning centers. In the vicinity of 895°C low-angle colony boundaries and intracolonial lattice defects (point defects, dislocations and stacking faults) are the dominant pinning centers. The large volume fraction of low-angle boundaries is due to the combination of strong basal texture and low colony size. At higher deformation temperatures both the low-angle colony boundaries and those due to the decomposition of the Bi2212 phase near the melting point causing an increasing number of nonsuperconducting particles, lead to a further increase in the critical current density.

4. BI2212/MG$_{1-x}$CU$_x$O COMPOSITE

4.1. Results

4.1.1. XRD Patterns and Texture

Figure 14 shows XRD patterns of the composite in initial sintered and deformed at 915°C states. Along with the Bi2212 phase one can reveal a small amount of phases Mg$_{1-x}$Cu$_x$O, (Sr,Ca)$_{4-y}$Bi$_2$O$_z$ and (Sr,Ca)$_{14}$Cu$_{24}$O$_{41}$ in the initial sintered state. As it was shown in Reference [28], formation of the Mg$_{1-x}$Cu$_x$O solid solution initiated by the interaction of the nanocrystalline MgO powder with Bi2212 at 850°C. Over the whole temperature range understudy the XRD patterns of the deformed samples almost completely correspond to the phase Bi2212. The preservation of the phase Bi2212 until 915°C is connected with increasing

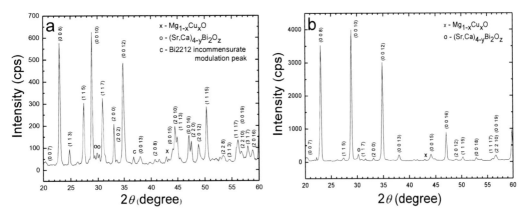

Figure 14. XRD patterns of the Bi2212/Mg$_{1-x}$Cu$_x$O samples (a) as-sintered at 855°C and deformed by torsion under pressure at (b) 915°C.

Table 4. Deformation temperature (T_d), Lotgering factor (F), mean length (L) and mean thickness (H) of Bi2212 colonies, mean size of individual $Mg_{1-x}Cu_xO$ particles (d_{MgCuO}) and volume fraction of the total amount of nonsuperconducting phases (V_{ns}) of samples under study

Sample No.	$T_d(^\circ C)$	F	L (μm)	H (μm)	d_{MgCuO} (μm)	V_{ns} (%)
38S	Undeformed	0.48	2.61±0.19	0.90±0.12	0.18±0.05	24
44	815	0.90	2.65±0.21	0.68±0.07	0.19±0.01	24
38	855	0.90	3.50±0.05	0.87±0.05	0.24±0.02	25
39	865	0.94	3.46±0.18	0.83±0.06	0.20±0.02	26
40	875	0.97	3.64±0.28	0.80±0.05	0.33±0.01	27
41	895	0.98	3.78±0.22	0.78±0.04	0.41±0.02	30
42	905	0.98	4.39±0.09	0.85±0.06	0.41±0.03	34
43	915	0.97	4.12±0.27	0.84±0.04	0.44±0.05	35

melting temperature of this phase due to the quasi-hydrostatic pressure applied [18]. Moreover, deformation resulted in formation of a strong texture that was testified by the increased intensity of (001) type peaks. The dependence of the factor F on the deformation temperature is shown in Table 4. In the undeformed sample $F = 0.48$. It is seen that rather strong texture ($F = 0.90$) is formed already at 815°C. Over the range 815-855°C the growth of texture is weak. At temperatures above 855°C there occurs further growth of texture. The maximum value $F = 0.98$ is observed at deformation temperatures 895 and 905°C. Above 905°C the factor F decreases slightly.

4.1.2. Microstructure

At all deformation temperatures there remains a microstructure typical for the phase Bi2212 which consists of colonies of plate-like grains. Deformation leads to formation of a distinct metallographic texture with the orientation of colony axis [001] parallel to the axis of compression (Figure 15). Microstructure is rather uniform on characteristic sizes: the mean colony length (L) and thickness (H) in the center and the periphery do not differ more than 5%. The averaged colony sizes are shown in Table 4.

In the undeformed sample of the composite in addition to $Mg_{1-x}Cu_xO$ there are two secondary phases, namely $(Sr, Ca)_{4-y}Bi_2O_z$ and $(Sr, Ca)_{14}Cu_{24}O_{41}$. In the process of deformation no new phases appear.

The size of $Mg_{1-x}Cu_xO$ particles grows quite substantially (in 9 times) already at sintering and their average size reaches 0.18 μm. Particles have round shape and their distribution within the sample interior is rather uniform (Figure 15, Table 4). During deformation the growth of particles continues, but rather weakly up to 865°C. Above 865°C the growth rate of particles increases considerably and at the temperature 915°C the value $d_{MgCuO} = 0.44$ μm. The appropriate study has shown that the size of $Mg_{1-x}Cu_xO$ particles changes mainly during deformation and no in the process of further recovery annealing at 850°C.

Table 4 shows the dependence of the total amount of $Mg_{1-x}Cu_xO$, Bi-free and Cu-free secondary phases V_{ns} on deformation temperature. Until 875°C the value V_{ns} changes very weakly and not exceeds 30%. The appreciable increase in the V_{ns} value takes place above 895°C due to partial decomposition of Bi2212 phase [40]. At 915°C their amount achieves 35%. The particles of Bi-free and Cu-free phases are mainly of an anisotropic shape. They are

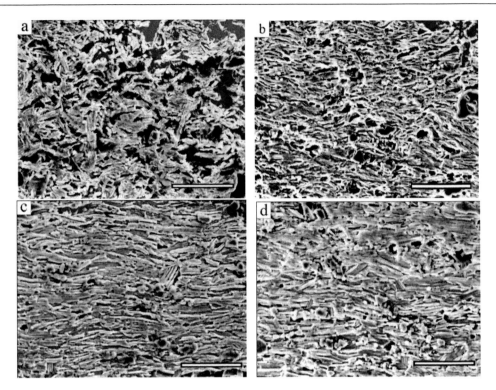

Figure 15. SEM micrographs of the Bi2212/Mg$_{1-x}$Cu$_x$O samples (a) as-sintered at 855°C and deformed by torsion under pressure at (b) 815°C, (c) 895°C and (d) 915°C. The axis of compression is vertical. Bar = 10 μm. Grey submicron particles protruding on the surface of the metallographic section are the Mg$_{1-x}$Cu$_x$O.

elongated perpendicular to the compression axis and characterized by an essential size spread. In the temperature range 815 – 895°C the average size of the particles is equal to 7 – 9 μm along the compression axis and 12 – 14 μm in the direction normal to the compression axis. Above 895°C the particle size increases and at 915°C their size is 9 and 20 μm along and perpendicular to the compression axis, respectively.

4.1.3. Superconducting Properties

The character of the dependence of the critical current density on the deformation temperature $J_c(T_d)$ at 4.2 K and 30 K and the magnetic field 1.5 T is no monotonous (Figure 16). The curve $J_c(T_d)$ has two distinct enhancements: at 875 and 915°C. The maximum is observed in the sample deformed at 915°C. In spite of the non-monotonous character of the curve $J_c(T_d)$ the value J_c in the deformed samples is much higher than in the undeformed one. Note that with increasing temperature from 4.2 K to 30 K the difference between minimum and maximum values of $J_c(T_d)$ increases essentially. If at 4.2 K the difference between maximum and minimum is no more than 25% then it is about 300% at 30 K. The dependence $B_{irr}(T_d)$ at 30 K also has two enhancements: 1.4 T at 875°C and 1.5 T at 915°C (Figure 17).
Figure 18 shows the dependence of the mean effective activation energy on the deformation temperature $\langle E \rangle(T_d)$. The dependence $\langle E \rangle(T_d)$ has two decrements: at 865°C and 895 – 905°C and three enhancements: at 815, 875 and 915°C. The maximum value (31 meV) is achieved at 915°C.

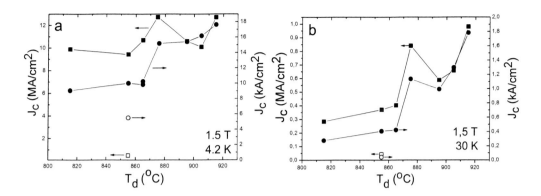

Figure 16. Dependence of J_c at 1.5 T on deformation temperature T_d of composite Bi2212/Mg$_{1-x}$Cu$_x$O at (a) 4.2 K and (b) 30 K, J_c is interpreted as both intra-(full square) and inter-(full circle) granular current density, unfilled symbols correspond to undeformed sample.

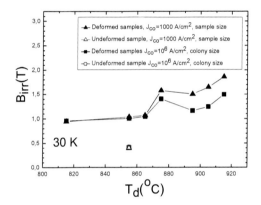

Figure 17. Dependence of irreversibility field B_{irr} at 30 K determined for J_c criterion of $1·10^3$ and $1·10^6$ A/cm^2 on deformation temperature (T_d) of composite Bi2212/Mg$_{1-x}$Cu$_x$O, J_c is interpreted as intra-(square) and inter-(triangle) granular current density.

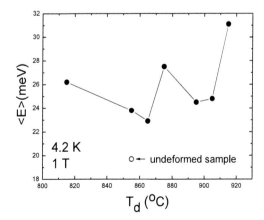

Figure 18. Dependence of mean effective activation energy $\langle E \rangle$ at 1 T and 4.2 K of composite Bi2212/Mg$_{1-x}$Cu$_x$O on deformation temperature T_d.

Table 5. The values of the superconducting transition onset (T_c), midpoint (T_c^{mid}) and critical current density (J_c) at 60 K and 77 K for deformed samples Bi2212/Mg$_{1-x}$Cu$_x$O

Sample No.	T_d (°C)	T_c (K)	T_c^{mid} (K)	J_c at 60K (A/cm^2)	J_c at 77K (A/cm^2)
44	815	90	33	160	0
38	855	92	45	490	10
39	865	92	45	430	10
40	875	92	48	760	10
41	895	92	55	1350	20
42	905	89	55	1800	120
43	915	90	56	3100	250

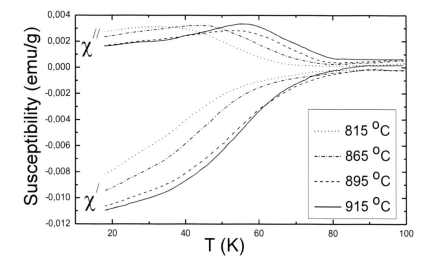

Figure 19. Temperature dependence of real χ' and imaginary χ'' parts of the AC magnetic susceptibility for the Bi2212/Mg$_{1-x}$Cu$_x$O samples deformed at 815°C, 865°C, 895°C and 915°C, the excitation AC-field amplitude $H_0 = 100$ Oe.

The temperatures of the onset (T_c) and the middle (T_c^{mid}) of superconducting transition for all deformed samples are shown in Table 5, and the temperature dependence of magnetic susceptibility of some of them is shown in Figure 19. It is seen that T_c values of the samples are rather neighbor (T_c = 89 – 92 K), whereas T_c^{mid} values depend on deformation temperature and increase with increasing deformation temperature. For example, in the sample deformed at 815°C a value of T_c^{mid} is about 33 K while in the sample deformed at 915°C this value is 56 K. Since the T_c^{mid} value is proportional to the width of the superconducting transition (ΔT_c) this means that the ΔT_c value decreases with increasing deformation temperature. Values of transport J_c at 60 K and 77 K correlate with T_c^{mid} and also increase with deformation temperature (Table 5).

4.2. Discussion

4.2.1. Superconducting Properties at 4.2 K and 30 K

Let us consider the influence of particles $Mg_{1-x}Cu_xO$ on current-carrying capacity at 4.2 and 30 K of the deformed samples. In the undeformed composite the $Mg_{1-x}Cu_xO$ particle size is rather small. However, the energy of pinning exceeds the corresponding value for the undoped Bi2212 sample only by 1.5 meV (19.5 against 18 meV) [40]. Neighbor values were obtained in the melted composites Bi2212 with submicron inclusions $SrZrO_3$, $SrHfO_3$, Sr_2CaMoO_6 and Sr_2CaWO_6, in which due to doping the value of pinning energy $\langle E \rangle$ is increased by no more than 3 meV [64]. Such a small increase in $\langle E \rangle$ is likely to be typical for undeformed Bi2212 base composites with submicron impurities of secondary phases.

When deformation is applied the situation is different. The $Mg_{1-x}Cu_xO$ particles being obstacles for sliding contribute to accumulation of high density of lattice defects (dislocations and stacking faults) in their vicinity. That is why the essential growth of $\langle E \rangle$ (up to 26 meV) and the rather high J_c in the sample deformed at 815°C (see Figure 18) are attributed to the joint influence of particles $Mg_{1-x}Cu_xO$ and lattice defects on flux pinning. Note that deformation of the undoped Bi2212 sample at 815°C does not lead to such strong growth of the pinning energy where $\langle E \rangle = 19 - 21$ meV [40].

With increasing deformation temperature there occurs an increase in the particle size and the distance between particles. It is known that the size of the stable dislocation substructure formed around the particles is conditioned by the geometry of disperse particles (mainly by the distance between the particles) [31]. That is why the increase of the distance between the particles decreases the density of dislocations connected with them. Moreover, regardless of the presence of particles, there occurs an acceleration of recovery processes within the grain interior, grain boundaries and intercolonial boundaries with increasing deformation temperature. As a result, at the temperature 865°C the value $\langle E \rangle$ drops to 23 meV, that is comparable with the value $\langle E \rangle = 22$ meV in the undoped Bi2212 sample, having basal texture of the same sharpness ($F = 0.94$) and neighbor colony sizes [40]. Thus, the influence of the $Mg_{1-x}Cu_xO$ particles and the lattice defects connected with them on flux pinning becomes negligible at the deformation temperatures above 865°C.

In spite of the high superconducting properties of the sample deformed at 815°C, the properties are maximal in the samples deformed at 875 and 915°C. In these samples the role of $Mg_{1-x}Cu_xO$ particles is not so important already since other centers of flux pinning are dominant.

The former studies of microstructure and superconducting properties of the deformed undoped Bi2212 ceramics allowed establishing that there exist two enhancements of superconducting properties: at 895 and 940°C [40]. The first peak is explained by the joint action of low-angle colony boundaries and intracolonial lattice defects (point defects, dislocations and stacking faults), while the second one by the joint action of low-angle boundaries and fine particles of the non-superconducting phases occurring in the process of decomposition of the Bi2212 phase near the melting temperature. The large contribution of low-angle boundaries is connected with formation of strong basal plane texture ($F > 0.93$) [40]. The analogous enhancements are observed in the Bi2212/$Mg_{1-x}Cu_xO$ composite. However their temperatures are slightly lower i.e. 875 and 915°C. This testifies that similar to the undoped Bi2212 ceramics the same pinning centers are responsible for formation of two

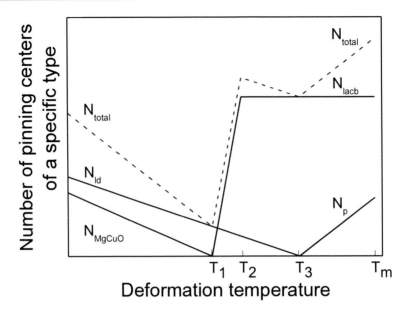

Figure 20. Scheme illustrating the dependence of the number of pinning centers of a specific type on deformation temperature of composite Bi2212/Mg$_{1-x}$Cu$_x$O. N_{ld} is the number of pinning centers in the system of the intracolonial lattice defects (point defects, dislocations and stacking faults), N_p is the number of pinning centers in the system of the particles of nonsuperconducting phases formed during decomposition of the Bi2212 phase near the melting point, N_{lacb} is the number of pinning centers in the system of the low-angle colony boundaries, N_{MgCuO} is the number of pinning centers in the system of Mg$_{1-x}$Cu$_x$O particles, N_{total} (dashed line) denotes the total number of pinning centers, T_1 is the temperature at which $F \sim 0.90$ (FWHM ~ 9°) is achieved, T_2 is the temperature at which the texture reaches a high and relatively constant value ($F = 0.97 - 0.98$), T_3 and T_m is the temperature of decomposition onset and macroscopic melting of Bi2212 phase, respectively.

enhancements identified. The data obtained can be explained in terms of the scheme proposed earlier for Bi2212 ceramics (see Section 3.2.2, Figure 13) and are shown in Figure 20 with the only difference being that the composite Bi2212/Mg$_{1-x}$Cu$_x$O contains an additional pinning center, namely particles Mg$_{1-x}$Cu$_x$O. The scheme is based on the simplified assumption that there exist four main types of pinning centers of the magnetic flux in the composite: (i) intracolonial lattice defects, (ii) low-angle colony boundaries, (iii) non-superconducting phase particles occurring at decomposition of the Bi2212 phase, and (iv) Mg$_{1-x}$Cu$_x$O particles.

The shape of line N_{total} resembles the temperature dependences of J_c, B_{irr} and especially $\langle E \rangle$ and allows one to explain the temperature dependence of the superconducting properties consistently. The scheme shows that both maxima (at 875°C and 915°C) appear when no one, but at least two types of pinning centers operate equally. The peak at 875°C forms when the parameter F becomes rather high and stable ($F = 0.97$). At this temperature the density of lattice defects is still sufficiently high. That is why the maximum at 875°C is caused by the action of low-angle colony boundaries and intracolonial lattice defects. The minimum at 895 – 915°C is connected with the fact that within this temperature range, due to the recovery inside of the colonies, the density of lattice defects is rather low and mainly only low-angle colony boundaries contribute to pinning. Above 905°C the amount of particles increases sharply, but the texture is high ($F = 0.97$), and the parameter L is minimal. Therefore the maximum at 915°C is due to the action of low-angle colony boundaries and increasing number of nonsuperconducting particles occurring at decomposition of the Bi2212 phase.

Note that values of J_c and B_{irr} at 4.2 K and 30 K in the best sample of Bi2212/Mg$_{1-x}$Cu$_x$O composite ($T_d = 915°C$) are less by a factor of 1.5 – 2 as compared with the undoped Bi2212 material deformed at the same temperature and having almost the same values of the factor F and the colony size [40]. There may be two explanations, namely: (i) superconducting properties of the Bi2212 phase are partially weaken due to Cu depletion which results from its interaction with MgO particles at temperatures above 850°C [28], and/or (i) the regime of annealing after deformation (950°C for 96 h in air) is no optimal for Bi2212/Mg$_{1-x}$Cu$_x$O-composite.

4.2.2. Superconducting Properties at 60 K and 77 K

Let us consider features of superconducting transition and transport properties at 60 and 77 K of the deformed composite. As it has been established in Section 3, in the undoped Bi2212 sample after deformation at 815°C and recovery annealing the factor $F = 0.76$, and a value of J_c at 60 K and 77 K is 800 and 40 A/cm^2, respectively [40]. After similar deformation and annealing conditions a stronger texture ($F = 0.90$) is formed in the composite. The colony size is approximately the same but a value of J_c at 60 K and 77 K is much less and equals about 160 and 0 A/cm^2, respectively. It testifies that after deformation at 815°C and annealing the intercolony boundaries in the composite have lower intergranular J_c than in the undoped sample, though the mean effective pinning energy at 4.2 K is much higher. This phenomenon can be explained by the following. The starting material for the composite was prepared by mechanical mixing of two powders (Bi2212 and MgO) and after sintering the most of Mg$_{1-x}$Cu$_x$O particles dispose at colony boundaries. As shown in [40] the main mechanism of deformation in Bi2212 is intercolonial sliding. That is why during deformation of the composite the particles preventing the sliding are stress concentrators. The less is the particle size the larger is the amount of boundaries where they dispose and the larger is the distortion they initiate. That is why in the composite the level of internal stresses in the vicinity of colony boundaries is higher than in the undoped Bi2212 ceramics, and the lower is the deformation temperature the higher is the stress.

High-T_c superconductors are polyatomic crystalline bodies. Atoms of different chemical elements in ideal defectless polyatomic solid bodies free of stress fields are distributed orderly. Defects being sources of stress fields can initiate local disordering of ideal arrangement of atoms in polyatomic solids, since different atoms in the stress fields are distinguished in their behavior. In this case there are the dilatation stresses that exert the most essential influence on spatial distribution of atoms characterized by different atomic volumes in elastic-stressed crystals. Large (fine) atoms tend to shift toward the areas where tensile (compressive) stresses are present [65]. Internal interfaces are flat defects able to create spatial-heterogeneous stresses, in particular, dilatation stresses and that is why they can cause local changes of stoichiometry which essentially weakens intergranular J_c. It has been shown that at low-angle boundaries the edge lattice dislocations are sources of dilatation stresses which lead to local changes of oxygen stoichiometry [66]. In the deformed composite the colony boundaries in the vicinity of particles Mg$_{1-x}$Cu$_x$O contain high density of dislocations. That is why a high value of ΔT_c, and a low value of J_c can be explained by local violation of the oxygen stoichiometry in the vicinity of the colony boundaries under the effect of stresses from $Mg_{1-x}Cu_xO$ particles. The dislocation tangles, trapped by particles are conventionally very stable and are preserved till high temperatures [31]. That is why it is not wonder that even such long (96 h) annealing at 850°C does not lead to formation of narrow superconducting transition. To decrease ΔT_c and increase J_c at 77 K in the samples obtained at low deformation temperatures another regime of post deformation annealing is likely to be required to relieve internal stresses more completely.

As the deformation temperature increases the $Mg_{1-x}Cu_xO$ particles grow. Moreover, with increasing temperature the processes of recovery of dislocation structure in Bi2212 phase accelerate. Because of that with increasing deformation temperature the negative influence of particles on intercolonial current decreases.

Similar as a matter of fact phenomenon connected with the negative influence of excessive stresses on the oxygen sublattice of the superconducting phase was observed in deformed composite Y123/Y211 [34, 35]. After deformation and recovery annealing one can often observe strongly stressed areas in the vicinity of Y211 particles, and the oxygen amount in these areas is less than in the ones far from the particles. It is manifested in the formation of a tweed type ortho-2 structure [34] and the absence of twins of tetra-ortho phase transformation [35]. Compressive stresses are likely to operate in these areas which force out the oxygen atoms to the areas with tensile stresses.

Thus, the influence of $Mg_{1-x}Cu_xO$ particles on superconducting properties of the deformed samples has a dual character. On the one hand, they essentially increase the pinning energy and critical current density at low temperatures. On the other hand, during process of deformation the particles located at colony boundaries become sources of uncompensated stresses, distort oxygen stoichiometry near the boundaries and increase ΔT_c. However, unfortunately, strong texture does not form in that temperature region where $Mg_{1-x}Cu_xO$ particles are sufficiently small and effective as centers of flux pinning (up to $T_d = 865$°C). We think that for restricting the negative aspects of doping and achieving higher properties of the composites it is necessary to do the following. Firstly, other methods of preparing composites are required, for example, sol-gel technique providing more uniform distribution of particles in the material. Secondly, one should try to obtain strong texture ($F \geq 0.93$) at low deformation temperatures ($T_d < 865$°C) via refining the initial microstructure and changing regimes of deformation.

4.3. Conclusions

[1] Strong basal plane texture ($F = 0.90$) forms already at 815°C and remains at this level till 855°C. Above 855°C the value of F starts to grow again and the most strong texture ($F = 0.97 - 0.98$) forms during deformation within the temperature range $T_d = 875 - 915$°C. The matrix Bi2212 phase preserves plate-like shaped colonies within the whole deformation temperature range under study. The mean size of colonies weakly depends on the temperature of deformation.

[2] The noticeable contribution of $Mg_{1-x}Cu_xO$ particles to the flux pinning at 4.2 K and 30 K is observed only after low deformation temperatures ($T_d = 815 - 865$°C) when they are in a fine disperse state. The contribution of particles to the flux pinning is first of all conditioned by lattice defects (dislocations and stacking faults) trapped by

the particles. As the deformation temperature increases the size of $Mg_{1-x}Cu_xO$ particles and the distance between them increase while the density of defects trapped by them decreases. Besides as the deformation temperature increases the recovery processes accelerate that also promotes reduction of dislocation density. That is why above 865°C the joint contribution of particles and trapped lattice defects to the flux pinning becomes insignificant.

[3] The $Mg_{1-x}Cu_xO$ particles and the lattice defects connected with them exert a dual effect on superconducting properties of the composite. On the one hand, they essentially increase low-temperature (4.2 K and 30 K) properties, especially the pinning energy. On the other hand, during deformation the particles located at colony boundaries initiate high stresses that weakens intercolonial J_c and increases ΔT_c. The highest value of ΔT_c is observed in the sample deformed at 815°C. As the deformation temperature increases the value ΔT_c decreases while J_c at 60 K and 77 K increases, respectively.

[4] At low temperatures (4.2 K and 30 K) the composite displays the best superconducting properties in two samples obtained at 875°C and 915°C, respectively. The main pinning centers in the sample deformed at 875°C are low-angle colony boundaries and intracolonial lattice defects (point defects, dislocations and stacking faults). The low-angle colony boundaries and particles of non-superconducting phases arising from decomposition of the Bi2212 phase near the melting point dominate as a flux pinning centers in the sample deformed at 915°C.

5. UNDOPED BI2212 CERAMICS AND BI2212/SR$_2$CAWO$_6$ COMPOSITE FROM SOL-GEL PRECURSORS

5.1. Introduction

The influence of W and Mo impurities and phases on their base on the superconducting properties of Bi2212 and Bi2223 has been studied significantly [64, 67–71]. However, the addition of W and Mo into Bi-Sr-Ca-Cu-O matrix is of much interest since the substitution of Bi results in increasing concentration of holes because the valency of Mo and W is higher than that of Bi [69]. This phenomenon cardinally differs from the one occurring in the case of analogous heterovalent partial substitution of Ca by Y in Bi2212 and Y by Pr in Y123. Zhang et al. [69] assume that it is due to the fact that the Bi-O planes dispose far from the Cu-O planes, so the Mo and W in the Bi sites could not fill or localize the holes. At the same time they promote increasing oxygen content [69].

Makarova et al. [64] considered the influence of additions W, Zr and Hf in Bi2212 on formation of phases within the temperature range 850 – 900°C. It has been established that at doping there consequently occurs formation of phases Sr_2CaMO_6 ($M = Mo$, W) and $SrAO_3$ ($A = Zr$, Hf), being chemically compatible with Bi2212. In composites Bi2212/0.25Sr$_2$CaMO$_6$ and Bi2212/0.5SrAO$_3$ prepared via melt processing of sol-gel precursor the pinning energy of disperse phases at 5 K increases in row Sr_2CaMoO_6 – $SrHfO_3$ – $SrZrO_3$ – Sr_2CaWO_6. Moreover, in composite Bi2212/Sr$_2$CaWO$_6$ there forms the narrowest superconducting transition and values J_c and B_{irr} at 60 K are also higher than in other composites.

Naturally, further studies of composite Bi2212/Sr$_2$CaWO$_6$ are highly appreciated. It is interesting an attempt to force pinning of magnetic vortices on particles Sr$_2$CaWO$_6$ due to severe hot plastic deformation since at deformation the particles can trap dislocations [31], which, in turn, are strong centers for pinning of the magnetic flux [72]. The efficiency of such a method has been demonstrated in experiments on torsion straining under pressure composite Bi2212/Mg$_{1-x}$Cu$_x$O, when such deformation led to the increase of the pinning energy concerned with particles from 19.5 meV to 26 meV at 4.2 K (see Figure 18). This section is devoted to the study of the influence of torsion under pressure on microstructure, texture and superconducting properties of undoped Bi2212 and composite Bi2212/Sr$_2$CaWO$_6$, the initial oxide precursor of which was processed by the sol-gel method.

5.2. Results

5.2.1. XRD Patterns and Texture

Figure 21 shows diffraction patterns of non-deformed samples of Bi2212 and composite Bi2212/Sr$_2$CaWO$_6$ and the ones deformed at the temperature 915°C. The deformed samples

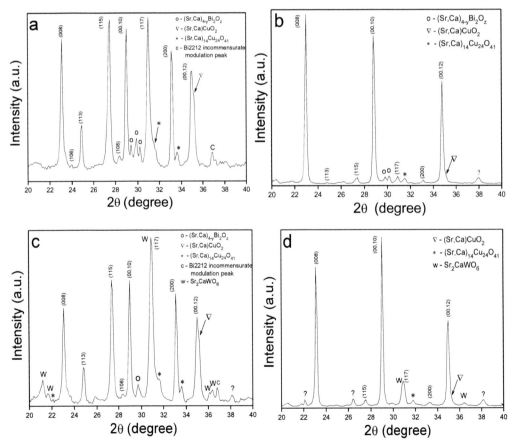

Figure 21. XRD patterns of the sol-gel (a, b) undoped Bi2212 and (c, d) composite Bi2212/Sr$_2$CaWO$_6$, (a, c) as sintered at 855°C and (b, d) deformed by torsion under pressure at 915°C.

almost completely consist of the Bi2212 phase. Deformation led to sharpening of basal plane texture in both of the materials that is confirmed by the increased intensity of the (00l) peaks (Figure 21 b, d). With increasing deformation temperature the degree of basal texture non-monotonously increases from $F = 0.74$ at $T_d = 815°C$ up to $F = 0.94 – 0.95$ at $T_d = 915°C$ (Table 6). Pressure also exerts an essential influence on texture. In particular, due to the increase in pressure by twice (from 15 to 30 MPa) at the temperature 895°C the factor F has grown from 0.87 up to 0.94 in the undoped Bi2212 ceramics (Table 6).

Cu-free phase (Sr, Ca)$_{4-y}$Bi$_2$O$_z$ and two Bi-free phases (Sr,Ca)CuO$_2$ and (Sr,Ca)$_{14}$Cu$_{24}$O$_{41}$ are present in the undeformed sol-gel Bi2212. In addition to these phases there is phase Sr$_2$CaWO$_6$ in the composite. No other W-containing phases are revealed in the samples of composite under study. During deformation the specified phases are preserved though their relative intensity decreases because of increasing intensity of (00l) peaks of matrix phase Bi2212.

5.2.2. Microstructure

In both materials the colonies of the Bi2212 phase preserve their plate-like shape at all regimes of deformation under study (Figures 22 and 23). In the undoped Bi2212 over the

Table 6. Parameters of microstructure and texture of samples of sol-gel undoped Bi2212 and composite Bi2212/Sr₂CaWO₆

Sample No.	T_d (°C)	T_c, (K)	$d_\perp$, (µm)	$d_\parallel$, (µm)	F	V_{ns}, (%)	L_{sp}, (µm)	H_{sp}, (µm)	d_w, (µm)
Undoped Bi2212									
92S	Undeformed	88	1.21	1.11	0.26	16	48	30	
101	815	94	1.55	0.65	0.74	18	33	10	
100	855	95	2.02	0.87	0.73	20	50	20	
102	865	94	2.16	0.99	0.89	19	44	14	
96	875	94	2.45	1.51	0.81	17	40	12	
92	895	95	3.53	1.69	0.87	19	41	18	
98	895, $P = 30$ MPa	95	3	1.4	0.94	16	31	13	
95	905	94	4.1	1.65	0.90	17	40	12	
97	915	95	3	1.14	0.94	17	37	12	
Composite Bi2212/Sr₂CaWO₆									
93S	Undeformed	88	1.47	0.90	0.29	28	31	15	0.30
108	815	84	1.37	0.76	0.74	30	31	12	0.34
107	865	92	2.11	0.91	0.89	36	32	15	0.34
105	875	94	2.10	0.87	0.88	31	33	12	0.41
93	895	90	2.65	1.01	0.87	38	29	10	0.43
104	905	91	2.78	1.14	0.95	39	30	13	0.46
106	915	90	2.17	0.96	0.95	37	34	18	0.60

Notes: T_d is the temperature of deformation, T_c is the onset of transition temperature, $d_\perp$ and $d_\parallel$ is the mean colony size of Bi2212 phase in the perpendicular direction and along the axis of compression, respectively,(F is Lotgering's factor, V_{ns} is the total volume fraction of Cu-free, Bi-free and Sr$_2$CaWO$_6$ (for composite) secondary phases, L_{sp} and H_{sp} is the mean size of Cu-free and Bi-free phases in the perpendicular direction and along the axis of compression, respectively, d_w is the mean size of separate particles Sr$_2$CaWO$_6$.

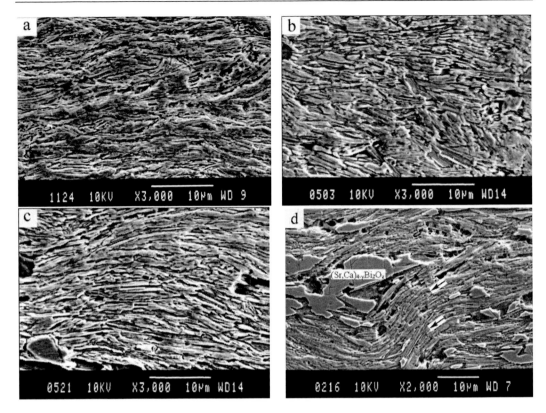

Figure 22. Microstructure of deformed samples of sol-gel undoped Bi2212: (a) $T_d = 815°C$, (b) $T_d = 915°C$, (c) and (d) $T_d = 895°C$ ($P = 30$ MPa, sample No 98); (a,b,c) far from coarse particles of secondary phases, (d) near the coarse particle $(Sr,Ca)_{4-y}Bi_2O_z$; the arrows show change of the flow direction of Bi2212 colonies near the particle; the axis of compression is vertical.

deformation temperature range 815 – 865°C the colony size grows rather slowly (Figure 22 and Table 6). Beginning from 865°C the colony size starts to grow more rapidly and achieves maximum at 905°C. Above 905°C the colony size decreases. The increase in pressure from 15 MPa up to 30 MPa at $T_d = 895°C$ (sample No 98) also leads to some decrease in the colony size (Table 6). Unlike Bi2212 the growth of the colony size in the composite until 905°C is more uniform. However, similar to Bi2212 the colony size also decreases slightly above 905°C (Table 6).

In the undoped Bi2212 until 915°C the total volume fraction of Bi-free and Cu-free phases (V_{ns}) does not exceed 20% (Table 6). The particles of secondary phases are rather coarse and in deformed samples have an elongated shape, as a rule (Figure 22 d). The mean particle size in all samples is $H_{sp} = 16$ μm and $L_{sp} = 40$ μm (Table 6). In the process of deformation the coarse particles cause some disturbances in the flow of matrix Bi2212 phase and locally weaken its texture (Figure 22 d).

In the composite, due to the presence of particles Sr_2CaWO_6, the total volume fraction of secondary phases is higher than in the undoped Bi2212 by 18% on average (Table 6). Until 915°C the value V_{ns} varies weakly and is about 28 – 39%. The volume fraction of Bi-free and Cu-free phases is approximately similar to that in the undoped Bi2212. The particles of these phases have an elongated shape and their size is nearly similar to that in the undoped Bi2212 (Table 6).

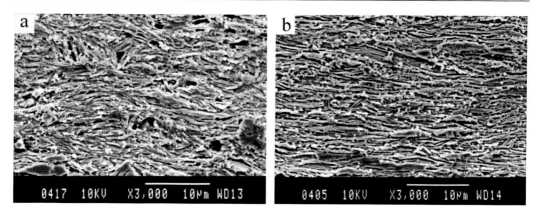

Figure 23. Microstructure of deformed samples of sol-gel composite Bi2212/Sr$_2$CaWO$_6$ far from coarse particles of secondary phases: (a) $T_d = 815°C$, (b) $T_d = 915°C$; the axis of compression is vertical.

Let us consider particles Sr$_2$CaWO$_6$. Within the temperature range 815 – 895°C about 20 – 25% of particles Sr$_2$CaWO$_6$ constitute conglomerates while the rest ones are separate particles. Within the temperature range 905 – 915°C the amount of particles Sr$_2$CaWO$_6$ constituting conglomerates increases up to 30 – 35%. The mean conglomerate size varies weakly with temperature and does not exceed 2 µm. After sintering the mean size of separate particles Sr$_2$CaWO$_6$ is about 0.30 µm (Table 6). With increasing deformation temperature from 815 up to 865°C the particle size grows weakly. Above 865°C the rate of particle growth increases essentially and at 915°C their size achieves 0.60 µm.

5.2.3. Superconducting Properties

Figures 24 and 25 show the dependence of critical current density on deformation temperature $J_c(T_d)$ at 4.2 K and 30 K at 1.5 T for the both materials. It is seen that J_c in the undoped B2212 is higher than in the composite.

In undoped Bi2212 the dependences of intragranular and intergranular J_c on deformation temperature T_d are not correlated throughout the temperature range of deformation, so we consider them separately. Both at 4.2 K and 30 K the dependence of intragranular J_c on deformation temperature has a distinct "V-type" shape. With increasing T_d above 815°C a value of intragranular J_c decreases, achieves minimum at 895°C and then increases until 915°C. The rise of pressure by twice at the temperature 895°C increases J_c at 4.2 K and 30 K up to the level of the sample deformed at 915°C. Intergranular J_c at 4.2 and 30 K weakly depends on T_d up to $T_d = 895°C$, but then one increases. The rise of pressure by twice at 895°C noticeably increases intergranular J_c at 4.2 K and 30 K, too.

In composite Bi2212/Sr$_2$CaWO$_6$ at 4.2 K as T_d increases from 815 up to 895°C the tendency to a decrease in intragranular J_c is also observed though it is less prominent than in undoped Bi2212. Above 895°C a value of intragranular J_c increases steadily. At 30 K a decrease in intragranular J_c within the range 815 – 895°C is more distinct than at 4.2 K. Similar to 4.2 K the increase in intragranular J_c is observed above 895°C. Intergranular J_c at 4.2 K weakly depends on T_d up to $T_d = 895°C$, but then one increases. At 30 K intergranular J_c has a distinct "V-type" shape with minimum at $T_d = 895°C$.

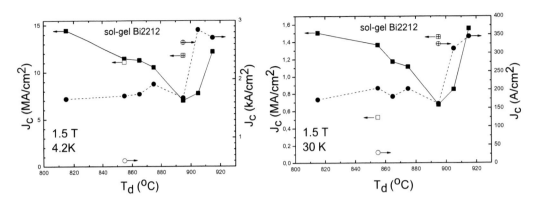

Figure 24. Dependence of J_c at 1.5 T on deformation temperature T_d of sol-gel Bi2212 at 4.2 K (a) and 30 K (b), J_c is interpreted as both intra-(full square) and inter-(full circle) granular current density, unfilled symbols correspond to undeformed sample.

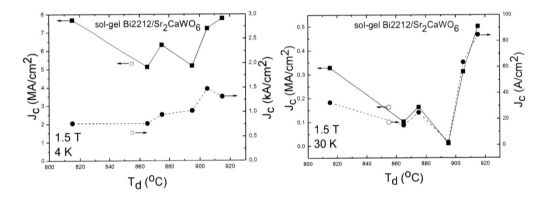

Figure 25. Dependence of J_c at 1.5 T on deformation temperature T_d of sol-gel composite Bi2212/Sr$_2$CaWO$_6$ at 4.2 K (a) and 30 K (b), J_c is interpreted as both intra-(full square) and inter-(full circle) granular current density, unfilled symbols correspond to undeformed sample.

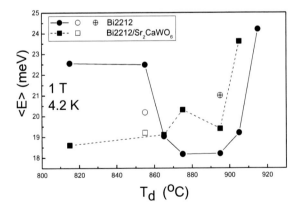

Figure 26. Dependence of mean effective activation energy $\langle E \rangle$ at 1 T and 4.2 K on deformation temperature T_d of sol-gel undoped Bi2212 and composite Bi2212/Sr$_2$CaWO$_6$. Undeformed samples (sintered at 855°C) are denoted by open symbols. The crossed circle at T_d = 895°C corresponds to the sample No 98 of Bi2212 deformed under pressure P = 30 MPa.

The activation energy of flux pinning behaves as follows (Figure 26). A clear minimum of $\langle E \rangle$ is observed in undopped Bi2212 deformed at 895°C. The rise in pressure by twice at 895°C increases $\langle E \rangle$ from 18.2 up to 21 meV. In composite Bi2212/Sr$_2$CaWO$_6$ within the range $T_d = 815 - 895$°C the pinning energy is very weak (less than 20.5 meV) and does not at all depend on deformation temperature. However, above 895°C a value of $\langle E \rangle$ grows significantly and achieves 23.6 meV at 905°C.

5.3. Discussion

In both materials, undoped Bi2212 and composite Bi2212/Sr$_2$CaWO$_6$ (both are designated as sol-gel materials), at the regimes similar to those before applied to Bi2212 [40] and composite Bi2212/Mg$_{1-x}$Cu$_x$O [73], processed by solid-state synthesis (designated as solid-state materials), no strong texture has been formed. We think that it is caused by the presence of large particles of Cu-free and Bi-free phases in sol-gel materials. Large particles initiate disturbances in the flow of the matrix Bi2212 phase and weaken locally texture (see Figure 22 d). For example, if in the solid-state Bi2212 and the composite Bi2212/Mg$_{1-x}$Cu$_x$O the mean particle size of the phases identified is about $7 - 10$ μm along and $10 - 15$ μm perpendicular the axis of compression then in the sol-gel materials their sizes being much larger are about $14 - 16$ and $30 - 40$ μm, respectively. The reasons responsible for formation of such coarse particles in sol-gel materials are not considered herein being a subject of another study.

Note another rather unexpected result, namely, the superconducting properties of sol-gel composite Bi2212/Sr$_2$CaWO$_6$ are weaker than those of sol-gel undoped Bi2212. It is difficult to clearly specify the reason of it. There may be two reasons: (i) tungsten having dissolved partially in phase Bi2212 worsens its superconducting properties and (ii) the regime of post-deformation annealing (850°C, 96 h) is not optimal.

As we have noted earlier (see Figures 13 and 20) [40, 73], superconducting properties of solid-state Bi2212 and the composite Bi2212/Mg$_{1-x}$Cu$_x$O can be explained on the basis of supposition that in dependence on deformation temperature the various types of pinning centers can operate. By reasoning from the data of microstructure, texture and phase composition let us determine defects responsible for typical dependences $J_c(T_d)$ for sol-gel undoped Bi2212 and composite Bi2212/Sr$_2$CaWO$_6$.

5.3.1. Sol-Gel Undoped Bi2212

Though methods of X-ray diffraction, light microscopy and SEM did not reveal any increase in the volume fraction of secondary phases above 895°C, nevertheless assume that in the range above this temperature there occurs decomposition of Bi2212 phase with precipitation of fine particles of non-superconducting phases. The noticeable change in V_{ns} is observed only at much higher deformation temperatures ($T_d = 930$°C, $\alpha = 90°$, $V_{ns} = 25\%$) and/or at essentially high strain values ($T_d = 905$°C, $\alpha = 180°$, $V_{ns} = 30\%$) [40]. The assumption on precipitation of very fine particles fits well with the fact that above 895°C the growth of Bi2212 colonies is retarded. It is known that fine particles retard efficiently the migration of internal interfaces [74]. That is why for estimating the critical current density of

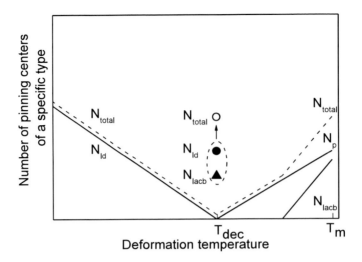

Figure 27. Scheme illustrating the dependence of the number of pinning centers of a specific type on deformation temperature of sol-gel undoped ceramics Bi2212; N_{ld} is the number of pinning centers in the system of the intracolonial lattice defects, N_p is the number of pinning centers in the system of the particles of nonsuperconducting phases formed during decomposition of the Bi2212 phase near the melting point, N_{lacb} is the number of pinning centers in the system of the low-angle colony boundaries, T_{dec} and T_m is the temperature of decomposition onset and macroscopic incongruent melting of Bi2212 phase, respectively, points denoted N_{lacb} (full triangle) and N_p (full circle) at T_{dec} correspond to number of pinning centers of an appropriate type for sample No 98, N_{total} (dashed line and open circle at T_{dec}) denotes the total number of pinning centers.

samples deformed above 895°C, it is necessary to take into account the contribution of fine particles of secondary phases forming during decomposition of the Bi2212 phase to pinning of magnetic flux.

So, depending on microstructure three types of defects can operate as centers of flux pinning in sol-gel Bi2212. These defect types are: (i) intracolonial defects, (ii) particles precipitating from decomposition of phase Bi2212, and (iii) low-angle colony boundaries. The contribution of low-angle boundaries to flux pinning can be essential only if $F \geq 0.94$ [40, 73]. In the majority of sol-gel Bi2212 samples (except samples 97 and 98) $F < 0.94$, that is the amount of low-angle boundaries is small and they do not make any contribution to flux pinning [40]. Figure 27 shows the scheme of pinning which explains the intragranular $J_c(T_d)$ dependence in sol-gel Bi2212. The whole temperature range of deformation is divided conventionally into two regions: low temperature and high temperature. The boundary between two regions is at the temperature of the onset of Bi2212 phase decomposition (T_{dec}), namely, at 895°C. At low deformation temperatures (up to T_{dec}) the level of texture is low ($F < 0.94$), that is why only intracolonial lattice defects make main contribution to flux pinning (line N_{ld}). With increasing deformation temperature the density of lattice defects reduces, that is why line N_{ld} decreases. Above T_{dec} decomposition of Bi2212 phase begins and particles of non-superconducting phases start to contribute to flux pinning (line N_p). With increasing deformation temperature line N_p grows since the number of particles increases. Besides, above T_{dec} the texture becomes stronger ($F = 0.94$), that is why at 915°C both the particles of secondary phases and the low-angle colony boundaries make contribution to flux pinning (line N_{lacb} grows).

The increase in pressure by twice at $T_d = 895°C$ (sample No 98) makes the texture stronger ($F = 0.94$) and there occurs contribution from low-angle boundaries (point N_{lacb} at T_{dec}). Moreover, the colony sizes $d_\parallel$ and $d_\perp$ decrease by $15 - 17\%$ (see Table 6). But the decrease in the colony size occurs due to formation of transverse subgrain boundaries which, in turn, are formed due to activation of intracolonial dislocation slip [40]. That is why the increase in pressure leads to increase in the density of intracolonial lattice defects, too (point N_{ld} at T_{dec}).

Let us explain dependences of inter- and intragranular $J_c(T_d)$ basing on the analysis of operating centers of flux pinning. The dependencies intraganular $J_c(T_d)$ and $\langle E \rangle(T_d)$ at 4.2 K and 30 K are clearly correlated throughout the temperature range of deformation. It testifies that dependences of intragranular $J_c(T_d)$ are defined by the appropriate centers of flux pinning.

Intergranular $J_c(T_d)$ does not correlates with the dependence $\langle E \rangle(T_d)$ up to $T_d = 895°C$. This can be explained as follows. Up to 895°C the texture is weak ($F < 0.94$), therefore the majority of intercolony boundaries are weak links. Obviously, in this case the intracolonial pinning centers do not affect the intergranular current density. Therefore, in the temperature range $T_d = 815 - 895°C$ intergranular current is almost independent of deformation temperature. Above $T_d = 895°C$ due to increased texture the number of weak links is reduced, therefore the intergranular J_c begins to depend on intracolonial flux pinning, and intergranular J_c increases.

Note one more interesting result. At 4.2 K the values of intragranular J_c and $\langle E \rangle$ in the samples deformed above 855°C (except the sample with $T_d = 915°C$) are less than in the undeformed one. At 30 K this effect is not observed, namely, superconducting properties of all deformed samples are higher than those of undeformed ones. A similar decrease in current carrying capacity of hot-deformed ceramics was observed by Dou et al [20]. The authors [20] compared superconducting properties of three Ag/Bi2223 tapes processed by different methods: (i) undeformed (rolled samples), (ii) deformed at room temperature (cold-pressed), and (iii) subjected to hot deformation at 800°C and pressure 15 MPa (hot-pressed). It has been established that the normalized pining force density for the cold-pressed tape is higher than for the as-rolled tape, the peak of which is in turn higher than that for the hot-pressed tape, consistent with the $J_c - H$ and irreversibility line measurement results. In our case the relatively high superconducting properties at 4.2 K of the undeformed sample of Bi2212 can be explained by the following. It is well known that after high-temperature deformation (especially after superplastic deformation) the density of intragranular defects is very low [75-77]. Two circumstances are responsible for this. Firstly, at high temperatures recovery processes are initiated within the grain interior. Secondly, at high temperatures the diffusion accelerates and the internal interfaces become active sinks for lattice defects. That is why low superconducting properties of the samples deformed in the temperature range $855 - 905°C$ may be explained by the fact that high-temperature deformation decreases the amount of intracolonial defects existing previously in the undeformed sample. The fact that at 30 K this effect is not observed testifies that in the undeformed sample there are mainly such defects which pining energy rapidly decreases with increasing temperature. It is evidently point defects and microcracks which were introduced into the material in the process of preparation of sol-gel powder and cold pressing prior to sintering.

5.3.2. Sol-Gel Composite Bi2212/Sr_2CaWO_6

Attention is drawn to two unusual results. Firstly, the superconducting properties of the composites are lower than undoped Bi2212. Secondly, unlike the undoped Bi2212 the mean effective activation energy of the composite weakly depends on T_d up to 895°C and almost corresponds to the $\langle E \rangle$ value of the samples of undoped Bi2212 with low density of intracolonial defects ($T_d = 865 - 905$°C). It is obvious that the density of intracolonial defects in the composite, as well as in the undoped Bi2212, reduces noticeably with increasing deformation temperature from 815 to 895°C. However, unlike the Bi2212 the decrease in $\langle E \rangle$ with increasing T_d is not observed in the composite, i.e. intracolonial defects play a less role in formation of properties of the composite than in the undoped Bi2212. Evidently, in the composite the regions of weakened superconductivity are extended rather far from interphase boundaries Bi2212/Sr_2CaWO_6, that is why both the particles Sr_2CaWO_6 and their trapped lattice dislocations do not exert a noticeable contribution to the flux pinning.

Low flux pinning energy and weak texture are responsible for weak dependence of intra- and intergranular currents on deformation temperature up to 895°C. The growth of J_c and $\langle E \rangle$ above $T_{dec} = 895$°C is connected with increasing both the density of particles occurring from decomposition of phase Bi2212 and number of low-angle boundaries ($F > 0.94$).

5.4. Conclusions

[1] In the sol-gel undoped Bi2212 and composite Bi2212/Sr_2CaWO_6 the volume fraction of Bi2212 phase does not undergo significant changes during torsion under pressure in the temperature range 815 – 915°C. At the same time relatively not strong basal plane texture ($F \leq 0.95$) is formed. In the process of deformation Bi2212 phase colonies preserve their plate-like shape. Their size grows until 905°C, and above this temperature it decreases. The decrease in the colony size above 905°C is most likely connected with the dragging of Bi2212 colony boundaries by fine particles of non-superconducting phases formed during decomposition of Bi2212 phase near the melting point.

[2] Undeformed samples of both materials contain coarse particles of Bi-free and Cu-free phases. Deformation doesn't influence almost their size and amount. The sizes of these particles are almost similar in both materials and in average they are about 14 – 16 μm along and 30-40 μm perpendicular to the compression axis. Such coarse particles initiate disturbances in the flow of matrix Bi2212 phase and prevent formation of strong texture.

[3] Particles Sr_2CaWO_6 have a relatively high tendency to coarsening at the stage of powder processing and sintering. After sintering at 855°C for 24 h the mean size of separate Sr_2CaWO_6 particles is about 0.30 μm. With increasing deformation temperature from 815°C to 915°C the size of separate Sr_2CaWO_6 particles grows from 0.34 to 0.60 μm.

[4] The sol-gel composite Bi2212/Sr_2CaWO_6 has lower superconducting properties than the sol-gel undoped Bi2212. This may be due to two reasons: (i) tungsten dissolved partially in phase Bi2212 decreases its superconducting properties, (ii) the regime of post-deformation annealing (850°C, 96 h) is no optimal.

[5] The dependence of superconducting properties of sol-gel undoped Bi2212 and composite $Bi2212/Sr_2CaWO_6$ on deformation temperature can be explained consistently by the scheme of flux pinning proposed earlier for explanation of superconducting properties of solid-state Bi2212 and composite $Bi2212/Mg_{1-x}Cu_xO$.

CONCLUSION

For the first time a systematic investigation of Bi2212 ceramics deformed by torsion under quasi-hydrostatic pressure has been carried out. The investigation included the evolution of microstructure, texture and phase composition with deformation temperature, load, twist angle and twist rate. It has been established that the method applied may essentially improve the superconducting properties of the material. Important advantages of the method of torsion under pressure are the ability to deform to high strains in the conditions of quasi-hydrostatic pressure. Applying low quasi-hydrostatic pressure (1 – 10 MPa) allows increase the onset temperature for incongruent melting of phase Bi2212 by 50 – 60°C. Bi2212-base materials exhibit a strongly non-monotonic dependence of the superconducting properties (J_c, B_{irr}, $\langle E \rangle$) on deformation temperature. In the undoped ceramic Bi2212 as well as in the composites $Bi2212/Mg_{1-x}Cu_xO$, $Bi2212/Sr_2CaWO_6$ the samples, deformed near melting temperature, have best superconducting properties. High properties are formed due to large volume fraction of low-angle colony boundaries and high density of nonsuperconducting particles formed during decomposition of Bi2212 phase near melting temperature. It was established that the superconducting properties of the composites are lower than the undoped Bi2212. It can be connected with two circumstances, namely: (i) partial deterioration of Bi2212 phase by the doped oxide, and (ii) non-optimal regime of the annealing after deformation applied to composites. The regime of annealing of the composites after deformation requires further optimization. The introduction of small particles of inert oxides (e.g. MgO) into initial charge in order to enhance flux pinning has not met expectations. This is due to the fact that the maximum texture is formed near the melting point, where the doped particles grow to large sizes and lose their effectiveness as pinning centers of magnetic flux. The results obtained can be used to produce rings with high superconducting properties, which will find use in various cryogenic electrical devices.

ACKNOWLEDGMENTS

This work was partly supported by the Austrian ÖAD (project I.15/2001), Russian Foundation for Basic Research (project 01-03-02003 BNTS_a) and Program for Basic Researches No 8 of OEMMPU RAS.

REFERENCES

[1] Miao, H.; Kitaguchi, H.; Kumakura, H.; et al. *Physica C* 1998, *vol. 303*, 81 - 90.
[2] Kumakura, H.; Togano, K.; Maeda, H.; et al. *Appl. Phys. Lett.* 1991, *vol. 58*, 2830 - 2832.

[3] Palstra, T.T.M.; Batlogg, B.; Schneemeyer, L.F.; et al. *Phys. Rev. Lett.* 1988, *vol. 61*, 1662 - 1665.

[4] Gammel, P.L.; Schneemeyer, L.F.; Waszczak, J.V.; et al. *Phys. Rev. Lett.* 1988, *vol. 61*, 1666 - 1669.

[5] Clem, J.R. *Phys. Rev.* 1991, *vol.43*, 7837 - 7846.

[6] Larbalestier, D.C. *IEEE Trans. on Appl. Supercond.* 1997, *vol.7*, 90 - 97.

[7] Metlushko, V.V.; Guntherodt, G.; Wagner, P.; et al. *Appl. Phys. Lett.* 1993, *vol 63*, 2821 - 2823.

[8] Pan, V.M.; Kasatkin, A.L.; Svetchnikov, V.L.; et al. *Cryogenics.* 1993, *vol. 33*, 21 - 27.

[9] Zhang, Y.; Mironova, M.; Lee, D.F.; et al. *Jpn. J. Appl. Phys.* 1995, *vol.34*, 3077 - 3081.

[10] Bagnall, K.E.; Grigorieva, I.V.; Steeds, J.W.; et al. *Supercond. Sci. Technol.* 1995, *vol. 8*, 605 - 612.

[11] Miller, D.J.; Sengupta, S.; Hettinger, J.D.; et al. *Appl. Phys. Lett.* 1992, *vol. 61*, 2823 - 2825.

[12] Chu, C.Y.; Routbort, J.L.; Chen, Nan; et al. *Supercond. Sci. Technol.* 1992, *vol. 5*, 306 - 312.

[13] Garnier, V.; Caillard, R.; Sotelo, A.; et al. *Physica C.* 1999, *vol. 319*, 197 – 208.

[14] Caillard, R.; Garnier, V.; Desgardin, G.; *Physica C.* 2000, *vol. 340*, 101 - 111.

[15] Bridgeman, P.W. *Studies in Large Plastic Flow and Fracture*; McGraw-Hill: New York, 1952; pp 1 – 362.

[16] Imayev, M.F.; Kabirova, D.B.; Korshunova, A.N.; et al. In: *Proc. of the 4th International Conference on Recrystallization and Related Phenomena.* The Japan Institute of Metals, Tokio, Japan, 1999, pp 899 – 903.

[17] Daminov, R.R.; Imayev, M.F.; Reissner, M.; et al. *Physica C.* 2004, *vol. 408-410*, 46 - 47.

[18] Imayev, M.F.; Daminov, R.R.; Popov, V.A.; et al. *Physica C.* 2005, *vol. 422*, 27-40.

[19] Majewski, P. *Appl. Supercond.* 1995, *vol. 3*, 289-301.

[20] Dou, S.X.; Wang, X.L.; Guo, Y.C.; et al. *Supercond. Sci. Technol.* 1997, *vol. 10*, A52-A67.

[21] Kazin, P.E.; Tretyakov Yu.D. *Russian Chemical Reviews.* 2003, *vol. 72*, 849-865.

[22] Jin, S.; Tiefel, T.; Sherwood, R.; at al. *Phys.Rev. B.* 1988, *vol.37*, 7850 - 7853.

[23] Lee, D.F.; Selvamanickam, V.; Salama, K. *Physica C.* 1992, *vol. 202*, 83-96.

[24] Chakrapani, V.; Balkin, D.; McGinn P. *Applied Superconductivity.* 1993, *vol. 1*, 71-80.

[25] Kazin, P.E.; Jansen, M.; Larrea, A.; et al. *Physica C.* 1995, *vol. 253*, 391-400.

[26] Huang, S.L.; Dew-Hughes, D.; Zheng, D.N.; et al. *Supercond. Sci. Technol.* 1996, *vol. 9*, 368-373.

[27] Pavard, S.; Villard, C.; Bourgault, D.; et al. *Supercond. Sci. Technol.* 1998, *vol. 11*, 1359-1366.

[28] Kazin, P.E.; Tretyakov, Y.D.; Lennikov V.V.; et al. *J. Mater. Chem.* 2001, *vol.11*, 168-172.

[29] Yamaguchi, K.; Murakami, M.; Fujimoto, H.; et.al. *Jpn. J. Appl. Phys.* 1990, *vol. 29*, L1428-L1431.

[30] Wang, R.K.; Ren, H.T.; Xiao, L.; et al. *Supercond. Sci. Technol.* 1990, *vol. 3*, pp 344-346.

[31] Martin, J.W. *Micromechanisms in Particle-Hardened Alloys*; Cambridge University Press: New York, 1980; pp 1 - 201.

[32] Rodgers, D.; White, K.; Selvamanickam, V.; et al. *Supercond. Sci. Technol.* 1992, *vol. 5*, 640-644.

[33] Zhang, Y.; Selvamanickam, V.; Lee, D.F.; et al. *Jpn. J. Appl. Phys.* 1994, *vol.33*, 3419-3423.

[34] Mironova, M.; Lee, D.F.; Selvamanickam, V.; et al. *Philosophical Magazine A.* 1995, *vol. 71*, 855-870.

[35] Vilalta, N.; Sandiumenge, F.; Rabier, J.; et al. *Philosophical Magazine A.* 1997, *vol. 76*, 837-855.

[36] Vilalta, N.; Sandiumenge, F.; Rodriguez, E.; et al. *Philosophical Magazine B.* 1997, *vol. 75*, 431-441.

[37] Ullrich, M.; Leenders, A.; Krelaus, J.; et al. *Materials Science and Engineering B.* 1998, *vol. 53*, 143-148.

[38] Pavard, S.; Bourgault, D.; Villard, C.; et al. *Physica C.* 1999, *vol. 316*, 198-204.

[39] Kazin, P.E.; Poltavets, V.V.; Tretyakov, Y.D.; et al. *Physica C.* 1997, *vol. 280*, 253-265.

[40] Imayev, M.F.; Daminov, R.R.; Reissner, M.; et al. *Physica C.* 2007, *vol. 467*, 14-26.

[41] Lotgering, F.K. *J. Inorg. Nucl. Chem.* 1959, *vol. 9*, 113 - 123.

[42] Saltykov, S. A. *Stereometric Metallography*; Metallurgiya: Moscow, 1970; pp 1 - 375 (in Russian).

[43] Bean, C.P. *Phys. Rev. Lett.* 1962, *vol. 8*, 250 - 253.

[44] Anderson, P.W.; Kim, Y.B. *Rev. Mod. Phys.* 1964, *vol.36*, 39 - 43.

[45] Kazin, P.E.; Os'kina, T.E.; Tretyakov, Yu.D. *Appl. Supercond.* 1993, *vol. 1*, 1007 - 1013.

[46] Liu, H.; Liu, L.; Zhang, Y.; et al. *J. of Mat. Sci.* 1999, *vol. 34*, 6099-6105.

[47] Riley, G.N.; Malozemoff, A.P.; Li, Q.; et al. *JOM.* 1997, No. 11, 24-27 and 60.

[48] Gubernatorov, V.V.; Sokolov, B.K.; Vladimirov, L.R.; et al. *Rus. Doklady Physics.* 1999, vol. 44, 75 - 77.

[49] Poirier, J.P. *Plasticité á Haute Température des Solides Cristallins*; Eyrolles: Paris, 1976; pp 1 - 320.

[50] Song, C.; Liu, F.; Gu, H.; et al. *J. of Mater. Sci.* 1991, *vol. 26*, 11 - 16.

[51] von Mises R. *Z. Angew. Math. Mech.* 1928, *vol. 8*, 161 - 185.

[52] Wassermann, G.; Grewen, J. *Texturen Metallischer Werkstoffe*; Springer Verlag: Berlin, 1962; pp 1-654.

[53] Imayev, M.F.; Kaibyshev, R.O.; Musin, F.F.; et al. *Mater. Sci. Forum.* 1994, *vol. 170-172*, 445 - 451.

[54] Kondo, N.; Sato, E.; Wakai, F. *J. Am. Ceram. Soc.* 1998, *vol. 81*, 3221 - 3227.

[55] Zhengping, Xi; Lian, Zhou. *Supercond. Sci. Technol.* 1994, *vol. 7*, 908 - 912.

[56] Cahn, W.; Hillig, W.B.; Sears, G.W. *Acta Metall.* 1964, *vol. 12*, 1421 - 1437.

[57] Imayev, M.F.; Kazakova, D.B.; Gavro, A.N.; et al. *Physica C.* 2000, *vol. 329*, 75 - 87.

[58] Andersen, L. G.; Poulsen, H.F.; Abrahamsen, A.B.; et al. *Supercond. Sci. Technol.* 2002, *vol. 15*, 190 - 201.

[59] Demianczuk, D.W.; Aust, K.T. *Acta Metall.* 1975, *vol. 23*, 1149 - 1162.

[60] Fridman, E.M.; Kopezkii, C.V.; Shvindlerman, L.S. *Z. Metallkd.* 1975, *vol. 66*, 533 - 539.

[61] Trcka, M.; Reissner, M.; Varahram, H.; Steiner, W.; Hauser, H. *Physica C.* 2000, *vol. 341-348, Pt 3*, 1487-1488.

[62] Nakamura, N.; Gu, G.D.; Takamuku, K.; et al. *Appl. Phys. Lett.* 1992, *vol. 61*, 3044 - 3046.

[63] Diaz, A.; Mechin, L.; Berghuis, P.; et al. *Phys. Rev. Lett.* 1998, *vol. 80*, 3855 - 3858.

[64] Makarova, M.V.; Kazin, P.E.; Tretyakov, Yu. D.; et al. *Physica C.* 2005, *vol. 419*, 61-69.

[65] Girifalco, L.A.; Welch, D.O. *Point Defects and Diffusion in Strained Metals*; Gordon and Breach: New York, 1967; pp 1-312.

[66] Kraevski, A. Yu.; Ovid'ko, I. A. *Rus. Physics of the Solid State.* 2000, *vol. 42*, 1218-1221.

[67] Mao, Z.; Wang, H.; Dong, Y.; et al. *Physica C.* 1990, *vol. 170*, 35-40.

[68] Abram'an, P.B.; Avag'an, A.A.; Gevorg'an, S.G.; at al. *Rus. Sverkhprovodimost: Phyz. Chim. Tech.* 1990, *No. 3*, 698-707.

[69] Zhang, H.; Wu, K.; Feng, Q.R.; et al. *Physics Letters A.* 1992, *vol.169*, 214-218.

[70] Tatsumisago, M.; Inoue, S.; Tohge, N.; et al. *J. Mat. Sci.* 1993, *vol.28*, 4193-4196.

[71] Han, S.H.; Cheng, C.H.; Dai, Y.; et al. *Supercond. Sci. Technol.* 2002, *vol. 15*, 1725-1727.

[72] Campbell, A.M.; Evetts J.E. *Critical Currents in Superconductors*; Taylor and Francis Ltd: London, 1972; pp 1-243.

[73] Reissner, M.; Daminov, R.R.; Imayev M.F.; et al. In *Proc. of the 6th European Conference on Applied Superconductivity* (EUCAS-2003). 2003, pp 2195-2201.

[74] Smith, C.S. *Trans. AIME.* 1948, *vol. 175*, 15 - 51.

[75] Ball, A.; Hutchison, M.M. *Met. Sci.* 1969, *vol. 3*, 1-7.

[76] Lee, E.U.; Underwood, E.E. *Met. Trans.* 1970, *vol. 1*, 1399-1406.

[77] Lee, D. *Met. Trans.* 1970, *vol. 1*, 309-312.

In: Ferroelectrics and Superconductors
Editor: Ivan A. Parinov

ISBN: 978-1-61324-518-7
©2012 Nova Science Publishers, Inc.

Chapter 2

ORIENTATION EFFECTS IN NOVEL COMPOSITES BASED ON RELAXOR-FERROELECTRIC SINGLE CRYSTALS

V. Yu. Topolov[,1], A. V. Krivoruchko[≠1, 2], P. Bisegna[‡3] and S. V. Glushanin[£4]*

[1]Department of Physics, Southern Federal University, Rostov-on-Don, Russia
[2]Don State Technical University, Rostov-on-Don, Russia
[3]Department of Civil Engineering, University of Rome "Tor Vergata", Rome, Italy
[4]Scientific Design & Technology Institute "Piezopribor",
Southern Federal University, Rostov-on-Don, Russia

ABSTRACT

Single crystals of perovskite-type relaxor-ferroelectric solid solutions $(1 - x)Pb(A_{1/3}Nb_{2/3})O_3 - xPbTiO_3$ (A = Mg or Zn) with so-called "engineered domain structures" and compositions near the morphotropic phase boundary exhibit outstanding electromechanical properties at room temperature. In the last decade, these single crystals have often been used to manufacture advanced piezo-active composites. The high performance of these modern composites consists in high piezoelectric activity, sensitivity, considerable hydrostatic parameters, etc. The present chapter is devoted to the analysis of interconnections between the electromechanical properties of the single-crystal component and the single crystal/polymer composite at various volume fractions of the components therein. Results on the performance of the single crystal/polymer composites are discussed for the $2 - 2$ and $1 - 3$ composites with the regular structure and specific orientations of the main crystallographic axes of the single-crystal component. The emphasis is placed on the effect of the orientation of the main crystallographic axes and the anisotropy of electromechanical properties of the single-crystal components on the piezoelectric coefficients, electromechanical coupling factors and figures of merit of

---

[*]E-mail: vutopolov@sfedu.ru.
[≠]E-mail: kolandr@yandex.ru.
[‡]E-mail: bisegna@uniroma2.it.
[£]E-mail: glushanin@inbox.ru.

the composites. Examples of the performance of the composites based on the single-domain and polydomain single crystals with the regular arrangement of components are reported and compared. Extremum values of the effective parameters of the composite at the variable orientation of the main crystallographic axes of the single-crystal component are compared to the related parameters of the single-crystal component, and important advantages of the composite are described in this context. The role of the anisotropy of the electromechanical properties, poling direction, microgeometry, and other factors in forming the considerable hydrostatic piezoelectric response, piezoelectric anisotropy of the composite and its electromechanical coupling factors is discussed. The transition from $1 - 3$ to $0 - 3$ connectivity and related changes in the effective parameters are considered. Advantages and high-performance potential of the composites based on single crystals of $(1 - x)Pb(A_{1/3}Nb_{2/3})O_3 - xPbTiO_3$ are of value for modern piezotechnical applications, including active elements of sensors, actuators, transducers, hydrophones, acoustic antennae, etc.

1. INTRODUCTION

The interest in advanced composites based on relaxor-ferroelectric single crystals (SCs) of $(1 - x)Pb(Mg_{1/3}Nb_{2/3})O_3 - xPbTiO_3$ (PMN $- xPT$) or $(1 - x)Pb(Zn_{1/3}Nb_{2/3})O_3 - xPbTiO_3$ (PZN $- xPT$) with the perovskite-type structure is mainly concerned with remarkable piezoelectric and dielectric properties of these SCs close to the morphotropic phase boundary [1–5]. The aforementioned solid solutions combine the physical properties of the relaxor $[Pb(Mg_{1/3}Nb_{2/3})O_3$ or $Pb(Zn_{1/3}Nb_{2/3})O_3]$ and "normal" ferroelectric $(PbTiO_3)$ components, and the proximity of the morphotropic phase boundary becomes the important factor that strongly influences the behavior of these systems in the wide temperature range, under external electric and mechanical fields, under certain poling conditions, etc. The outstanding characteristic of the PMN $- xPT$ and PZN $- xPT$ SCs with engineered domain structures is a very high piezoelectric activity: for instance, the longitudinal piezoelectric coefficient of SCs $d_{33} \sim (10^2 - 10^3)$ pC/N [1, 2, 4–7], while in ferroelectric ceramics the typical value is $d_{33} \sim (10 - 10^2)$ pC/N [8]. To achieve high piezoelectric activity of SC samples, the relaxor-ferroelectric solid solutions are engineered by compositional adjustment with a corresponding decrease in Curie temperature, and this method enables one to attain extreme values of some parameters in the relatively easy way. The presence of various heterophase and engineered domain structures [1, 3–6] considerably influences the electromechanical (i.e., piezoelectric, dielectric and elastic) properties, electromechanical coupling of the SCs and the anisotropy of their properties. Numerous experimental data suggest that the morphotropic phase boundary at room temperature is located near the molar concentration $x = 0.30 - 0.33$ in PMN $- xPT$ [9–11] and $x = 0.08 - 0.10$ in PZN $- xPT$ [12]. The performance of the PMN $- xPT$ and PZN $- xPT$ SCs has been studied in the single-domain [13–15] and polydomain [7, 14] states and at certain poling directions of the SCs [16, 17]. Examples of the orientation dependence of the electromechanical properties of the single-domain PMN $- 0.33PT$ [18, 19], PZN $- 0.045PT$ [6] and PZN $- 0.09PT$ [20] SCs show that the high performance at the specific orientation of the crystallographic axes represents an advantage over the performance of the conventional poled ferroelectric ceramics [8].

In the last decade intensive studies of the novel composites based on the aforementioned SCs were carried out to obtain important information on the effective properties and related parameters of the composites, their optimum microgeometric characteristics, volume fractions of components, and so on. One can mention the following composites studied using different experimental and theoretical methods: PZN – 0.08PT/polymer [21], PMN – 0.33PT/epoxy [22, 23], PMN – 0.30PT/epoxy [24] (1 – 3 connectivity[1]), PMN – xPT/copolymer of vinylidene fluoride – trifluoroethylene [25] at x = 0.30, 0.33 and 0.42 (0 – 3 connectivity), PZN – 0.07PT/polymer [26], PMN – 0.33PT/polyvinylidene fluoride [27], PMN – 0.29PT/polyvinylidene fluoride [28] (2 – 2 connectivity), etc. Moreover, changes in the electromechanical properties and their anisotropy in the novel composites were studied in connection with changes in the domain structure or the poling direction of the PMN – xPT and PZN – xPT SCs [26 – 28]. The aim of the present chapter is to give a systematic description of the orientation effects in the two-component composites based on either PMN – xPT or PZN – xPT and to show advantages that are caused by the orientation effects studied and are important for modern piezotechnical applications. Below we consider a series of examples of the SC/polymer composites wherein the orientation effects play the important role and lead to large extreme values of the parameter series. In this context we analyze physical and crystallographic factors that promote the high performance of the novel composites.

2. ORIENTATION EFFECTS AT THE ROTATION OF THE SPONTANEOUS POLARIZATION VECTOR IN THE SINGLE-CRYSTAL COMPONENT

In this section we discuss the performance of the 2 – 2 SC/polymer composites wherein the SC component is either single-domain or polydomain. Below we demonstrate how the PMN – xPT and PZN – xPT SCs improve this performance and provide a series of extreme points of important effective parameters of the composites. The main emphasis is made on the volume-fraction and orientation dependences of the effective parameters that are of practical interest.

2.1. 2 – 2 PZN – 0.12PT / Polymer Composite and Its Performance

As is known from experimental data, there are complete sets of room-temperature electromechanical constants (Table 1) measured on the single-domain PMN – 0.42PT [15] and PZN – 0.12PT [14] SCs with $4mm$ symmetry. Both these compositions are located in the vast tetragonal region of the (x, T) diagrams [9 – 12] of the studied solid solutions. We see the difference between the piezoelectric coefficients $d_{ij}^{(1)}$ of PMN – 0.42PT and PZN – 0.12PT SCs, for which the order-of-magnitude of $d_{ij}^{(1)}$ remains 10^2 pC/N (Table 1).[2] The single-domain PZN – 0.12PT SC exhibits a more pronounced elastic anisotropy in comparison to that of the single-domain PMN – 0.42PT SC. According to data shown in Table 1, ratios

$$s_{33}^{(1),E} / s_{11}^{(1),E} = 2.71; \ s_{11}^{(1),E} / |s_{13}^{(1),E}| = 1.10 \tag{1}$$

---

[1]Connectivity indices $\alpha - \beta$ and $\alpha - \beta - \gamma$ are written in terms of papers [30, 31].
[2]Hereafter we use superscript "(1)" to denote electromechanical constants of the SC component.

Table 1. Elastic compliances $s_{ab}^{(n),E}$ (in 10^{-12} Pa^{-1}), piezoelectric coefficients $d_{ij}^{(n)}$ (in pC/N)

and relative dielectric permittivity $\varepsilon_{pp}^{(n),\sigma}/\varepsilon_0$ of single-domain SC and polymer components at room temperature

Components	$s_{11}^{(n),E}$	$s_{12}^{(n),E}$	$s_{13}^{(n),E}$	$s_{33}^{(n),E}$	$s_{44}^{(n),E}$	$s_{66}^{(n),E}$
PZN – 0.12PT SC [14]	20.1	– 4.6	– 18.2	54.5	19.5	17.2
PMN – 0.42PT SC [15]	9.43	– 1.68	– 6.13	19.21	35.09	12.5
Polyurethane [29]	400	– 148	– 148	400	1100	1100
Components	$d_{31}^{(n)}$	$d_{33}^{(n)}$	$d_{15}^{(n)}$	$\varepsilon_{11}^{(n),\sigma}/\varepsilon_0$	$\varepsilon_{33}^{(n),\sigma}/\varepsilon_0$	
PZN – 0.12PT SC [14]	– 207	541	653	10000	750	
PMN – 0.42PT SC [15]	– 91	260	131	8627	660	
Polyurethane [29]	0	0	0	3.5	3.5	

are valid for the single-domain PZN – 0.12PT SC at room temperature. At the same time, the above-mentioned single-domain SCs are characterized by the almost equal anisotropy of the piezoelectric coefficients $d_{3j}^{(1)}$ as follows from Table 1,

$$d_{33}^{(1)}/|d_{31}^{(1)}| = 2.61 \ (\text{PZN} - 0.12\text{PT}); \ d_{33}^{(1)}/|d_{31}^{(1)}| = 2.86 \ (\text{PMN} - 0.42\text{PT}). \tag{2}$$

The distinctions shown in Equations (1) can influence the piezoelectric and hydrostatic parameters and the anisotropy of the electromechanical properties of the composite based on the single-domain SC, however such an effect was not yet studied in detail. An example of the composite, wherein the orientation effect and the anisotropy of the electromechanical properties of the SC component play the important role, is considered below.

2.1.1. Model Concepts and Averaging Procedure

It is assumed that the composite represents a system of the parallel-connected SC and polymer layers which form the regular laminar structure (Figure 1) with 2 – 2 connectivity. The crystallographic axes X, Y, and Z of the tetragonal single-domain SC in the initial state ($\alpha = 0$ or $\beta = 0$, see insets 1 and 2 in Figure 1) are parallel to the following perovskite unit-cell directions: $X \parallel [100] \parallel OX_1$, $Y \parallel [010] \parallel OX_2$ and $Z \parallel [001] \parallel OX_3$. In this case the spontaneous polarization vector in each SC layer is $P_s^{(1)} \parallel OX_3$. We consider rotations of the $P_s^{(1)}$ vector around one of the co-ordinate axes – either OX_1 or OX_2 (see insets 1 and 2 in Figure 1), on assumption that all the SC layers in the composite sample have the same orientation of $P_s^{(1)}$. The system of SC cuts with a fixed orientation of the crystallographic axes X, Y, and Z is to be prepared before manufacturing the 2 – 2 composite sample. The single-domain state of the SC layers (cuts) in the sample can be stabilized under bias [13, 15]. In each polydomain SC layer, the rotation of the effective spontaneous polarization vector $P_s^{(1)*} = m_d P_{s,1} + (1 - m_d) P_{s,2}$ in the $(X_2 O X_3)$ plane is caused by changes in the volume fraction m_d of the laminar 90° domains (inset 3 in Figure 1), and these changes depend on an external electric field that is preliminary applied to the SC layers.

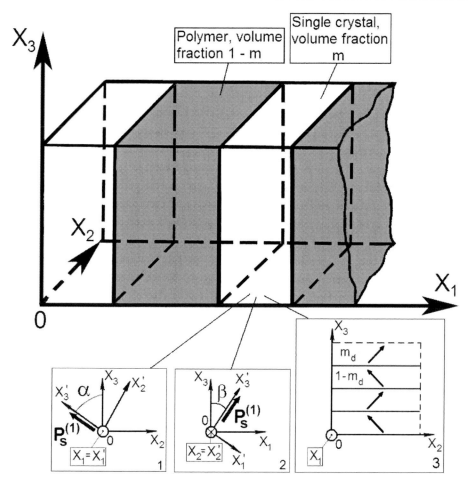

Figure 1. Schematic of the 2 – 2 parallel-connected SC/polymer composite. $(X_1X_2X_3)$ is the rectangular co-ordinate system of the composite sample, α and β are angles of rotation of the spontaneous polarization vector $\boldsymbol{P}_s^{(1)} \parallel OX_3'$ of the single-domain SC layer around the OX_1 axis (inset 1) or around the OX_2 axis (inset 2). In inset 3 the 90° domain structure in the SC layer is schematically shown, where m_d and $1 - m_d$ are volume fractions of the 90° domains in the SC layer, and their spontaneous polarization vectors are shown with arrows.

The effective electromechanical properties of the 2 – 2 composite are studied within the framework of the matrix approach [31, 32] that is applied to piezo-active composite materials with planar microgeometry. The 9 × 9 matrix of the effective properties of the composite

$$\parallel C^* \parallel = \begin{pmatrix} \parallel s^{*E} \parallel & \parallel d^* \parallel^T \\ \parallel d^* \parallel & \parallel \varepsilon^{*\sigma} \parallel \end{pmatrix} \qquad (3)$$

is written in the rectangular co-ordinate system $(X_1X_2X_3)$ in terms of $\parallel s^{*E} \parallel$ (6 × 6 matrix of elastic compliances at constant electric field), $\parallel d^* \parallel$ (3 × 6 matrix of piezoelectric charge coefficients) and $\parallel \varepsilon^{*\sigma} \parallel$ (3 × 3 matrix of dielectric permittivities measured at constant stress), where superscript "T" denotes the transposed matrix. The $\parallel C^* \parallel$ matrix in Equation (3) is determined from averaging the electromechanical properties of components on the volume fraction m and given by

$$\| C^* \| = [\| C^{(1)} \| \cdot \| M \| m + \| C^{(2)} \| (1 - m)] \cdot [\| M \| m + \| I \| (1 - m)]^{-1}, \tag{4}$$

where $\| C^{(1)} \|$ and $\| C^{(2)} \|$ are matrices of the electromechanical properties of SC and polymer, respectively, $\| M \|$ is concerned with the electric and mechanical boundary conditions [11] at interfaces $x_1 = \text{const}$ (Figure 1), and $\| I \|$ is the identity 9×9 matrix. The $\| C^{(n)} \|$ matrices from Equation (4) have a structure similar to that shown in Equation (3).

Elements of the $\| C^{(1)} \|$ matrix are written taking into account the orientation of $\boldsymbol{P}_s^{(1)}$ in the single-domain SC layer (insets 1 and 2 in Figure 1) or the volume fraction m_d of the 90° domains (inset 3 in Figure 1). For example, in case of the single-domain SC layers, the elements of $\| C^{(1)} \|$ are represented in the tensor form as $\varepsilon_{np}^{(1),\sigma} = r_{nk} r_{pl} (\varepsilon_{kl}^{(1),\sigma})_T$ (dielectric properties, 2nd rank), $d_{ifg}^{(1)} = r_{it} r_{fu} r_{gv} (d_{tuv}^{(1)})_T$ (piezoelectric properties, 3rd rank) and $s_{pqvw}^{(1),E} = r_{pc} r_{ql} r_{vh} r_{wn} (s_{clhn}^{(1),E})_T$ (elastic properties, 4th rank), where r_{nk} is the element of the matrix that describes the rotation of the $\boldsymbol{P}_s^{(1)}$ vector and the co-ordinate axes, and subscript "T" means that the electromechanical constant is given in the main crystallographic axes, i.e., taken directly from Table 1. The above mentioned rotation is shown in inset 1 in Figure 1 [then we have in the general form $r_{nk} = r_{nk}(\alpha)$] and in inset 2 in Figure 1 [then we have $r_{nk} = r_{nk}(\beta)$].

The $\| M \|$ matrix from Equation (4) is concerned [31] with the boundary conditions at the layer interfaces and represents as follows: $\| M \| = \| W_1 \|^{-1} \cdot \| W_2 \|$, where

$$\|W_n\| = \begin{pmatrix}
1 & 0 & 0 & 0 & 0 & 0 & 0 & 0 & 0 \\
s_{12}^{(n),E} & s_{22}^{(n),E} & s_{23}^{(n),E} & s_{24}^{(n),E} & s_{25}^{(n),E} & s_{26}^{(n),E} & d_{12}^{(n)} & d_{22}^{(n)} & d_{32}^{(n)} \\
s_{13}^{(n),E} & s_{23}^{(n),E} & s_{33}^{(n),E} & s_{34}^{(n),E} & s_{35}^{(n),E} & s_{36}^{(n),E} & d_{13}^{(n)} & d_{23}^{(n)} & d_{33}^{(n)} \\
s_{14}^{(n),E} & s_{24}^{(n),E} & s_{34}^{(n),E} & s_{44}^{(n),E} & s_{45}^{(n),E} & s_{46}^{(n),E} & d_{14}^{(n)} & d_{24}^{(n)} & d_{34}^{(n)} \\
0 & 0 & 0 & 0 & 1 & 0 & 0 & 0 & 0 \\
0 & 0 & 0 & 0 & 0 & 1 & 0 & 0 & 0 \\
d_{11}^{(n)} & d_{12}^{(n)} & d_{13}^{(n)} & d_{14}^{(n)} & d_{15}^{(n)} & d_{16}^{(n)} & \varepsilon_{11}^{(n),\sigma} & \varepsilon_{12}^{(n),\sigma} & \varepsilon_{13}^{(n),\sigma} \\
0 & 0 & 0 & 0 & 0 & 0 & 0 & 1 & 0 \\
0 & 0 & 0 & 0 & 0 & 0 & 0 & 0 & 1
\end{pmatrix},$$

$n = 1$ is related to SC, and $n = 2$ is related to polymer. $\| C^{(1)} \|$ is a function of m and α (at rotation shown in inset 1 in Figure 1), m and β (at rotation shown in inset 2 in Figure 1), or m and m_d (for the polydomain SC layer shown in inset 3 in Figure 1). The effective electromechanical properties of the composite [see matrix $\| C^* \|$ from Equation (4)] are determined on an assumption that wavelengths of acoustic waves propagated are considerably longer than the thickness of each layer in the composite sample (Figure 1).

Based on matrix elements of $\| C^* \|$ from Equation (4), we study the volume-fraction and orientation dependences of the hydrostatic piezoelectric coefficients as

$$d_h^* = d_{31}^* + d_{32}^* + d_{33}^*; \; g_h^* = g_{31}^* + g_{32}^* + g_{33}^*; \; e_h^* = e_{31}^* + e_{32}^* + e_{33}^*, \tag{5}$$

and hydrostatic electromechanical coupling factor (ECF) of the composite as

$$k_h^* = d_h^* / \sqrt{s_h^{*E} \varepsilon_{33}^{*\sigma}}. \tag{6}$$

The piezoelectric coefficients g_{ij}^* from Equations (5) are determined from the matrix $\| g^* \| = \| d^* \| \| \varepsilon^{*\sigma} \|^{-1}$, the piezoelectric coefficients e_{ij}^* from Equations (5) are determined from the matrix $\| e^* \| = \| d^* \| \| s^{*E} \|^{-1}$ in agreement with conventional formulae [33] for a piezoelectric medium. Hydrostatic compliance s_h^{*E} of the composite at $E = $ const [see Equation (6)] is written in terms of elements of $\| s^{*E} \|$ as follows: $s_h^{*E} = \sum_{a=1}^{3} \sum_{b=1}^{3} s_{ab}^{*E}$. We add that the hydrostatic parameters Φ_h^* from Equations (5) and (6) characterize the performance of the 2 – 2 composite sample (Figure 1) with electrodes oriented parallel to the (X_1OX_2) plane.

2.1.2. Hydrostatic Parameters of the Composite Based on PZN – 0.12PT

The effective electromechanical properties of the polydomain SC layer (inset 3, Figure 1), the composite as a whole [see Equation (4)] and its hydrostatic parameters [see Equations (5) and (6)] are calculated using the full sets of experimental electromechanical constants from Table 1. Examples of the calculated volume-fraction and orientation dependences of the hydrostatic parameters Φ_h^* of the composite based on the single-domain PZN – 0.12PT SC are shown in Figures 2 and 3. The presence of the SC component with $4mm$ symmetry and the polymer component with ∞mm symmetry enables us to establish the periodic dependence of the hydrostatic parameters on the rotation angles α and β. For any value of m from the range $0 < m < 1$, the equalities $\Phi_h^*(m, \alpha) = \Phi_h^*(m, 180° - \alpha)$ and $\Phi_h^*(m, \beta) = \Phi_h^*(m, -\beta)$ hold, where $0° \leq \alpha < 90°$ and $0° \leq \beta < 90°$. These ranges are taken with regard for the tetragonal symmetry of the single-domain SCs [14, 15].

Changes in the orientation of the $P_s^{(1)}$ vector of the SC layer in the (X_2OX_3) plane (inset 1 in Figure 1) mean changes in projections of $P_s^{(1)}$ on the OX_2 and OX_3 axes, along which both the components of the composite are distributed continuously. Such a mode of rotation of the

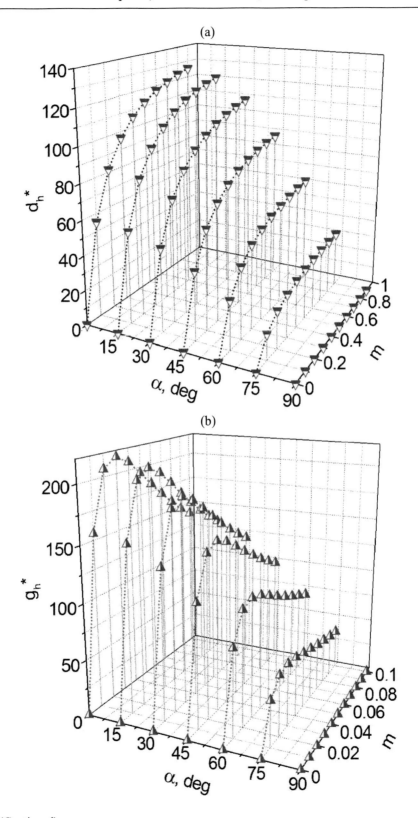

Figure 2 (Continued).

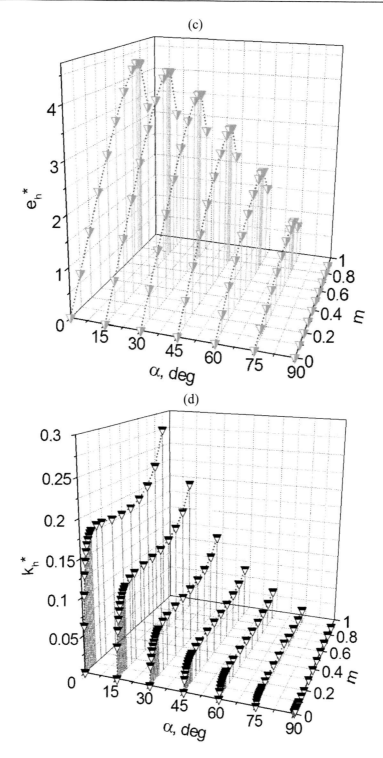

Figure 2. Hydrostatic piezoelectric coefficients (a – c) and ECF (d) of the PZN – 0.12PT SC/polyurethane composite at the rotation of the spontaneous polarization vector $P_s^{(1)}$ of the single-domain SC layer around the OX_1 axis (see inset 1 in Figure 1): (a) d_h^* (in pC/N), (b) g_h^* (in mV·m/N), (c) e_h^* (in C/m^2), and (d) k_h^*.

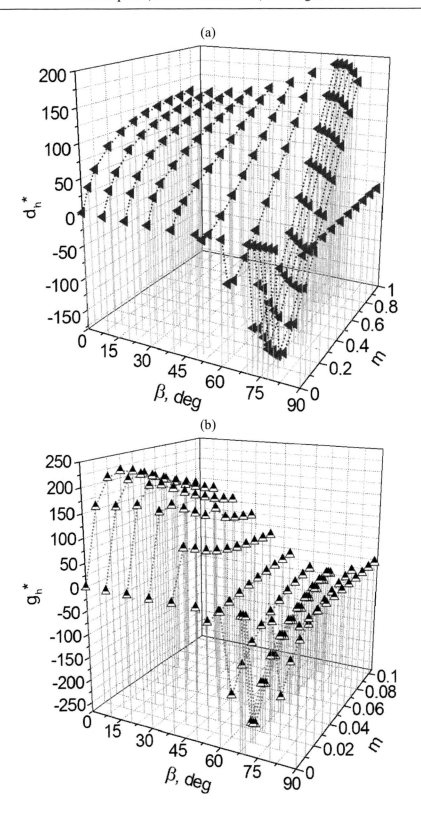

Figure 3 (Continued).

Orientation Effects in Novel Composites ... 55

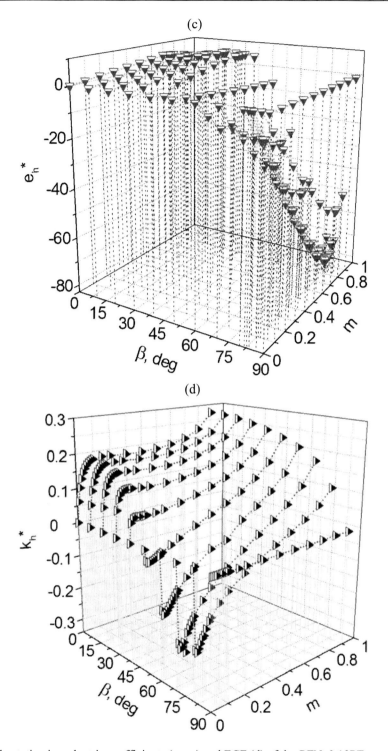

Figure 3. Hydrostatic piezoelectric coefficients (a – c) and ECF (d) of the PZN–0.12PT SC/polyurethane composite at the rotation of the spontaneous polarization vector $P_s^{(1)}$ of the single-domain SC layer around the OX_2 axis (see inset 2 in Figure 1): (a) d_h^* (in pC/N), (b) g_h^* (in mV·m/N), (c) e_h^* (in C/m^2), and (d) k_h^*.

$P_s^{(1)}$ vector leads to a fairly smooth dependence of the hydrostatic parameters Φ_h^* on α (Figure 2). It is seen that the hydrostatic piezoelectric coefficient d_h^* of the composite (Figure 2, a) remains less than $d_h^{(1)}$ of the single-domain SC. A combination of the piezoelectric (d_{ij}^*) and dielectric ($\varepsilon_{pp}^{*\sigma}$) properties gives rise to the pronounced maxima of g_h^* at $\alpha =$ const (Figure 2, b), however these maxima take place at the relatively small volume fraction m. In Figure 2, b we omit the volume-fraction range $0.1 < m < 1$ where g_h^* monotonically decreases at $\alpha =$ const. Combining the piezoelectric (d_{ij}^*) and elastic (s_{ab}^{*E}) properties, one can attain the non-monotonic volume-fraction behavior of the hydrostatic piezoelectric coefficient e_h^* (Figure 2, c), however values of e_h^* near the local maximum points are relatively small. The complicated shape of the surface of $k_h^*(m, \alpha)$ (Figure 2, d) is accounted for by the active influence of the dielectric properties at $0 < m < 0.1$ and of the elastic properties at $0.4 < m < 0.9$. As in the case of d_h^*, we see that $k_h^*(m, \alpha) < k_h^{(1)}$, i.e., there is a correlation between $k_h^*(m, \alpha)$ and $d_h^*(m, \alpha)$ in certain ranges of m and α, and this correlation stems from Euation (6).

The related composite based on the polydomain SC (see inset 3 in Figure 1) is characterized by the hydrostatic parameters varying in the fairly narrow ranges (Table 2). The polydomain SC layers have the fixed orientations and volume fractions (m_d and $1 - m_d$) of the $90°$ domains over the whole composite sample, and a change in m_d means a rotation of the effective spontaneous polarization vector $P_s^{(1)*}$ in the (X_2OX_3) plane. As in the aforementioned case of the single-domain SC layer, this rotation does not lead to considerable hydrostatic parameters. In our evaluations, the $90°$ domain walls of the polydomain SC layer are assumed to be motionless. The $90°$ domain-wall displacements [34] in the SC layer provide a contribution into electromechanical constants of SC. Our estimations show that this domain-wall contribution might give rise to increasing the upper bounds on Φ_h^* (Table 2) by about $1.5 - 2$ times, however here we did not consider this contribution in details.

Table 2. Lower and upper bounds on hydrostatic parameters $\Phi_h^*(m, m_d)$ of the PZN – 0.12PT SC/polyurethane composite with polydomain SC layers (see inset 3 in Figure 1) at $0 < m < 1$ and $m_d =$ const

Φ_h^*	$m_d = 0.1$ and $m_d = 0.9$	$m_d = 0.3$ and $m_d = 0.7$	$m_d = 0.5$
d_h^* (in pC/N)	$0 < d_h^* < 89.8$	$0 < d_h^* < 89.8$	$0 < d_h^* < 89.8$
g_h^* (in mV·m/N)	$0 < g_h^* < 10.6$	$0 < g_h^* < 4.68$	$0 < g_h^* < 1.89$
e_h^* (in C/m^2)	$-0.816 < e_h^* < 0.763$	$0 < e_h^* < 7.81$	$0 < e_h^* < 10.5$
k_h^*	$0 < k_h^* < 0.115$	$0 < k_h^* < 0.115$	$0 < k_h^* < 0.115$

Varying the angle β means the crossing of the spontaneous polarization vector $\boldsymbol{P}_s^{(1)}$ of the single-domain SC layer and the interface x_1 = const (see inset 2 in Figure 1). This mode of rotation of $\boldsymbol{P}_s^{(1)}$ becomes favorable to achieve large values of the studied hydrostatic parameters $\Phi_h^*(m, \beta)$ (Figure 3). Along with the aforementioned combination of the electromechanical properties, the anisotropy of elastic compliances $s_{ab}^{(1),E}$ of SC promotes increasing $|\Phi_h^*|$ in different volume-fraction ranges. Of particular interest is the hydrostatic piezoelectric coefficient $e_h^*(m, \beta)$ (Figure 3, c) with the deep absolute minimum. Undoubtedly, values of $|e_h^*| \approx 77$ C/m^2 near the absolute minimum of $e_h^*(m, \beta)$ (Figure 3, c) enable us to regard the studied 2 – 2 composite as a novel promising piezoelectric material for hydroacoustic applications. To the best of our knowledge, in the studied ferroelectric ceramic/polymer composites, the typical e_h^* values do not exceed 14 C/m^2, and values of $e_h^* \approx 10$ C/m^2 are peculiar to conventional ferroelectric ceramics at room temperature [31].

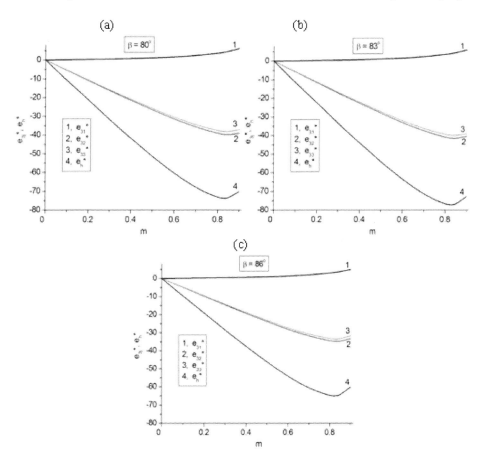

Figure 4. Correlation between piezoelectric coefficients e_{3j}^* and e_h^* (in C/m^2) of the PZN–0.12PT SC/polyurethane composite near absolute minimum of e_h^*. The spontaneous polarization vector $\boldsymbol{P}_s^{(1)}$ of the single-domain SC layer is oriented as shown in inset 2 in Figure 1, the rotation angle β = 80° (a), 83° (b) and 86° (c).

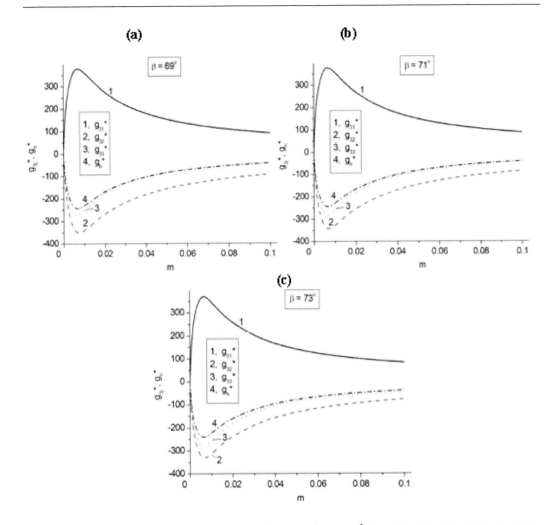

Figure 5. Correlation between piezoelectric coefficients g_{3j}^* and g_h^* (in mV·m/N) of the PZN–0.12PT SC/polyurethane composite near absolute minimum of g_h^*. The spontaneous polarization vector $P_s^{(1)}$ of the single-domain SC layer is oriented as shown in inset 2 in Figure 1, the rotation angle $\beta = 69°$ (a), 71° (b) and 73° (c).

There is an important correlation between the volume-fraction dependences of the hydrostatic piezoelectric coefficients and the piezoelectric coefficients that contribute into the hydrostatic response of the composite (see Equations (5) and Figures 4 and 5). Graphs in Figure 4 suggest that, near the absolute minimum of e_h^*, the piezoelectric coefficients e_{3j}^* from Equations (5) obey conditions:

$$e_{32}^* \approx e_{33}^* ; |e_{33}^*| \gg e_{31}^*. \qquad (7)$$

The validity of conditions (7) means that the interfaces $x_1 = $ const (see Figure 1) weaken the electromechanical interaction between the SC layers and strongly influences the

Orientation Effects in Novel Composites … 59

Table 3. Extreme values of hydrostatic parameters $\Phi_h^*(m, \beta)$ of 2–2 composites with single-domain SC layers (see inset 2 in Figure 1) at $0 < m < 1$ and $0° \leq \beta \leq 90°$

Para-meter	Single-domain PZN – 0.12PT SC/polyurethane composite		Single-domain PMN – 0.42PT SC/polyurethane composite	
	Absolute minima	Absolute maxima	Absolute minima	Absolute maxima
d_h^*	– 161 pC/N at $m = 0.090$ and $\beta = 78°$	214 pC/N at $m = 0.972$ and $\beta = 72°$	– 69.9 pC/N at $m = 0.081$ and $\beta = 78°$	130 pC/N at $m = 0.968$ and $\beta = 72°$
g_h^*	– 261 mV·m/N at $m = 0.005$ and $\beta = 71°$	215 mV·m/N at $m = 0.028$ and $\beta = 0°$	– 821 mV·m/N at $m = 0.005$ and $\beta = 70°$	197 mV·m/N at $m = 0.017$ and $\beta = 0°$
e_h^*	– 77.2 C/m^2 at $m = 0.825$ and $\beta = 83°$	7.78 C/m^2 at $m = 0.768$ and $\beta = 35°$	– 28.6 C/m^2 at $m = 0.725$ and $\beta = 79°$	10.2 C/m^2 at $m = 0.931$ and $\beta = 0°$
k_h^*	– 0.237 at $m = 0.020$ and $\beta = 74°$	The largest value corresponds to SC $(m = 1)$	– 0.100 at $m = 0.025$ and $\beta = 74°$	0.144 at $m = 0.063$ and $\beta = 0°$

anisotropy of e_{3j}^*. Contrary to e_{3j}^* from Figure 4, the piezoelectric coefficients g_{3j}^* from Figure 5 obey conditions:

$$g_{31}^* \approx |g_{32}^*|; \ g_{33}^* \approx g_h^*. \tag{8}$$

In our opinion, main distinctions between conditions (7) and (8) are associated with the different volume-fraction ranges of validity (e.g., large m values for e_{3j}^* and small m values for g_{3j}^*) and with features of the combination of the properties in the $2 - 2$ parallel-connected composite. In case of e_{3j}^*, the combination of the piezoelectric and elastic properties plays the leading role whereas behavior of g_{3j}^* is accounted for by the combination of the piezoelectric and dielectric properties. Graphs in Figure 5 suggest that the large values of $|g_h^*|$ can be attained at volume fractions of SC $m \approx 0.04 - 0.07$, and this range is to be taken into consideration when manufacturing the composite sample with high piezoelectric sensitivity. Our analysis of behavior of the piezoelectric coefficients near the absolute maxima (Figures 4 and 5) enables us to conclude that not only the elastic and dielectric anisotropy, but also the interfaces $x_1 = $ const in the composite sample (Figure 1) influence the validity of conditions (7) and (8) in certain volume-fraction and orientation ranges.

2.1.3. Comparison of Data on 2 – 2 Composites with Tetragonal Single-Domain Layers

Our comparison of the performance of the composites based on single-domain SCs of PZN – 0.12PT and PMN – 0.42PT (Table 3) suggests that the main difference between the hydrostatic parameters of these composites is associated with the piezoelectric coefficients of the SCs (Table 1). In addition, the hydrostatic piezoelectric coefficient g_h^* (see Table 3 and

Figures 2, b and 3, b) at small volume fractions ($0 < m < 0.1$) strongly depends on the dielectric properties of components, while the hydrostatic piezoelectric coefficient e_h^* mainly depends on the anisotropy of the elastic properties of SC. It is clear that the anisotropy of elastic compliances $s_{ab}^{(1),E}$ of the single-domain PMN – 0.42PT SC is less pronounced (Table 1) and therefore does not lead to the very large $|e_h^*|$ values in the composite based on this SC. The extreme values of g_h^* are achieved at small volume fractions m, and the order-of-magnitude of g_h^* does not differ from that known from papers [25 – 29, 31] on the SC/polymer and ceramic/polymer composites.

2.2. 2 – 2 PMN – 0.33PT / Polymer Composite and its Performance

Like PZN – 0.12PT, the PMN – 0.33PT SC was studied both in the single-domain and polydomain states. After poling along the [001] direction, the polydomain SC is characterized by the distinct engineered domain structure that comprises four non-180° domain types of the rhombohedral phase ($3m$ symmetry). The macroscopic symmetry of the SC sample with the engineered domain structure is $4mm$ [6, 7]. The corresponding full sets of electromechanical constants measured at room temperature (Table 4) are given in the main crystallographic axes. The mutual arrangement of the main crystallographic axes and the cubic perovskite unit-cell axes in the single-domain PMN – 0.33PT SC is shown in Figure 6. In this section we consider some features of the hydrostatic piezoelectric response of the 2–2 composite based on the single-domain PMN – 0.33PT SC. The arrangement of the layers in the composite sample is shown in Figure 1. It is assumed that the spontaneous polarization vector $\boldsymbol{P}_s^{(1)} \parallel OX_3'$, as shown in insets 1 and 2 of Figure 1, and the rotation of the vector with respect to the composite sample system ($X_1 X_2 X_3$) is described in terms of one of the angles either α or β. The averaging procedure described in Section 2.1.1 is also applied to the studied 2 – 2 composite.

Among the hydrostatic parameters Φ_h^* to be analyzed, of particular interest are the piezoelectric coefficients from Equations (5), ECF from Equation (6) and squared figure of merit that is written in terms of d_h^* and g_h^* as follows:

$$(Q_h^*)^2 = d_h^* g_h^*. \tag{9}$$

The parameter $(Q_h^*)^2$ from Equation (9) is used [31] to describe the sensor signal-to-noise ratio of the piezo-active composite and to characterize its piezoelectric sensitivity in hydrophones and other hydroacoustic applications. In general, the hydrostatic parameters are represented as $\Phi_h^*(m, \alpha)$ or $\Phi_h^*(m, \beta)$ for the rotation modes shown in insets 1 and 2 of Figure 1, respectively.

Table 4. Elastic compliances $s_{ab}^{(n),E}$ (in 10^{-12} Pa^{-1}), piezoelectric coefficients $d_{ij}^{(n)}$ (in pC/N) and relative dielectric permittivity $\varepsilon_{pp}^{(n),\sigma}/\varepsilon_0$ of PMN – 0.33PT SC and polyvinylidene fluoride (PVDF) at room temperature

Components	$s_{11}^{(n),E}$	$s_{12}^{(n),E}$	$s_{13}^{(n),E}$	$s_{14}^{(n),E}$	$s_{33}^{(n),E}$	$s_{44}^{(n),E}$	$s_{66}^{(n),E}$
PMN – 0.33PT single-domain SC [13]	62.2	– 53.8	– 5.6	– 166.2	13.3	511.0	232.0
PMN – 0.33PT polydomain SC poled along [001] [7]	69.0	– 11.1	– 55.7	0	119.6	14.5	15.2
PVDF with $P_r^{(2)} \uparrow\uparrow OX_3$ [35] [a]	333	– 148	– 87.5	0	299	$1.90 \cdot 10^4$	943

Components	$d_{15}^{(n)}$	$d_{22}^{(n)}$	$d_{31}^{(n)}$	$d_{33}^{(n)}$	$\varepsilon_{11}^{(n),\sigma}/\varepsilon_0$	$\varepsilon_{33}^{(n),\sigma}/\varepsilon_0$
PMN – 0.33PT single-domain SC [13]	4100	1340	– 90	190	3950	640
PMN – 0.33PT polydomain SC poled along [001] [7]	146	0	– 1330	2820	1600	8200
PVDF with $P_r^{(2)} \uparrow\uparrow OX_3$ [35] [a]	– 38	0	10.42	– 33.64	7.513	8.431

[a] $P_r^{(2)}$ is the remanent polarization vector of the polymer component.

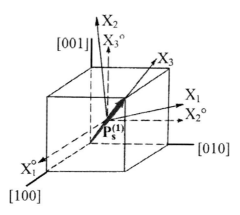

Figure 6. Orientation of the spontaneous polarization vector $P_s^{(1)}$ in the single-domain PMN – 0.33PT SC (3m symmetry) with respect to the principal OX_j and perovskite $OX_j°$ axes. [100], [010] and [001] are perovskite unit-cell directions.

Behavior of $\Phi_h^*(m, \alpha)$ and $\Phi_h^*(m, \beta)$ near absolute extreme points is shown in Figure 7. For the 2–2 composite based on the rhombohedral PMN – 0.33PT SC, the rotation of $P_s^{(1)}$ around the OX_1 axis leads to the larger values of $|d_h^*|$, $|k_h^*|$ and $(Q_h^*)^2$ (Figure 7, a – c).

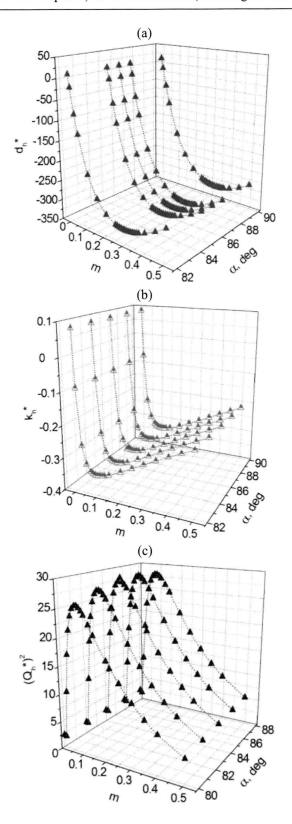

Figure 7 (Continued).

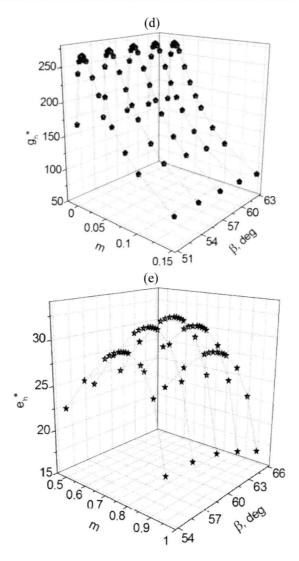

Figure 7. Hydrostatic piezoelectric coefficients (a, d and e), ECF (b) and squared figure of merit (c) of the 2 – 2 single-domain PMN – 0.33PT SC/PVDF composite in the vicinity of extreme points: (a) d_h^* (in pC/N), (b) k_h^*, (c) $(Q_h^*)^2$ (in 10^{-12} Pa^{-1}), (d) g_h^* (in mV·m/N), and (e) e_h^* (in C/m^2). The polymer component is poled so that its remanent polarization $P_r^{(2)} \uparrow\downarrow OX_3$, and the rotation of the spontaneous polarization vector $P_s^{(1)}$ of the single-domain SC layer takes place around either the OX_1 or OX_2 axis (see insets 1 and 2 in Figure 1).

The rotation of $P_s^{(1)}$ around the OX_2 axis is favorable to attain the larger piezoelectric coefficients g_h^* and e_h^* (Figure 7, d and e). It is seen that the maximum values of g_h^* at β = const are comparable to those calculated for the composite based on PZN – 0.12PT (Figure 3, b) while the piezoelectric coefficient e_h^* remains less than $|e_h^*|$ in Figure 3, c. In our opinion, such a difference in the e_h^* values is accounted for by the elastic anisotropy of the SC component. The piezoelectric and elastic anisotropy of the PMN – 0.33PT SC becomes favorable to obtain a series of extreme in the narrow regions of α (Figure 7, a – c) or β

(Figure 7, d and e). To the best of our knowledge, the similar orientation behavior of the hydrostatic piezoelectric coefficient e_h^* was not mentioned in the literature, especially in papers on the piezo-active 2 – 2 composites, and can be of value for hydroacoustic applications (hydrophones, antennae, etc.).

2.3. 2 – 2 PMN – 0.29PT / Polymer Composite and Its Performance

In the present section we analyze the effect of the polarization orientation on the hydrostatic piezoelectric coefficients d_h^* and e_h^* in the parallel-connected 2 – 2 composite based on the polydomain [011]-poled PMN – 0.29PT SC with macroscopic $mm2$ symmetry. The single-domain composition PMN – 0.29PT SC is characterized by the rhombohedral distortion of the perovskite unit cell and $3m$ symmetry. In the polydomain SC poled along [011] of the perovskite unit cell, one can observe the laminar 71° domain structure, the considerable anisotropy of elastic and piezoelectric properties (Table 5) and high piezoelectric activity [36]. Below we discuss some features of the hydrostatic piezoelectric response of the 2 – 2 composite based on the [011]-poled PMN – 0.29PT SC.

It is assumed that the layers of the 2 – 2 composite based on PMN – 0.29PT [28] are parallel-connected and form the regular structure (Figure 1). Following Reference [35], we orient the main crystallographic axes X, Y, and Z of the [011]-poled PMN – 0.29PT SC as follows: $X \parallel [0\bar{1}1]$, $Y \parallel [100]$ and $Z \parallel [011]$ (inset 1 in Figure 8), where the Miller indices [hkl] are written with respect to the perovskite unit cell. Spontaneous polarization vectors of the non-180° domains (Figure 8) are denoted as $P_{s,1}$ and $P_{s,2}$. The orientation of X, Y, and Z is characterized by rotation angles α, β and γ (insets 2 – 4 in Figure 8), where an equality $\alpha = \beta = \gamma = 0°$ corresponds to the following arrangement: $X \parallel OX_1$, $Y \parallel OX_2$ and $Z \parallel OX_3$. The 71° domains in SC are characterized by equal volume fractions and the spontaneous polarization vector of each SC layer is $P_s^{(1)} = (P_{s,1} + P_{s,2})/2$. We consider rotations of the main crystallographic axes of SC around one of the co-ordinate axes, OX_1, OX_2 or OX_3 (Figure 8), so that the spontaneous polarization vectors $P_{s,1}$ and $P_{s,2}$ remain situated either over the (X_1OX_2) plane or at this plane [28]. Taking into account the mutual orientations of the $P_{s,i}$ vectors and the OX_j axes (Figure 8), we determine the following ranges [28] in which the orientation angles are varied:

$$ -\arcsin(1/\sqrt{3}) \leq \alpha \leq \arcsin(1/\sqrt{3});\ -45° \leq \beta \leq 45°;\ 0° \leq \gamma \leq 360°. \tag{10} $$

Moreover, in the presence of the orthorhombic SC and isotropic polymer components, the periodic dependence of the hydrostatic parameters Φ_h^* from Equations (5) and (6) on α, β and γ from conditions (10) is observed. For example, the hydrostatic piezoelectric coefficients from Equations (5) obey conditions:

$$ \Phi_h^*(m, \alpha) = \Phi_h^*(m, -\alpha);\ \Phi_h^*(m, \beta) = \Phi_h^*(m, -\beta);\ \Phi_h^*(m, \gamma) = \Phi_h^*(m, 180° - \gamma) \tag{11} $$

Table 5. Room-temperature elastic compliances $s_{ab}^{(1),E}$ (in 10^{-12} Pa^{-1}), piezoelectric coefficients $d_{ij}^{(1)}$ (in pC/N) and relative dielectric permittivity $\varepsilon_{pp}^{(1),\sigma}/\varepsilon_0$ of PMN – 0.29PT SC poled along [011] of the perovskite unit cell [35]

$s_{11}^{(1),E}$	$s_{12}^{(1),E}$	$s_{13}^{(1),E}$	$s_{22}^{(1),E}$	$s_{23}^{(1),E}$	$s_{33}^{(1),E}$
18.0	−31.1	8.4	112	−61.9	49.6
$s_{44}^{(1),E}$	$s_{55}^{(1),E}$	$s_{66}^{(1),E}$	$d_{31}^{(1)}$	$d_{32}^{(1)}$	$d_{33}^{(1)}$
14.9	69.4	13.0	610	−1883	1020
$d_{15}^{(1)}$	$d_{24}^{(1)}$	$\varepsilon_{11}^{(1),\sigma}/\varepsilon_0$	$\varepsilon_{22}^{(1),\sigma}/\varepsilon_0$	$\varepsilon_{33}^{(1),\sigma}/\varepsilon_0$	
1188	167	3564	1127	4033	

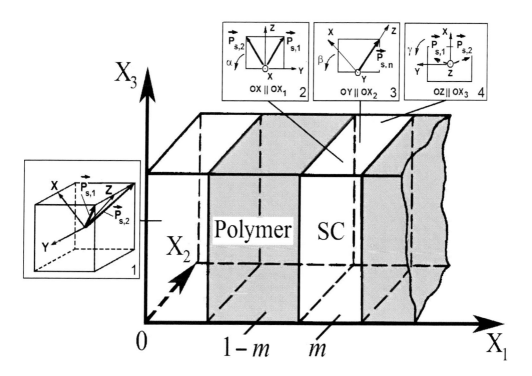

Figure 8. Schematic of the 2 – 2 parallel-connected composite, arrangement of the spontaneous polarization vectors $P_{s,1}$ and $P_{s,2}$ of domains in the [011]-poled polydomain SC (inset 1) and modes of rotation of $P_{s,1}$ and $P_{s,2}$ in SC layers (insets 2 – 4). ($X_1X_2X_3$) is the rectangular co-ordinate system, m and $1 - m$ are volume fractions of SC and polymer, respectively. X, Y and Z are main crystallographic axes of the polydomain SC, and α, β and γ are rotation angles. It is assumed that $n = 1$ and 2 (inset 3).

at volume fractions $0 < m < 1$. Taking into account conditions (10) and (11), we considered the dependences $\Phi_h^*(m, \alpha)$, $\Phi_h^*(m, \beta)$ and $\Phi_h^*(m, \gamma)$, where Φ_h^*'s are the hydrostatic parameters from Equations (5), (6) and (9).

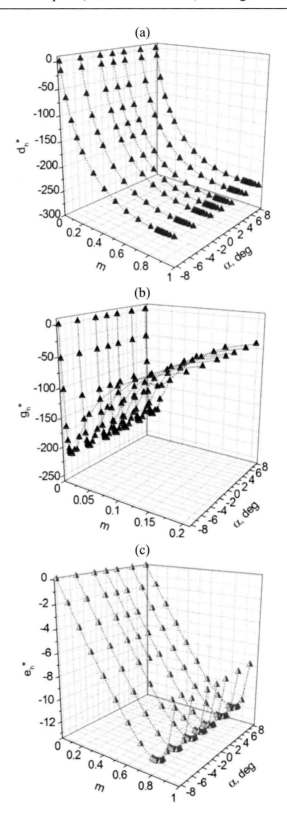

Figure 9 (Continued).

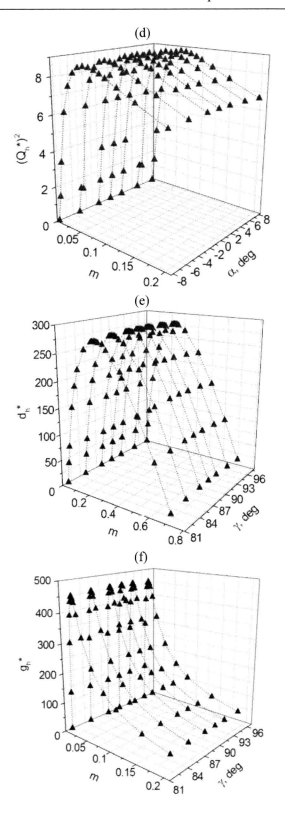

Figure 9 (Continued).

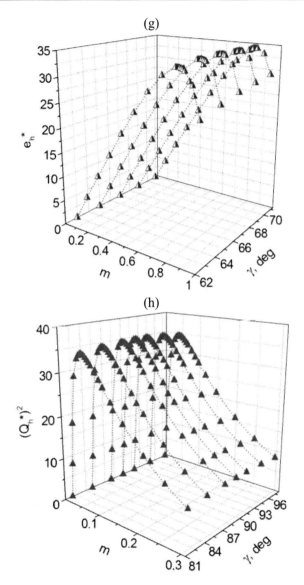

Figure 9. Hydrostatic piezoelectric coefficients (a – c and e – g) and squared figures of merit (d and h) of the 2 – 2 polydomain PMN – 0.29PT SC/polyurethane composite in the vicinity of extreme points: (a), (e) d_h^* (in pC/N), (b), (f) g_h^* (in mV·m/N), (c), (g) e_h^* (in C/m^2), and (d), (h) (Q_h^*)^2 (in 10^{-12} Pa^{-1}). Modes of rotation of the spontaneous polarization vectors $P_{s,1}$ and $P_{s,2}$ of domains in the [011]-poled PMN – 0.29PT SC are shown in Figure 8.

2.4. Effective Parameters and Advantages of 2 – 2 Single Crystal/Polymer Composites

The examples of the volume-fraction and orientation dependence of the hydrostatic parameters of the 2 – 2 composites (Sections 2.1 – 2.3) enable us to regard these composites as promising modern piezo-active materials. The main advantage consists in the fact that absolute values $|\Phi_h^*|$ can be more than those related to the conventional ferroelectric ceramic/polymer composites [31, 37 – 39] with 2 – 2 and other connectivities. For example, typical maximum values of the hydrostatic piezoelectric coefficients achieved in the 2 – 2 composites based on ceramics are 60 – 80 pC/N (for d_h^*) and 150 – 300 mV·m/N (for g_h^*) [37], however max g_h^* is observed at volume fractions of the ceramic component $m_c \approx 0.01$. The typical value of $(Q_h^*)^2$ in the 1 – 3 ceramic/polymer composite is about $3 \cdot 10^{-12}$ Pa^{-1} [39]. The experimental value of max $(Q_h^*)^2 \approx 5 \cdot 10^{-11}$ Pa^{-1} was achieved in the 2 – 2 PZT-type ceramic/polymer composite at a specific orientation of the ceramic layers [38] with respect to the poling direction. Taking into account the remarkable electromechanical properties of SCs and the orientation effect in the presence of the anisotropic and highly piezo-active component, we attain the hydrostatic piezoelectric coefficients $|d_h^*| > 200$ pC/N (Figures 3, a, 7, a, 9, a, and 9, e), $|g_h^*| > 200$ mV·m/N (Figures 2, b, 3, b, 7, d, 9, b, and 9, f), 33 C/m$^2 < |e_h^*|$ < 78 C/m^2 (Figures 3, c, 7, e, and 9, g), the ECF $0.20 < |k_h^*| < 0.35$ (Figures 2, d, 3, d and 7, b), and the squared figure of merit $(Q_h^*)^2 \sim 10^{-11}$ Pa^{-11} (Figures 7, c and 9, h). Undoubtedly, the results represented in Sections 2.1 – 2.3 indicate the potential of the novel 2–2 composites with specified orientations of the SC crystallographic axes for a variety of hydroacoustic and piezotechnical applications. The high piezoelectric performance of these materials is of value for specialists using the relaxor-ferroelectric SC as a component in the novel 2 – 2 composites and optimizing their effective piezoelectric properties, related hydrostatic parameters, ECFs, figures of merit, etc.

3. ORIENTATION EFFECTS AT THE CHANGE IN THE POLING DIRECTION OF POLYMER COMPONENTS

Below we discuss the performance of the SC/polymer composites with 1 – 3 and 0 – 3 connectivity patterns. A possibility of the subsequent poling of the ferroelectric SC and polymer components, especially in structures with a continuous polymer matrix [40–42], opens up new possibilities of controlling the orientation effect in the 1 – 3 and 0 – 3 composites. The orientation effect, concerned with the change in the orientation of the remanent polarization vector in the matrix of the 1 – 3 SC/polymer composite, was first studied [43] on assumption that the coercive field of ferroelectric polymer is considerably higher than the coercive field of SC. In Sections 3.1 – 3.3 we consider examples of the composites with components whose coercive fields satisfy the aforementioned condition.

3.1. 1 – 3 and 0 – 3 Single Crystal/Polymer Composites: Modeling of Properties

It is assumed that the 1 – 3 FC/polymer composite consists of a system of very long SC rods distributed regularly in a continuous polymer matrix (Figure 10, left part). The rods have shape of circular cylinder with fixed radius and are aligned along the poling axis OX_3 of the rectangular co-ordinate system ($X_1X_2X_3$). The centers of the sections of these cylinders constitute a square lattice in the (X_1OX_2) plane, i.e., a square pattern in the arrangement of the SC component is considered. The SC and polymer components are characterized by volume fractions m and $1 - m$, respectively. The main crystallographic axes in each rod are oriented as follows: $X \parallel OX_1$, $Y \parallel OX_2$ and $Z \parallel OX_3$. The 1 – 3 composite as a whole is poled along the OX_3 axis so that the spontaneous polarization vector of each SC rod is $\boldsymbol{P}_s^{(1)} \uparrow\uparrow OX_3 \parallel Z$ and the remanent polarization vector of the polymer matrix is either $\boldsymbol{P}_r^{(2)} \uparrow\uparrow OX_3$ or $\boldsymbol{P}_r^{(2)} \uparrow\downarrow OX_3$.

A transition from 1 – 3 to 0 – 3 connectivity is observed if, for example, the SC rod is replaced by an aligned SC prolate inclusion (Figure 10, right part). The microgeometric similarity of the composite structures (the presence of the prolate inclusions instead of the long rods) suggests that this change in connectivity would not result in drastic changes in the effective parameters. The spheroidal shape of each SC inclusion in the 0–3 composite is described by the equation $(x_1 / a_1)^2 + (x_2 / a_2)^2 + (x_3 / a_3)^2 = 1$ in the rectangular co-ordinate system ($X_1X_2X_3$), where a_1, $a_2 = a_1$ and a_3 are semi-axes of the spheroid. Centers of symmetry of the SC inclusions occupy sites of a simple lattice with unit-cell vectors parallel to the OX_j axes. Each SC inclusion is characterized by the spontaneous polarization vector $\boldsymbol{P}_s^{(1)} \uparrow\uparrow OX_3 \parallel Z$. Like the 1 – 3 composite, the remanent polarization vector in the polymer matrix is oriented as either $\boldsymbol{P}_r^{(2)} \uparrow\uparrow OX_3$ or $\boldsymbol{P}_r^{(2)} \uparrow\downarrow OX_3$. Additionally, it is assumed that the SC inclusions have equal sizes over the whole composite sample. Hereafter the shape of each SC inclusion can be described in terms of the aspect ratio $\rho = a_1 / a_3$.

Electromechanical properties of the components in the studied 1 – 3 and 0 – 3 composites are described by elastic moduli $c_{ab}^{(n),E}$ measured at electric field $E = $ const, piezoelectric coefficients $e_{ij}^{(n)}$ and dielectric permittivities $\varepsilon_{pp}^{(n),\xi}$ measured at mechanical strain $\xi = $ const,

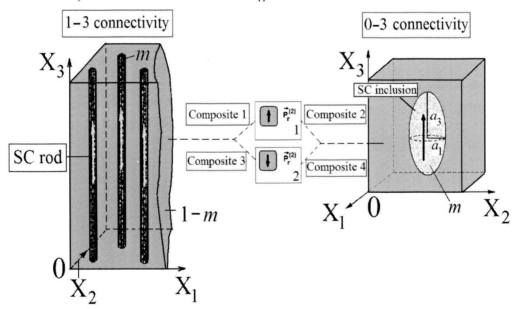

Figure 10. Schematic of the 1 – 3 and 0 – 3 composites based on SCs. The spontaneous polarization vector $\boldsymbol{P}_s^{(1)}$ of the SC rod (inclusion) is denoted by an arrow, the remanent polarization vector $\boldsymbol{P}_r^{(2)}$ of the ferroelectric polymer matrix is oriented as shown in insets 1 and 2. a_1 and a_3 are semi-axes of the spheroidal inclusion, m and $1 - m$ are volume fractions of SC and polymer, respectively.

where $n = 1$ (SC) or 2 (polymer). The electromechanical properties of the composites are regarded as effective (homogenized) properties in a longwave approximation, i.e., on assumption that the wavelength of an external acoustic field is to be much longer than the size of the rod or the largest semi-axis (a_3) of the inclusion.

In the last years the finite-element method (FEM) was developed for the SC/polymer piezo-active composites with a periodic arrangement of inclusions and high piezoelectric activity of the SC component (see, for instance, Reference [43]). In this method a comprehensive unit-cell model of the composite is put forward. In the present study the COMSOL package [44] is applied to obtain the volume-fraction dependence of the effective electromechanical properties of the $1 - 3$ and $0 - 3$ composites.

In case of the $1 - 3$ composite, a square polymer unit cell (in the X_1OX_2-plane) contains a circular SC inclusion, with radius adjusted to yield the appropriate volume fraction m. This unit cell is discretized using about 240,000 triangular elements. The unknown displacement field is interpolated using second-order Lagrange shape functions, leading to a problem with about 1,000,000 degrees of freedom. Periodic boundary conditions are enforced on the boundary of the unit cell [45], and the matrix $\| C^* \|$ of effective electromechanical constants of the composite is computed column-wise, performing calculations for diverse average strain and electric fields imposed to the structure. The matrix is represented in the general form as

$$\| C^* \| = \begin{pmatrix} \left\| c^{*E} \right\| & \left\| e^* \right\|^T \\ \left\| e^* \right\| & -\left\| \varepsilon^{*\xi} \right\| \end{pmatrix}, \tag{12}$$

where matrices of elastic moduli $\| c^{*E} \|$, piezoelectric coefficients $\| e \|$ and dielectric permittivities $\| \varepsilon^{*\xi} \|$ depend on m, and superscript T refers to the transposed matrix. After solving the equilibrium problem, the effective material constants from Equation (12) are computed by averaging the resulting local stress and electric-displacement fields over the unit cell.

By modeling the effective properties of the $0 - 3$ composite, we consider the unit cell, containing the spheroidal inclusion (Figure 10, right part) with radius adjusted to yield the appropriate volume fraction m, and this unit cell is discretized using tetrahedral elements. Their number, depending on the aspect ratio ρ of the spheroidal inclusion, varies from 700,000 to 1,600,000. The unknown displacement and electric-potential fields are interpolated using linear Lagrangian shape functions. The corresponding number of degrees of freedom varies from 500,000 to 1,200,000. Periodic boundary conditions are enforced on the boundary of the unit cell, and the matrix of effective constants of the composite is computed column-wise, performing calculations for diverse average strain and electric fields imposed to the structure. The Geometric Multigrid Iterative Solver (V-cycle, successive over-relaxation pre- and post-smoother, direct coarse solver) is employed. After solving the electro-elastic equilibrium problem of the piezo-active inclusion in the continuous piezo-active matrix, effective electromechanical constants of the $0 - 3$ composite are computed, by averaging the resulting local stress and electric-displacement fields over the unit cell. The matrix $\| C^* \|$ of the effective electromechanical properties of the $0 - 3$ composite is also given by Equation (12), and matrix elements depend on m and ρ.

Below for comparison we consider the effective parameters of the four related SC/polymer composites: $1 - 3$ with $P_r^{(2)} \uparrow\uparrow OX_3$ (*composite 1* in our subsequent notations, see also Figure 10), $0 - 3$ with $P_r^{(2)} \uparrow\uparrow OX_3$ (*composite 2*), $1 - 3$ with $P_r^{(2)} \uparrow\downarrow OX_3$ (*composite 3*), and $0 - 3$ with $P_r^{(2)} \uparrow\downarrow OX_3$ (*composite 4*). It is assumed that the $0 - 3$ composite comprises the prolate SC inclusions with $\rho = 0.1$.

3.2. Orientation Effects in Composites Based on Polydomain PMN – 0.33PT

Based on the full set of electromechanical constants from Equation (12) and taking into account conventional formulae [33] for the piezoelectric medium, we calculate the hydrostatic parameters from Equations (5) and (9) and the following ECFs:

longitudinal ECF

$$k_{33}^* = d_{33}^* / (\varepsilon_{33}^{*\sigma} s_{33}^{*E})^{1/2}, \tag{13}$$

lateral ECF

$$k_{31}^* = d_{31}^* / (\varepsilon_{33}^{*\sigma} s_{11}^{*E})^{1/2}, \tag{14}$$

thickness ECF

$$k_t^* = e_{33}^* / (\varepsilon_{33}^{*\varsigma} c_{33}^{*D})^{1/2}, \tag{15}$$

planar ECF

$$k_p^* = k_{31}^* [2 s_{11}^{*E} / (s_{11}^{*E} + s_{12}^{*E})]^{1/2}. \tag{16}$$

In this section we discuss features of the performance of the $1 - 3$ and $0 - 3$ composites based on the [001]-poled polydomain PMN $-$ 0.33PT SC. Among the ferroelectric polymer components, we choose copolymer of 75 / 25 mol. % copolymer of vinylidene fluoride and trifluoroethylene (VDF–TrFE) with known electromechanical properties (Table 6). This polymer component is characterized by unusual signs and large anisotropy of the piezoelectric coefficients $e_{3j}^{(2)}$ [47] so that at $P_r^{(2)} \uparrow\uparrow OX_3$, the relation sgn $e_{33}^{(2)} = -$ sgn $e_{31}^{(2)} < 0$ is valid.

Examples of the volume-fraction dependence of the effective parameters of the $1 - 3$ and $0 - 3$ composites are shown in Figure 11. It is seen that the orientation of the $P_r^{(2)}$ vector in the polymer matrix strongly influences the performance of the composite, especially at $m < 0.2$. It can be accounted for by the fact that at the relatively small volume fractions m, the elastic properties of the composite remain comparable to those of the polymer. This feature pre-determines the electromechanical interaction between the components, the piezoelectric effect and electromechanical coupling. The less favorable conditions for the electromechanical interaction take place in composite 2 wherein the inclusions are isolated and the polarization vector of polymer $P_r^{(2)}$ is antiparallel to the SC poling direction (Figure 10). At $m > 0.2$ we see minor differences between the parameters of composites 1 and 3 (Figure 11): due to the presence of the long SC rods with high piezoelectric activity, the role of the polymer matrix with low piezoelectric activity weakens in increasing the volume fraction of SC m.

Table 6. Elastic moduli $c_{ab}^{(2),E}$ (in 10^{10} Pa), piezoelectric coefficients $e_{ij}^{(2)}$ (in C/m²) and relative dielectric permittivity $\varepsilon_{pp}^{(2),\xi}/\varepsilon_0$ of 75 / 25 mol. % copolymer of vinylidene fluoride and trifluoroethylene at room temperature [47]

$c_{11}^{(2),E}$	$c_{12}^{(2),E}$	$c_{13}^{(2),E}$	$c_{33}^{(2),E}$	$e_{31}^{(2)}$	$e_{33}^{(2)}$	$\varepsilon_{33}^{(2),\xi}/\varepsilon_0$
0.85	0.36	0.36	0.99	0.008	−0.29	6.0

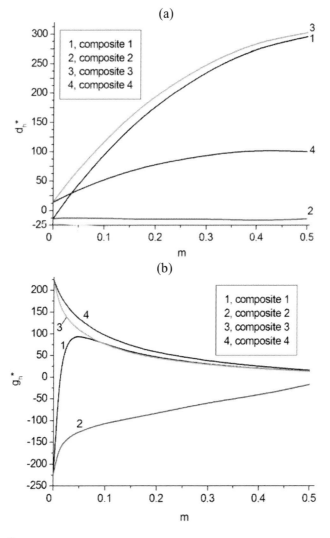

Figure 11 (Continued).

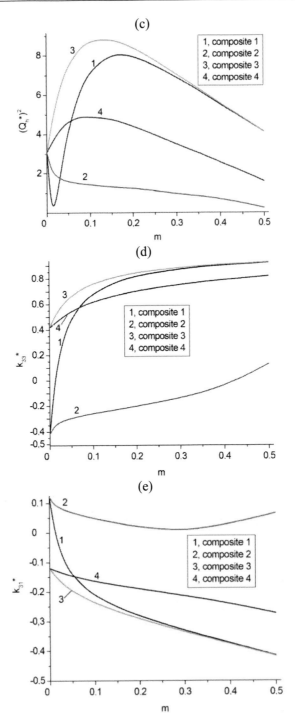

Figure 11. Hydrostatic piezoelectric coefficients (a and b), squared figure of merit (c) and ECFs (d and e) of the 1 – 3 and 0 – 3 polydomain PMN – 0.33PT SC/VDF–TrFE composites at different orientations of the remanent polarization vector $P_r^{(2)}$ in the polymer matrix: (a) d_h^* (in pC/N), (b) g_h^* (in mV·m/N), (c) $(Q_h^*)^2$ (in 10^{-12} Pa^{-1}), (d) k_{33}^*, and (e) k_{31}^*. Composites 1 – 4 are denoted as shown in Figure 10. In composites 2 and 4, spheroidal inclusions are characterized by the aspect ratio $\rho = 0.1$.

3.3. Orientation Effects in Composites Based on Single-Domain PMN – 0.42PT

In the present section we report the behavior of ECFs of the 1 – 3 single-domain SC/polymer composite based on single-domain PMN – 0.42PT SC. This SC represents the first tetragonal relaxor-based composition, for which the full set of electromechanical constants was measured [15] in the single-domain state (Table 1). A combination of considerable elastic and dielectric anisotropies as well as relatively high piezoelectric sensitivity make this SC attractive as a component in the advanced composites.

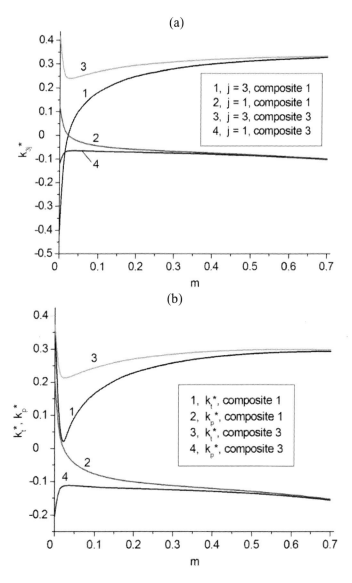

Figure 12. ECFs of the 1 – 3 single-domain PMN – 0.42PT SC/VDF–TrFE composite at different orientations of the remanent polarization vector $P_r^{(2)}$ in the polymer matrix: (a) k_{31}^* and k_{33}^*, and (b) k_t^* and k_p^*. Composites 1 and 3 are denoted as shown in Figure 10.

Volume-fraction dependences of ECFs from Equations (13) – (16) are shown in Figure 12. As in Figure 11, a and e (see curves 1 and 3 for 1 – 3 connectivity), the minor difference between values of ECFs is observed at $m > 0.2$. Curves 1 and 2 in Figure 12 suggest that the large anisotropy of ECFs (i.e., $|k_{33}^* / k_{31}^*| \gg 1$ and $|k_t^* / k_p^*| \gg 1$) is expected in the volume-fraction range of $0.05 < m < 0.1$, as the elastic properties of the polymer component and the composite are comparable.

To demonstrate the key role of the elastic anisotropy of the 1 – 3 composite, we consider interconnections between anisotropy factors $\zeta_{33\text{-}31} = |k_{33}^*/k_{31}^*|$ and $\zeta_{t\text{-}p} = |k_t^*/k_p^*|$. The anisotropy factors $\zeta_{33\text{-}31}$ and $\zeta_{t\text{-}p}$ of the composite strongly depend on ζ_d and elastic compliances s_{ab}^{*E} of the composite. Substituting from Equations (13) – (16), we derive that

$$\zeta_{33\text{-}31} = |k_{33}^* / k_{31}^*| = |d_{33}^* / d_{31}^*| (s_{11}^{*E} / s_{33}^{*E})^{1/2}. \tag{17}$$

Taking into account the equality $k_t^* \approx k_{33}^*$ (that is valid for the piezo-active 1 – 3 composite in the wide volume-fraction m range [31, 47], cf. Figures 12, a and 12, b), we write the anisotropy factor $\zeta_{t\text{-}p}$ as follows:

$$\zeta_{t\text{-}p} = |k_t^* / k_p^*| \approx |k_{33}^* / k_{31}^*| [(s_{11}^{*E} + s_{12}^{*E})/(2 s_{11}^{*E})]^{1/2} \approx |d_{33}^* / d_{31}^*| [(s_{11}^{*E} + s_{12}^{*E}) / (2 s_{33}^{*E})]^{1/2}. \tag{18}$$

Thus, Equations (17) and (18) illustrate the considerable dependence of the anisotropy factors $\zeta_{33\text{-}31}$ and $\zeta_{t\text{-}p}$ on the elastic properties of the composite. It is seen that Equations (17) and (18) comprise s_{11}^{*E} and s_{33}^{*E}, and it means that changes in the elastic response of the composite along the OX_3 axis (parallel to the SC rod, see Figure 10, left part) and along the OX_1 axis (perpendicular to the SC rod) would lead to distinct changes in the anisotropy of ECFs of the studied 1 – 3 composite.

By comparing data from Figures 11, d and e with data from Figure 12, a, we see that all the ECFs of composites 1 – 4 based on the polydomain PMN – 0.33PT SC change monotonically while k_{33}^* of composite 3 based on the single-domain PMN – 0.42PT SC exhibits a non-monotonic dependence on m. Such a feature is associated with the influence of the elastic anisotropy of the SC on the longitudinal piezoelectric response and ECF k_{33}^* as well as with the orientation effect that leads to the change in sgn $d_{33}^{(2)}$ and its influence on the electromechanical coupling of the composite. As a consequence, a competition of contributions from the components poled in opposite directions gives rise to unmonotonic $k_{33}^*(m)$ dependence in the volume-fraction range where elastic constants of the polymer component and the composite as a whole remain comparable. It should be added that the similar tendency is also observed in case of the $k_t^*(m)$ dependence predicted for composite 3 (cf. curve 3 in Figure 12, a and curve 3 in Figure 12, b).

The orientation effect plays the important role in forming the volume-fraction dependence of the piezoelectric coefficients e_{3j}^* of the $1-3$ composite (Table 7). The favorable result is achieved, as the SC and polymer components are poled in opposite directions. The $1-3$ composite structure promotes the large anisotropy of e_{3j}^* at different poling directions of the polymer matrix, i.e., condition $|e_{33}^* / e_{31}^*| \gg 1$ is valid for composites 1 and 3 in the wide m range (see data from Table 7).

Changes in the piezoelectric coefficient e_{33}^* (Table 7) represent an example of the sum effect [31, 48] at the determination of the piezoelectric response of the $1-3$ composite while the full sets of electromechanical constants, including the piezoelectric coefficients $e_{ij}^{(n)}$, are involved in the FEM calculation (see Section 3.1). This sum effect is also concerned with the proper microgeometry of the $1-3$ composite (Figure 10) that contains the system of the SC rods oriented parallel to the poling axis OX_3. We see that a difference between the values of e_{33}^* related to composites 1 and 3 at $m = \text{const}$ (Table 7) distinctly decreases at the volume fraction $m > 0.4$, i.e., as the effective piezoelectric coefficient e_{33}^* of the studied $1-3$ composite becomes comparable to $e_{33}^{(1)}$ of the SC component and the orientation effect in the composite weakens.

Table 7. Volume-fraction dependence[a] of the piezoelectric coefficients e_{33}^* (in C/m²) of the $1-3$ single-domain PMN $-$ 0.42PT SC/VDF–TrFE composite at different orientations of the remanent polarization vector $P_r^{(2)}$ in the polymer matrix

m	0.01	0.02	0.03	0.04	0.05	0.1
Composite 1[b]						
e_{31}^*	0.00659	0.00515	0.00368	0.00218	0.000649	-0.00747
e_{33}^*	-0.153	-0.0150	0.122	0.260	0.397	1.08
Composite 3[b]						
e_{31}^*	-0.00940	-0.0108	-0.0123	-0.0138	-0.0153	-0.0234
e_{33}^*	0.422	0.553	0.685	0.816	0.948	1.61
m	0.2	0.3	0.4	0.5	0.6	0.7
Composite 1[b]						
e_{31}^*	-0.0265	-0.0505	-0.0814	-0.123	-0.181	-0.271
e_{33}^*	2.46	3.82	5.19	6.55	7.89	9.22
Composite 3[b]						
e_{31}^*	-0.0423	-0.0660	-0.0967	-0.138	-0.196	-0.285
e_{33}^*	2.92	4.23	5.53	6.83	8.12	9.39

[a] FEM data.
[b] Composites 1 and 3 are denoted as shown in Figure 10.

Thus, the orientation effect concerned with the poling direction of the ferroelectric polymer matrix in the studied composites is important in the context of the hydrostatic piezoelectric response, the anisotropy of the piezoelectric properties and electromechanical coupling. The results shown in Figures 11 and 12 and Table 7 can be taken into consideration in electromechanical transducer, sensor and hydroacoustic applications of novel composite materials.

CONCLUSION

In the present chapter we analyzed the role of the orientation of the polarization vectors of components in forming the effective electromechanical properties and related parameters of the relaxor-ferroelectric SC/polymer composites. These composites are based on PMN – xPT and PZN – xPT SCs with high piezoelectric activity and a series of remarkable physical properties that strongly depend on the orientation of the crystallographic axes, domain structure, poling direction, composition, temperature, etc. Examples of the orientation effect studied in the chapter are related to the $2 - 2$, $1 - 3$ and $0 - 3$ composites. The effective electromechanical properties, hydrostatic parameters and ECFs of the SC/polymer composites were predicted and analyzed at different orientations of the main crystallographic axes of relaxor-ferroelectric SCs (single-domain and domain-engineered [001]- and [011]-poled samples) and at different poling directions of ferroelectric polymer. Modeling and property predictions were carried out within the framework of models related to the $2 - 2$, $1 - 3$ and $0 - 3$ connectivity patterns at the regular distribution of components over the composite sample. Calculations of the effective electromechanical properties and related parameters were performed using either the matrix approach ($2 - 2$ connectivity) or FEM ($1 - 3$ and $0 - 3$ connectivities). Changes in the effective parameters of the matrix composite at the transition from $1 - 3$ to $0 - 3$ connectivity were also considered.

The results point out important advantages of the studied $2 - 2$ and $1 - 3$ SC/polymer composites over the conventional ceramic/polymer composites. For instance, the large absolute values of the hydrostatic piezoelectric coefficients $|d_h^*|$, $|g_h^*|$ and $|e_h^*|$, the hydrostatic ECF $|k_h^*|$, the squared figure of merit $(Q_h^*)^2$, and considerable values of the anisotropy factors of ECFs ζ_{33-31} and ζ_{t-p} are achieved due to the specific orientations of the main crystallographic axes of the SC component (that can be either single-domain or polydomain). Hereby the anisotropy of the elastic and piezoelectric properties of SC also plays the important role. Extreme points of the hydrostatic parameters d_h^*, g_h^*, e_h^*, k_h^*, and $(Q_h^*)^2$ were found for the $2 - 2$ composites based on SCs from different symmetry classes and with different features of the anisotropy of the elastic and piezoelectric properties. Extreme points of the hydrostatic parameters d_h^*, g_h^* and $(Q_h^*)^2$ were found for the $1 - 3$ and $0 - 3$ composites with the fixed orientation of the remanent polarization vector of the polymer component. Examples of the large anisotropy of the piezoelectric coefficients e_{3j}^* in the $1 - 3$ composite were shown and discussed.

It is believed that the data represented in the chapter will be of interest for specialists considering the use of advanced relaxor-ferroelectric SCs in novel composites and optimizing the effective electromechanical properties, hydrostatic, transducer and other parameters of these materials suitable for modern piezotechnical applications.

ACKNOWLEDGMENTS

The authors would like to thank Dr. C. R. Bowen and Prof. Dr. R. Stevens (University of Bath, United Kingdom), Prof. Dr. A. E. Panich and Prof. Dr. I. A. Parinov (Southern Federal University, Russia), and Prof. Dr. L. N. Korotkov (Voronezh State Technical University, Russia) for their continued interest in the research problems.

REFERENCES

[1]   Park, S. - E.; Hackenberger, W. *Curr. Opin. Solid State Mater. Sci.* 2000, *vol. 6*, 11–18.

[2]   Cross, L. E. In: *Piezoelectricity. Evolution and Future of a Technology*; Heywang, W.; Lubitz, K.; Wersing, W.; Eds.; Springer Series in Materials Science *114*; Springer: Berlin, Heidelberg, 2008; pp. 131–156.

[3]   Davis, M. *J. Electrocer.* 2007, *vol. 19*, 23–45.

[4]   Liu, T.; Lynch, C. S. *Acta Mater.* 2003, *vol. 51*, 407–416.

[5]   Dammak, H.; Renault, A. - E.; Gaucher P.; Thi M. P.; Clavarin, G. *Jpn. J. Appl. Phys. Pt 1* 2003, *vol. 42*, 6477–6482.

[6]   Liu, T.; Lynch, C. S. *Continuum Mech. Thermodyn.* 2006, *vol. 18*, 119–135.

[7]   Zhang, R.; Jiang, B.; Cao, W. *J. Appl. Phys.* 2001, *vol. 90*, 3471–3475.

[8]   Xu, Y. *Ferroelectric Materials and Their Applications*; North-Holland: Amsterdam, London, New York, NY, Toronto, 1991, pp. 1–255.

[9]   Noheda, B.; Cox, D. E.; Shirane, G.; Gao, J.; Ye, Z. - G. *Phys. Rev. B* 2002, *vol. 66*, 054104 – 10 p.

[10] Singh, A. K.; Pandey, D. *Phys. Rev. B* 2003, *vol. 67*, 064102 – 12 p.

[11] Shuvaeva, V. A.; Glazer, A. M.; Zekria, D. *J. Phys.: Condens. Matter* 2005, *vol. 17*, 5709–5723.

[12] La-Orauttapong, D.; Noheda, B.; Ye, Z. - G.; Gehring, P. M.; Toulouse, J.; Cox, D. E.; Shirane, G. *Phys. Rev. B* 2002, *vol. 65*, 144101 – **7** p.

[13] Zhang, R.; Jiang, B.; Cao, W. *Appl. Phys. Lett.* 2003, *vol. 82*, 3737–3739.

[14] Guennou, M.; Dammak, H; Thi, M. P. *J. Appl. Phys.* 2008, *vol. 104*, 074102 – 6 p.

[15] Cao, H.; Hugo Schmidt, V.; Zhang, R.; Cao, W.; Luo, H. *J. Appl. Phys.* 2004, *vol. 96*, 549–554.

[16] Zhang, R.; Jiang, B.; Cao, W.; Amin, A. *J. Mater. Sci. Lett.* 2002, *vol. 21*, 1877–1879.

[17] Zhang, R.; Jiang, B.; Jiang, W.; Cao, W. *Appl. Phys.Lett.* 2006, *vol. 89*, 242908 – 3 p.

[18] Damjanovic, D.; Budimir, M.; Davis, M.; Setter, N. *Appl. Phys. Lett.* 2003, *vol. 83*, 527–529; Erratum *Ibid.*, 2490.

[19] Topolov, V. Yu. *J. Phys.: Condens. Matter* 2004, *vol. 16*, 2115–2128.

[20] Topolov, V. Yu. *Sensors Actuat. A* 2005, *vol. 121*, 148–155.

[21] Ritter, T.; Geng, X.; Shung, K. K.; Lopath, P. D.; Park, S. - E.; Shrout, T. R. *IEEE Trans. Ultrason., Ferroelec., Freq. Contr.* 2000, *vol. 47*, 792–800.

[22] Cheng, K. C.; Chan, H. L. W.; Choy, C. L.; Yin, Q.; Luo, H.; Yin, Z. *IEEE Trans. Ultrason., Ferroelec., Freq. Contr.* 2003, *vol. 50*, 1177–1183.

[23] Ren, K.; Liu, Y.; Geng, X.; Hofmann, H. F.; Zhang, Q. M. *IEEE Trans. Ultrason., Ferroelec., Freq. Contr.* 2006, *vol. 53*, 631–638.

[24] Wang, W.; He, C.; Tang, Y. *Mater. Chem. Phys.* 2007, *vol. 105*, 273–277.

[25] Topolov, V. Yu.; Krivoruchko, A. V.; Bowen, C. R. *Ferroelectrics* 2007, *vol. 351*, 145–152.

[26] Krivoruchko A. V.; Topolov, V. Yu. *J. Phys. D: Appl. Phys.* 2007, *vol. 40*, 7113–7120.

[27] Topolov, V. Yu.; Krivoruchko, A. V. *J. Appl. Phys.* 2009, *vol. 105*, 074105 – 7 p.

[28] Topolov, V. Yu.; Krivoruchko, A. V.; Bowen, C. R.; Panich, A. A. *Ferroelectrics* 2010, *vol. 400*, 410–416.

[29] Gibiansky, L. V.; Torquato, S. *J. Mech. Phys. Solids* 1997, *vol. 45*, 689–708.

[30] Newnham, R. E.; Skinner, D. P.; Cross, L. E. *Mater. Res. Bull.* 1978, *vol. 13*, 525–536.

[31] Topolov, V. Yu.; Bowen, C. R. *Electromechanical Properties in Composites Based on Ferroelectrics*. Springer: London, 2009; pp. 1–202.

[32] Topolov, V. Yu.; Bowen, C. R.; Glushanin, S. V. In: *Piezoceramic Materials and Devices*; Parinov, I. A., Ed.; Nova Science Publishers: New York, NY, 2010; pp. 71 – 112.

[33] Ikeda, T. *Fundamentals of Piezoelectricity*. Oxford University: Oxford, New York, NY, Toronto, 1990; pp. 1 – 263.

[34] Aleshin, V. I. *Zh. Tekhn. Fiziki* 1990, *vol. 60*, 179–183 (in Russian).

[35] Kar-Gupta, R.; Venkatesh, T. A. *Acta Mater.* 2007, *vol. 55*, 1093–1108.

[36] Wang, F.; Luo, L.; Zhou, D.; Zhao, X.; Luo, H. *Appl. Phys. Lett.* 2007, *vol. 90*, 212903 – 3 p.

[37] Grekov, A. A.; Kramarov, S. O.; Kuprienko, A. A. *Ferroelectrics* 1987, *vol. 76*, 43–48.

[38] Safari, A.; Akdogan, E. K. *Ferroelectrics* 2006, *vol. 331*, 153–179.

[39] Akdogan, E. K.; Allahverdi, M.; Safari, A. *IEEE Trans. Ultrason., Ferroelec., Freq. Contr.* 2005, *vol. 52*, 746–775.

[40] Nan, C. W.; Weng, G. J. *J. Appl. Phys.* 2000, *vol. 88*, 416–423.

[41] Chan, H. L.W.; Ng, P. K. L.; Choy, C. L. *Appl. Phys. Lett.* 1999, *vol. 74*, 3029–3031.

[42] Ng, K. L.; Chan, H. L. W.; Choy, C. L. *IEEE Trans. Ultrason., Ferroelec., Freq. Contr.* 2000, *vol. 47*, 1308–1315.

[43] Topolov, V. Yu.; Krivoruchko, A. V.; Bisegna, P.; Bowen, C.R. *Ferroelectrics* 2008, *vol. 376*, 140–152.

[44] COMSOL, Inc. *COMSOL Multiphysics™ User's Guide* 2010 (version 4.0a), *http://www.comsol.com*

[45] Bisegna, P.; Luciano, R. *J. Mech. Phys. Solids* 1997, *vol. 45*, 1329–1356.

[46] Hackbusch, W. *Multi-Grid Methods and Applications*. Springer: Berlin, 1985; pp. 1 – 396.

[47] Taunaumang, H.; Guy, I. L.; Chan, H. L. W. *J. Appl. Phys.* 1994, *vol. 76*, 484–489.

[48] Newnham, R. E. In *Concise Encyclopedia of Composite Materials*; Kelly, A.; Cahn, R. W.; Bever, M. B.; Eds.; Elsevier: Oxford, 1994; pp. 214–220.

In: Ferroelectrics and Superconductors
Editor: Ivan A. Parinov

ISBN: 978-1-61324-518-7
©2012 Nova Science Publishers, Inc.

Chapter 3

PYROELECTRIC EFFECT AT THERMALLY ACTIVATED PHASE TRANSITIONS IN FERROELECTRICS, FERROELECTRICS-RELAXORS AND ANTIFERROELECTRICS

Yu. N. Zakharov, A. G. Lutokhin and I. P. Raevski*
Research Institute of Physics, Southern Federal University, Rostov-on-Don, Russia

ABSTRACT

Both quasistatic and dynamic pyroelectric measurements have been used for studying the phase transitions of different types in perovskite ferroelectrics, ferroelectrics –relaxors and antiferroelectrics. Some new features on the x–T-phase diagrams of $PbZr_{1-x}Ti_xO_3$, $(1-x)PbFe_{1/2}Nb_{1/2}O_3 - xPbTiO_3$, $(1-x)PbMg_{1/3}Nb_{2/3}O_3 - xPbTiO_3$ solid solutions have been revealed. The possibilities of the polarization screening studies near the ferroelectric-paraelectric phase boundary in unipolar $BaTiO_3$ crystals by means of dynamic pyroelectric method are discussed. A dramatic influence of bias electric field on pyroelectric response was revealed for some $(1-x)PbMg_{1/3}Nb_{2/3}O_3 - xPbTiO_3$ ceramic compositions. These data are interpreted in the frame of an assumption of quasi-critical behavior of the pyroelectric response in relaxor ceramics.

1. POLARIZATION SCREENING MECHANISMS NEAR THE FERROELECTRIC-PARAELECTRIC PHASE TRANSITION IN THE UNIPOLAR BARIUM TITANATE CRYSTAL STUDIED BY DYNAMIC PYROELECTRIC EFFECT

Ferroelectric-paraelectric (FE-PE) phase transition in barium titanate and triglicine sulphate unipolar crystals had been studied by A. G. Chynoweth, by means of dynamic pyroelectric method. He discovered an anomalous behaviour of pyroelectric current near T_c

[1, 2]. However, Chynoweth's explanation of discovered pyrocurrent sign reversal cannot be considered as satisfactory because possible variations of polarization screening mechanism in crystal thickness and dynamics of this process were not taken into account. That is why we attempted to study more carefully the local pyroeffect in the region near a flat FE-PE phase boundary in the unipolar barium titanate crystal.

The samples were c-, a-c-domain and a-domain rectangular plates of $2 \times 3 \times 0.2$ mm^3 size cut from the virgin unipolar barium titanate crystals grown by Remeika method [3] from solution in KF flux. The plates were edge-electroded on edges 3×0.2 mm^2 perpendicular to a-domain direction. Heater and cooler were situated near electrodes, so they allowed the flat FE-PE phase boundaries formation. Their path was directed along plate length equaled to 2 mm with necessary speed and stabilization in necessary position of plate length by created temperature gradient and its variations. Crystals resistivity near T_C was $R = 10^{10}$ Ohm·m. One of plate faces was coated with non-conductive black paste and subjected to rectangular modulated 5 mW He-Ne laser heat pulses of pyroelectric probe with probe diameter of 20 μm. The heat pulse duration was 2 ms, repetition time was 50 ms (thermal diffusion length was 30 μm). Domain structure and phase boundary position were controlled simultaneously in reflected polarized light from opposite plate face. The above modulation regime was chosen in order to prevent flat boundary distortions and undesirable contributions to pyrocurrent connected with them. Absence of such distortions had been confirmed during experiments. The samples were short-circuited and pyrocurrent pulses were observed on oscillograph screen.

Two series of experiments were conducted. In the first one, phase boundary was formed and positioned in steady state in the appropriate place of plate length with successive heat probe scanning along plate in relation to the phase boundary. In the second one, the pyroelectric probe was positioned in the appropriate place of plate length with successive phase boundary scanning in relation to the pyroelectric probe. Typical dynamic local pyroelectric activity P_{dyn} distributions (in arbitrary units) versus probe distance to the phase boundary for a- and a-c-domain plates and for c-domain are shown in Figures 1, 2 and 3, respectively. We use here the term pyroactivity instead of pyrocoefficient because there exists a possibility of contributions to pyrocurrent other than contributions connected only with spontaneous polarization variations. This will be discussed later.

In Figure 1, local pyroactivity P_{dyn} distribution versus distance to the steady phase boundary and P_{dyn} variations in time are represented. The values of P_{dyn} decrease with successive sign reverse in the FE phase, achieves minimum of reversed P_{dyn} on phase boundary and returns in PE phase to P_{dyn} sign of FE phase remote of phase boundary. This P_{dyn} distribution is typical for a– and a–c–domain crystals. Note that when the phase boundary was at that length placed after it had been shifted from left to right (transition from PE to FE phase) and P_{dyn} distribution was similar except some amplitude variations.

In Figure 2 P_{dyn} distribution for a- and a-c-domain crystal plates versus distance to the moving boundary with probe near left electrode is present. Similar distributions without sign reverse were observed also in low external electric fields and on crystals poled in high electric fields. When a-and a-c-domain crystals were rapidly transfered to PE phase P_{dyn} signal decreased with time constant equal to Maxwell relaxation time $\varepsilon \varepsilon_0 R \sim 10^3$ s.

---

*delta-46@mail.ru.

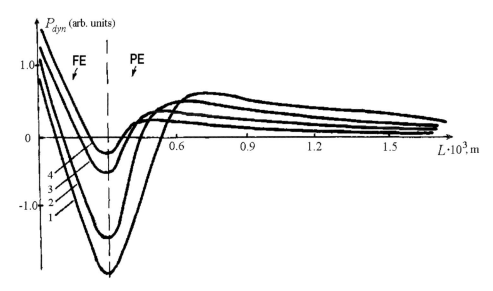

Figure 1. Local pyroactivity P_{dyn} distribution vs the distance to the steady phase boundary (1– immediately after stoppage; 2, 3 and 4 – after 1, 3 and 5 min, respectively).

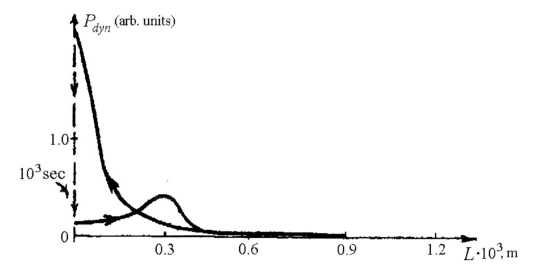

Figure 2. Local pyroactivity distribution vs the distance to the moving phase boundary (probe near left electrode, phase boundary shift directions are shown by arrows).

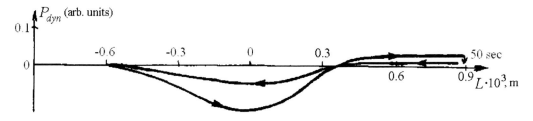

Figure 3. Local pyroactivity P_{dyn} distribution vs the distance to the moving phase boundary when probe is in the middle of plate length (c-domain crystal).

In Figure 3 P_{dyn} distribution versus distance to the moving boundary with probe in the middle of plate length is presented. This distribution is the typical one for all c-domain crystals with some variations in the distance to the boundary on which registration of pyroactivity begins, in most cases it was smaller than that shown in Figure 3. In all experiments with moving boundary its shift speed was equal to 60 microns per second.

Let us calculate dynamic pyrocurrent J for a plate with FE-PE phase boundary in two cases: (I) probe Δx is in FE region of length l, and (II) probe is in PE region of lenth $L - l$, where L is a plate length. Pyrocurrents in the plate regions in the case (I) are found as

i. in the probe region:
$$\frac{dP_s}{dt} + \frac{l}{4\pi}\frac{d}{dt}(\varepsilon_f E_f') \; ;$$

ii. in the FE region out of probe:
$$\frac{l}{4\pi}\varepsilon_f \frac{dE_f''}{dt} \; ;$$

iii. in the PE region out of probe:
$$\frac{l}{4\pi}\varepsilon_p \frac{dE_p}{dt} \; .$$

Currents are equal in all regions, thus

$$4\pi\frac{dP_s}{dt} + \frac{d}{dt}(\varepsilon_f E_f') = \varepsilon_f \frac{dE_f''}{dt} = \varepsilon_p \frac{dE_p}{dt} \; .$$

Short-circuit condition gives

$$\frac{dE_f}{dt}\Delta x + \frac{dE_f}{dt}(l - \Delta x) + \frac{dE_p}{dt}(L - l) = 0 \; .$$

Calculation for the case (II) may be conducted in a similar way. Then the pyrocurrents in the cases (I) and (II) are equal to

$$J_I = \frac{dP_s/dT + (l/4\pi)E_f(d\varepsilon_f/dT)}{L[(\varepsilon_f/\varepsilon_p) + (l/L)(1 - \varepsilon_f/\varepsilon_p)]}\Delta x\frac{dT}{dx} \; ;$$

$$J_{II} = \frac{\varepsilon_f}{\varepsilon_p}\frac{l}{4\pi}\frac{E_p'(d\varepsilon_p/dT)}{L[(\varepsilon_f/\varepsilon_p) + (l/L)(1 - \varepsilon_f/\varepsilon_p)]}\Delta x\frac{dT}{dx} \; .$$

Here ε_f and ε_p are permittivities in FE and PE phases, respectively; E_p, E_p', E_f', E_f'' are the electric fields in various plate regions.

It is obvious that pyrocurret in the PE region may exist only when there exists electric field no parallel to electrodes. If such electric field exists then in the FE and in PE phases there exists a contribution to pyrocurrent proportional to $E(d\varepsilon/dT)$ which is positive in the

FE phase and negative in the PE phase. When polarization variations are absent only this term will contribute to pyrocurrent. If E did not change sign in both phases then pyrocurrent must change sign because of $(d\varepsilon/dT)$ sign changing. We consider that just this case is shown in Figure 3 as here polarization direction is parallel to electrodes. Nevertheless, results shown in Figures 1 and 2 cannot be explained only on the basis of $E(d\varepsilon/dT)$ contribution to pyrocurrent. As the basis of existence of E in the vicinity of the FE-PE phase boundary, we consider two reasons, namely: (i) existence of unipolar field E_{un} which is due to the gradient of defects formed during crystal growth, then the spontaneous polarization formed near TC by means of concentration variations of antiparallel domains "screens" this field if one is high enough and forms unipolar polarization P_{un}; and (ii) existing field of unscreened boundary charge at phase boundary. Because other fields are absent (Figure 3) then we consider P_{dyn} sign reversal here as the evidence of the presence of E_{un} field directed parallel to crystal faces. When unipolar polarization is directed perpendicular to the phase boundary, then in accordance with [4, 5] in the low symmetry (FE) phase a-c- and c-domain twinning must take place. That is why boundary charge screening begins in the FE region (polarization in the region of domain twinning is close to zero), but twinning cannot screen boundary charge in whole (see Figure 1), and extra free charge from the PE region is needed. This free charge lags behind the boundary charge center which now is situated in the FE phase in distance of domain twinning of phase boundary. That is why electric field appears in the FE region (directed opposite to Pun) due to twinning and in the PE regions (in both E direction coincides with P_{un} direction in the FE phase). Then near phase boundary shown in Figure 1 the main contributions to P_{dyn} in various regions are: (i) in the FE phase remote of phase boundary $(dP_s/dT) < 0$, (ii) in the region of twinning $E(d\varepsilon_f/dT) > 0$, (iii) in the PE region $E(d\varepsilon p/dT) < 0$. Free charge lagging decreases with time and the region of domain twinning narrows. When there exists external electric field, for example contact field as in Figure 2, or polarization in crystal plate is rigidly fixed after plate poling in high electric field, then domain twinning becomes impossible or take place partially and the only screening source to be the free charge. Existence of this redistributed free charge may be confirmed by decrease of Pdyn with time constant equal to Maxwell relaxation time after rapid crystal transfer to PE phase (see Figure 2).

Thus sign reverse of P_{dyn} at heating near T_C with successive sign return in PE phase to initial one is an indicator of polarization screening due to domain twinning. Absence of such behavior of P_{dyn} is an indicator of screening by free charge.

2. ANOMALIES OF PYROELECTRIC AND DIELECTRIC PROPERTIES OF FERROELECTRIC $PbZr_{1-x}Ti_xO_3$ ($0.06 \leq X \leq 0.35$) CERAMICS AT THE $R3C \leftrightarrow R3M$ PHASE TRANSITION

Despite the large number of experimental studies devoted to the $PbZr_{1-x}Ti_xO_3$ (PZT) system, only several of them [6 – 8] report detailed X-ray diffraction data for the phase $x-T$ diagrams. In [7], such data are made it possible to find for the first time two tricritical points on the line of transitions to the paraelectric phase, at $x = 0.22$ and 0.55, which limit the region of second-order phase transitions. In [8], the range of existence of the monoclinic phase in

PZT was discovered and determined. We made an attempt to reproduce the fragment of the x–T diagram of the PZT binary system at $0.06 \leq x \leq 0.35$ based on the synchronous investigation of the temperature dependences of the pyroectric and dielectric properties.

The samples were prepared by the standard ceramic technology in the form of disks of 10 mm in diameter and of 1 mm thickness, with fired silver electrodes. The samples were polarized by applying a *dc*-electric field of 2×10^5 V/m after heating above the Curie point (T_C + 30°C) with subsequent cooling to 80°C for 1 h. Measurements of the temperature dependences of the relative permittivity $\varepsilon(T) = \varepsilon^T{}_{33}/\varepsilon_0$ (where $\varepsilon^T{}_{33}$ is the permittivity of the polarized sample, ε_0 is the permittivity of vacuum), dielectric loss tangent tan $\delta(T)$, and the dynamic pyroelectric current $i_{dyn}(T)$ were performed using the setup described in [9], modified by connecting to computer and using the L-CARD software [10]. The experimental technique included initial analysis of $\varepsilon(T)$ and tan $\delta(T)$ of each unpolarized sample using a CE5002 digital automatic bridge to determine the possibility of applying the capacitance meter method and then, after polarization, synchronous detection of $\varepsilon(T)$ and $i_{dyn}(T)$ using the setup [9,10]. The effect of the external *dc*-bias electric field E_b in the range from 0.5 up to 5×10^5 V/m on the temperature of the $R3c \leftrightarrow R3m$ phase transition was investigated using the measurement channel i_{dyn} and a block of accumulators.

Figure 4 shows the $\varepsilon(T)$ dependence for the sample with $x = 0.07$ upon heating (solid line) and cooling (dotted line). Jumps of $\varepsilon(T)$ in 100-magnified fragment of this dependence in the corresponding temperature range (from 30 up to 70°C) are shown in the inset. Upon repeated heating or cooling of a depolarized sample, the position of peaks of $\varepsilon(T)$ on the temperature scale for this fragment and near T_C retains. Similarly, $\varepsilon(T)$ curves were obtained for other polarized, depolarized, and unpolarized samples in the concentration range $0.06 \leq x \leq 0.15$. With an increase in concentration ($x > 0.15$), low-temperature peaks are absent in $\varepsilon(T)$ and tan $\delta(T)$, and the temperature hysteresis of $\varepsilon(T)$ near T_C decreases from $\Delta T = 12.5$°C at $x = 0.07$ down to $\Delta T = 2$°C at $x = 0.3$.

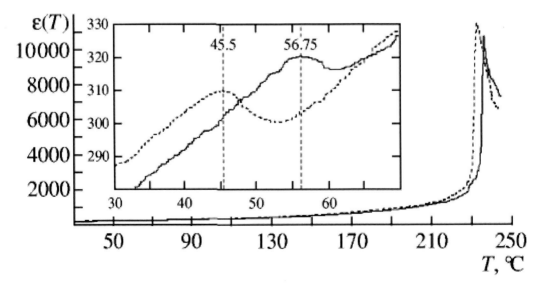

Figure 4. Temperature dependences $\varepsilon(T)$ for a PbZr$_{0.93}$Ti$_{0.07}$O$_3$ sample upon heating and cooling (solid and dashed lines, respectively).

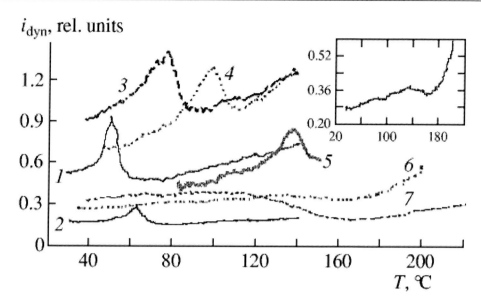

Figure 5. Dependences $i_{dyn}(T)$ for the PbZr$_{1-x}$T$_x$O$_3$ ferroelectric ceramic at (1) $x = 0.06$, (2) $x = 0.07$, (3) $x = 0.08$, (4) $x = 0.1$, (5) $x = 0.15$, (6) $x = 0.25$, and (7) $x = 0.3$.

Figure 5 shows the temperature dependences $i_{dyn}(T)$ for the samples with $0.06 \leq x \leq 0.35$ near the temperatures of the $R3c \leftrightarrow R3m$ phase transition. Figure 5 indicates that the low-temperature peak in $i_{dyn}(T)$ for compositions of the PZT binary system shifts to higher temperatures with an increase in the concentration x up to 0.25. However, with a further increase in x, this maximum smoothes out, exhibiting a high plateau for $x = 0.3$, whose extremum shifts to lower temperatures. A typical $i_{dyn}(T)$ dependence in the entire interval of existence of the polarized state, from T_{room} up to T_C for the PbZr$_{0.75}$Ti$_{0.25}$O$_3$ composition is shown in the inset into Figure 5.

Investigations of $i_{dyn}(T)$ under the action of the field E_b did not reveal any shifts of the $R3c \leftrightarrow R3m$ phase transition temperature upon heating and cooling. Further increase in $E_b > 5 \times 10^5$ V/m led to a sharp increase in the noise component in $i_{dyn}(T)$ and did not allow us to perform correct measurements within this study.

Figure 6 shows two x–T phase diagrams of PZT ($0.06 \leq x \leq 0.30$), which were obtained according to the pyroelectric and dielectric data for ceramics (□, +) and X-ray diffraction data for single crystals (○) [7]. They show good correlation in the heating mode. In addition, based on our data, the temperatures of the $R3m \rightarrow R3c$ phase transition were determined upon cooling the samples of ferroelectric ceramics (□).

The X-ray diffraction studies of the phase transition between the rhombohedral phases $R3m$ and $R3c$ in crystals [7] in the range $0.06 \leq x \leq 0.15$ revealed a small jump in the PZT unit-cell volume, corresponding to the first-order phase transition. For $x > 0.16$, stepwise changes in the volume and rhombohedral angle at the $R3m \rightarrow R3c$ phase transition have not been found. The transition point could be established only from the characteristic kink in the temperature dependence of the unit-cell edge length. It was noted in [7] that this fact cannot be explained only by diffusion of the phase transition due to the composition fluctuation, since the next portion of the $R3m \rightarrow R3c$ transition line in the phase diagram is flatter than the rising one, and the largest diffusion of the phase transition should be expected in the composition range $0.25 \leq x \leq 0.4$. In this context, it was assumed that the type of the $R3m \rightarrow$

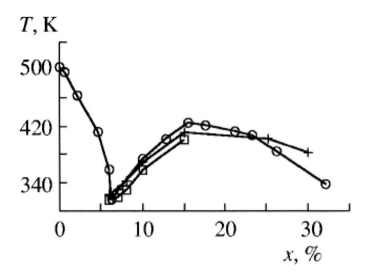

Figure 6. $x-T$ phase diagram of the binary ferroelectric ceramic PbZr$_{1-x}$Ti$_x$O$_3$, according to the data of pyroelectric and dielectric measurements of the ferroelectric ceramics upon heating (I) and cooling (□)compared with the X-ray diffraction data (○) for single crystals [7], obtained in the heating mode.

$R3c$ phase transition for $0.16 \leq x \leq 0.36$ is at least close to the second-order transition. This suggestion is justified well by the investigation of the dependences of the pyroelectric and dielectric properties in the temperature range from T_{room} up to T_C. In the compositions $0.06 \leq x \leq 0.15$, low-temperature peaks in $\varepsilon(T)$, tan $\delta(T)$, and $i_{dyn}(T)$ clearly manifest themselves. They correspond to the $R3m \rightarrow R3c$ phase transition with a hysteresis from $\Delta T = 12.5$ °C for $x = 0.07$ down to $\Delta T = 5$ °C for $x = 0.15$. Similar changes in ΔT are observed at T_C. In the compositions with x increased up to 0.20, the low-temperature peaks in $\varepsilon(T)$ and tan $\delta(T)$ diffuse with a decrease in the amplitude and ΔT.

In the compositions with $0.20 \leq x \leq 0.35$, low-temperature anomalies in $\varepsilon(T)$ and tan $\delta(T)$ are not detected, and, $\Delta T \sim 0$ at T_C for $x = 0.35$. This fact indicates that, for the PZT compositions in this range of concentrations x, the phase transitions in the $R3c \rightarrow R3m \rightarrow Pm3m$ line are close in type to the second-order phase transition. It should be noted that only pyroelectric measurements ($\gamma_{dyn}(T)$) make it possible to detect the changes in polarization that are related to the $R3c \leftrightarrow R3m$ phase transition in compositions with x up to 0.3. The investigation performed shows a high degree of correlation between the electric properties of single-crystal and polycrystalline samples of the PZT binary system.

3. TEMPERATURE DEPENDENCES OF PYROELECTRIC AND DIELECTRIC PROPERTIES OF PBZR$_{1-x}$TI$_x$O$_3$ ($0.37 \leq x \leq 0.57$) CERAMICS

This section presents the results of studying the pyroelectric and dielectric properties of solid-solution (SS) of the PZT ceramics the analysis of which is performed based on their real structure. Actuality of such investigations is due to the fact that the modern highly-effective

PZT-based ferro-piezoelectric materials are manufactured, mainly, in the form of ceramics – the objects having a compicated hierarchical structure with different types of imperfections.

The objects for studying were prepared by the usual ceramic technology from the high-purity oxides: PbO, TiO$_2$ and ZrO$_2$.

Discs of 10 mm diameter and of 1 mm thickness coated with Ag electrodes were poled in dc-electric field $E = (5 - 7) \times 10^6$ V/m at $T \sim 140$ °C for 40 min with the subsequent cooling down to 80°C. Registration of $i_{dyn}(T)$ was conducted at the frequency of 6.5 Hz by the sinusoidal modulation of the infrared radiation flow, and $i_{qu.st.}(T)$ was recorderd in the process of heating and cooling at a rate of 0.03 – 0.1 K/s synchronously for each sample. Simultaneously the temperature dependences of the relative dielectric permittivity $\varepsilon(T)$ were determined: $\varepsilon(T) = \varepsilon^T_{33}/\varepsilon_0$ (where ε^T_{33} is the dielectric permittivity of the poled sample and ε_0 is the permittivity of vacuum) with the help of ac-bridge E7-20. Moreover, the technique of studing involved a thermocycling of the samples according to the following scheme: heating from T_{room} up to T_1 → stabilization at T_1 during time t_1 → cooling down to $T_2 < T_1$ → stabilization at T_2 during time t_2. Each following cycle was accompanied by the increase of T_1 to the values no breaking the poled state. Registering the information and its processing were performed by using personal computer equipped by the system of data collection and the L-CARD software.

Figure 7 shows the temperature dependence $i_{dyn}(T)$ for PbZr$_{0.63}$T$_{0.37}$O$_3$ sample characterizing all stages of the thermally induced phase transitions (PT) $R3c \rightarrow R3m \rightarrow P4mm$ for the given x concentration established in Reference [10] by using X-ray studies. The PT boundaries are indicated by the dashed lines fixed at the $i_{dyn}(T)$ maxima. No anomalies were observed on the $\varepsilon(T)$ and $i_{qu.st.}(T)$ dependences.

With the aim to determine a repetition of the obtained results the repeated thermocycling with temperature fixation of the chosen isothermal range during 5 – 10 min was carried out. Figure 8a presents the fragments of the thermocycling process for the sample (with $x = 0.37$) with the recording $i_{qu.st.}(T)$ and $i_{dyn}(T)$. During heating and cooling $i_{dyn}(T)$ shows the stable

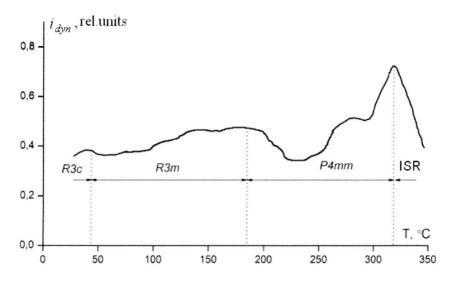

Figure 7. Temperature dependence of $i_{dyn}(T)$ for the PbZr$_{0.63}$T$_{0.37}$O$_3$ samples.

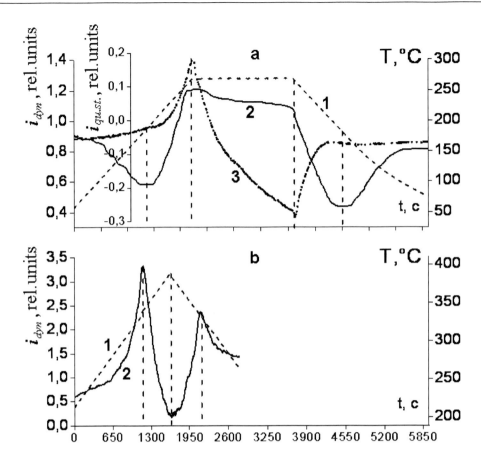

Figure 8. The plot of temperature variation: (1) $T(t)$, (2) $i_{dyn}(t, T)$, (3) $i_{qu.st.}(T)$ in the case of thermocycling of the sample with $x = 0.37$ up to $T = 250$ °C (a) and $T = 380$ °C (b).

minimum at $T \sim 180$ °C while at approaching $T \sim 250$ °C one relaxes slowly with increasing amplitude. At the same time, $i_{qu.st.}(T)$ increases with increasing temperature, at $T \sim$ const one inverts the sign during 5 min, and during 30 min it is reached the maximum value exceeding the initial one by a factor of 1.5 in modulus. The decrease of $i_{dyn}(T)$ with time at $T \sim$ const is caused by depolarization, whereas the behavior of $i_{qu.st.}(T)$ cannot be related only to the pyroelectric effect of ferroelectric polarization of the sample. Figure 8b shows a disappearance of the dynamic pyroelectric activitiy at $T = 380$ °C and its recovery at cooling. A study of dispersion of the relative dielectric permittivity was carried out in the temperature range of 370 – 400 °C and the results are present in Figure 9. Figure 9. shows the relaxor behavior of the samples of the PZT composition with $x = 0.37$ as evidenced by decreasing the magnitude of the relative dielectric permittivity maximum, its shift to the high-temperature range by 4 °C and, also, by the $\varepsilon(T)$ diffusion with increasing frequency of the measuring field. Similar, studies were conducted for the samples having the increased x concentration. An example of such a study is shown in Figure 10 for the PbZr$_{0.53}$Ti$_{0.47}$O$_3$ sample. As follows from the analysis of Figure 10, in the vicinity of T$_{1mh}$, there exist the anomalies of the above-mentioned dependences the manifestation of which will be discussed below.

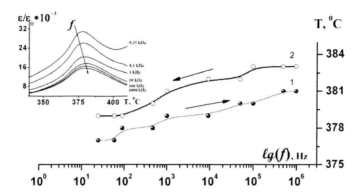

Figure 9. Temperature of max $\varepsilon/\varepsilon_0(T)$ in the vicinity of the Curie temperature versus the measuring field frequency $\lg f$ for the samples with $x = 0.37$ at heating (1), and cooling (2). The inset shows a general view of the dispersion of the $\varepsilon/\varepsilon_0(T)$ dependence.

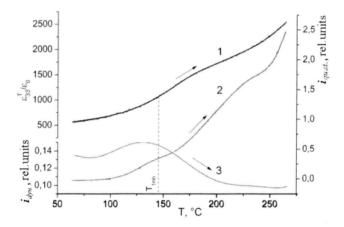

Figure 10. Temperature dependences: (1) $\varepsilon(T)$, (2) $i_{qu.st.}(T)$, and (3) $i_{dyn.}(T)$ at the heating of the PbZr$_{0.63}$Ti$_{0.37}$O$_3$ sample. T_{1mh} is the temperature interval of manifestation of the anomalies in the above dependence.

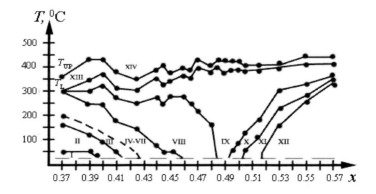

Figure 11. $x - T$ phase diagram of the PbZr$_{1-x}$Ti$_x$O$_3$ system in the compositional range studied. The phase are denoted as follows: (I) Rh$_1$; (II) Rh$_1$ +Rh$_2$; (III) Rh$_2$; (IV) Rb$_2$+Rh$_3$; (V) Rh$_3$; (VI) Rh$_3$+PSC$_1$; (VII) Rh$_3$+PSC$_1$+PSC$_2$; (VIII) Rh$_3$+PSC$_1$+PSC$_2$+T$_1$; (IX) PSC$_2$+T$_1$; (X) T$_1$; (XI) T$_1$+T$_2$; (XII) T$_2$; (XIII) ISR; (XIV) C.

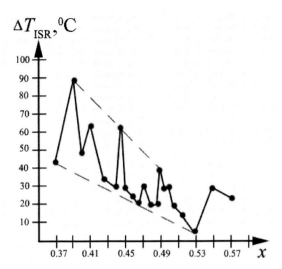

Figure 12. Width of the indistinct-symmetry region, ΔT_{ISR}, as a function of x.

According to Reference [11] from the data of X-ray studies, it follow that the phase diagram of the binary $PbZr_{1-x}Ti_xO_3$ system demonstrates the shape presented in Figure 11.

In all compositions with the increase of temperature from room to a certain (specific for each SS`) temperature T_L the distortion of noncubic (Rh, T) unit cell typical of $T > T_{LO}$ decreases ($c/a - 1 \to 0$, $a \to 900$), and at $T = T_L$ unit cell becomes practically cubic (C) (diffraction reflections are close to single ones, without any splittings peculiar to the low-symmetry phases). However at this temperature the $\varepsilon/\varepsilon_0$ maxima are not reached.

At the elevated temperature (below the Curie temperature) the X-ray studies revealed a region in which the crystal lattice parameters could not be clearly determined. It was called the indistinct-symmetry region (ISR). Figure 12 presents this region width as function of the concentration x.

Analysis of the data reported in Reference [11] and presented in Figures 11 and 12 shows that, at $0.37 \leq x \leq 0.57$, for SS of the binary PZT system, in the isothermal section of a rhombohedral (Rh) – tetragonal (T) PT there exist two intermediate phases identified (because of weak splittings) as pseudocubic (PSC1 and PSC2), and, also, indistinct-symmetry region the boundaries of which become closer as x increases. As a result, at $0.37 \leq x \leq 0.51$ in the isothermal sections the regions with different phase states of the PZT crystal lattice may exist separately or coexist. Variation in temperature and, also, the keeping at the direct current may lead to a breaking of the existing balance and a setting of a new energetic balance in such regions. This process is considerably complicated by the point defects (vacancies) caused by the heterophase compositions and the extended ones (crystallographic shear planes, imperfections connected with the infinitely adaptive TiO_2 structure, and spatial inhomogeneities of the ceramics manifesting themselves, especially, in the regions of phase transformations). Thus, the dependences $i_{qu.st.}(t, T)$ (see Figure 8a) integrating variations of all kinds of polarization and thermostimulated currents will require special methods for their interpretation. They may be the lower steps of cyclic heating the longer periods of keeping at the direct current or others but they will be possibly used in other study.

In Figure 10 the dependences $\varepsilon(T)$, $i_{dyn}(T)$ and $i_{qu.st.}(T)$ clearly illustrate a transition between phase states VIII into state IX in accordance with the PD [11] or $R3c \rightarrow P4mm$ phase transition for the PD [1]. For $i_{dyn}(T)$ it is a pronounced maximum, and for $\varepsilon(T)$ and $i_{qu.st.}(T)$ it is smooth step-like kinks on curves 1 and 2.

Thus a study of pyroelectric effect in the samples of ferroelectric materials in the dynamic and quasistatic modes enables one to determine with high precision the isothermal section in which there occur the phase variations and transformations and, also, to observe the relaxation and other physical phenomena caused by them. As a result, it becomes possible to get additional information about the state of ferroelectric polarization as compared to that obtained by the X-ray methods.

4. DIELECTRIC AND PYROELECTRIC PROPERTIES OF $(1-x)$PBFE$_{1/2}$NB$_{1/2}$O$_3 - x$PBTIO$_3$ $(0 \leq x \leq 0.08)$ FERROELECTRIC CERAMICS

Multiferroic lead ferroniobate $PbFe_{1/2}Nb_{1/2}O_3$ (PFN) is very promising base for designing new piezoelectric, electrostrictive, capacitor, pyroelectric, and posistor materials [12–16]. The preparation of PFN-based ceramics with low conductivity is a difficult problem [12, 14, 17]. In this context, the data on the phase transitions in PFN and its dielectric and other properties are rather contradictory. For example, along with the generally known phase transition at 370 − 380 K, which is assumed to be of the ferroelectric–paraelectric type, there are also data on the transition between two ferroelectric phases near ~ 355 K [17 − 21]. The room-temperature PFN structure was found to be rhombohedral in most studies [22], while in other studies it was identified as monoclinic [23]. According to the data of [19], the phase that is stable in the range of 355 − 370 K is tetragonal. Due to the high conductivity of the generally studied PFN samples, the intrinsic dielectric response is obscured by the Maxwell–Wagner relaxation and/or the relaxation related to oxygen defects [18, 23], while permittivity ε has a strong frequency dispersion. In this context, the anomaly of $\varepsilon(T)$ (corresponding to the transition near ~ 355 K) is observed in only PFN crystals and high-resistivity ceramic samples [17, 18, 20]. Among the PFN–based solid solutions, the $(1-x)$PbFe$_{1/2}$Nb$_{1/2}$O$_3-x$PbTiO$_3$ (PFN − xPT) system (promising for piezoelectronics [12, 23]) has been the one most thoroughly studied.

The $x - T$ phase diagram of these solid solutions has, however, been established only roughly; in particular, the morphotropic boundary between the rhombohedral (monoclinic) and tetragonal phases was localized only at room temperature [1, 12]. In this section, we analyze the temperature dependences of the dielectric and pyroelectric properties of PFN − xPT ceramic samples and determine (on the base of the experimental data) the temperature dependences of the position of the morphotropic boundary between the rhombohedral (monoclinic) and tetragonal phases.

PFN − xPT ceramic samples were obtained by solid-phase reactions from PbO of reagent grade and Fe_2O_3, Nb_2O_5, and TiO_2 of high purity grade. To increase the ceramics' resistivity, 1 − 2 wt % Li_2CO_3 of reagent grade was introduced into the charge [12, 17]. After powder synthesis at 850 − 900 °C during 4 h and grinding, the billets were pressed into disks of 10 mm in diameter and 2 − 4 mm thick. The samples were fired at 1050 − 1100 °C in tightly closed alumina crucibles on zirconia substrates without an atmosphere-forming charge. The

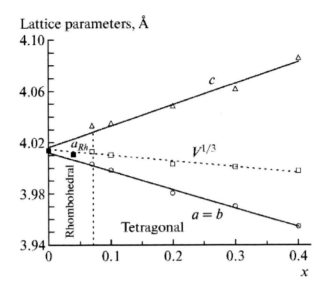

Figure 13. Concentration dependences of the unit-cell parameters of $(1-x)\text{PbFe}_{1/2}\text{Nb}_{1/2}\text{O}_3 - x\text{PbTiO}_3$ ceramic solid solutions at room temperature.

ceramics' density was 90 – 95% of the theoretical value. After grinding, electrodes were formed on the sample surfaces by the firing of a silver paste.

Permittivity ε was measured upon continuous cooling or heating at a rate of 2 – 3 K/min, using a computer-aided E7-20 immittance meter. The pyroelectric coefficient was measured upon continuous cooling or heating at a rate of 2 – 5 K/min by the quasistatic method [15, 17] and by the dynamic method, using a sinusoidal temperature modulation with a frequency of 2 – 3 Hz [24]. The dynamic pyroelectric coefficient γ was determined by comparing it against a reference (PCR-11 piezoelectric ceramic sample with known value of γ). X-ray diffraction studies were performed on DRON-7.0 diffractometer (CoKα radiation). The piezoelectric modulus d_{31} was measured by the standard resonance–antiresonance method.

X-ray diffraction studies of the compositions obtained showed that they all were single-phase and had a perovskite structure. The compositions with $x < 0.06 - 0.07$ had rhombohedral symmetry at room temperature, while the compositions with larger x values were characterized by tetragonal symmetry (see Figure 13). In [23], the symmetry of the PFN – xPT compositions with $x < 0.07$ was determined to be monoclinic. However, it is necessary to perform an X-ray diffraction analysis with synchrotron radiation in order to draw such a conclusion. With allowance for this, the obtained results are in agreement with the data in the literature [12, 19]. The concentration dependence of the pseudocubic-cell parameter ($\tilde{a} = V^{1/3}$) in the PFN – xPT system is linear (Figure 13). This suggests the formation of solid solutions over the whole concentration range under study.

The obtained PFN – xPT ceramics had a low dielectric loss due to doping with lithium (Figure 14, a). As a result, the frequency dispersion of ε in the studied samples was fairly low (Figure 14, b), and we may conclude that the dielectric properties were determined by the grain volume rather than the Maxwell–Wagner relaxation (related to the difference in the conductivities of the bulk and surface grain layers), as this generally occurs in highly conductive PFN ceramics [17, 18, 23].

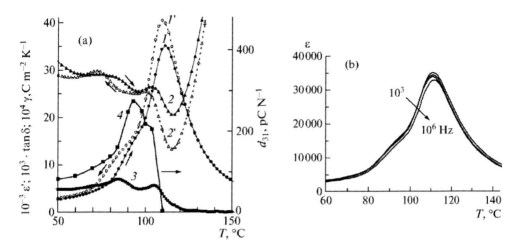

Figure 14. (a) Temperature dependences of (1, 1') permittivity ε, (2, 2') tan δ, (3) dynamic pyroelectric coefficient γ, and (4) piezoelectric coefficient d_{31} of PFN – 0.02PT ceramic samples, measured in the (1 – 4) heating and (1', 2') cooling modes; dependences 1', 2, and 2' were measured at a frequency of 10^3 Hz. (b) Temperature dependence of permittivity ε, measured at frequencies of 10^3, 10^4, 10^5, and 10^6 Hz (from top to bottom).

The temperature dependences of the real part of the permittivity, the loss tangent tan δ, the dynamic pyroelectric coefficient γ, and the piezoelectric coefficient d_{31} of the PFN – 0.02PT ceramic sample are shown in Figure 14. It can be seen that the temperature dependences of all these parameters have pronounced anomalies not only near the ferroelectric – paraelectric phase transition but also near the ferroelectric – ferroelectric transition. This allows us to determine the concentration dependence of the ferroelectric–ferroelectric phase transition in the PFN – xPT system. The temperature of this transition was found to decrease while the temperature hysteresis increased with an increase in the PbTiO₃ content (see Figure 15) This character of change in the ferroelectric–ferroelectric phase transition temperature for the PFN – xPT system is in agreement with the conclusion (drawn in [19] on the basis of X-ray diffraction studies of single crystals) that PFN passes from the cubic phase to the tetragonal phase upon cooling.

The $x - T$ phase diagram of the PFN – xPT system, plotted on the base of the dielectric and pyroelectric data [10], is shown in Figure 16. It can be seen that in the range of low ($x < 0.05$) PbTiO₃ contents, the dependence of the ferroelectric–ferroelectric transition temperature on x is relatively weak. At larger x, the rate of decrease rises with an increase in x, and at $x \approx 0.08$ the morphotropic phase boundary in the $x - T$ phase diagram becomes nearly vertical.

For the compositions with $x > 0.08$, no anomalies related to the ferroelectric–ferroelectric phase transition were observed in the temperature dependences of ε and the pyroelectric coefficient (down to 10 K). A change in the slope of the concentration dependence of the transition from the tetragonal phase to the cubic phase was observed in the range of $x \approx 0.07 - 0.08$. On the whole, the $x - T$ phase diagram of the $(1 - x)$PbFe₁/₂Nb₁/₂O₃ – xPbTiO₃ system resembles the corresponding diagram of the known system of $(1 - x)$PbMg₁/₃Nb₂/₃O₃ – xPbTiO₃ relaxor ferroelectric solid solutions [25].

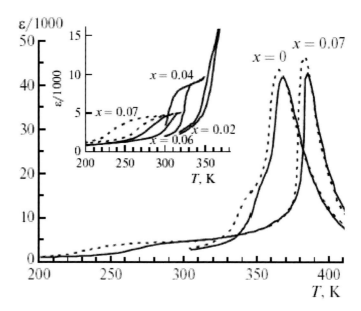

Figure 15. Temperature dependences of ε of the $(1-x)\text{PbFe}_{1/2}\text{Nb}_{1/2}\text{O}_3 - x\text{PbTiO}_3$ ceramic samples, measured at the frequency of 1 kHz upon heating (solid line) and cooling (broken line). The inset shows the anomalies of $\varepsilon(T)$ corresponding to the ferroelectric–ferroelectric phase transition for the compositions with different x values.

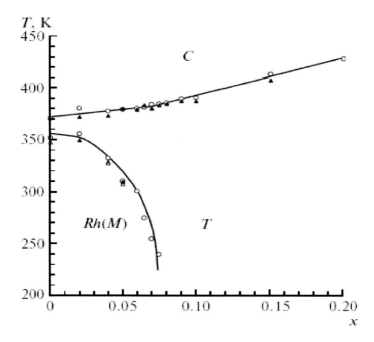

Figure 16. $x - T$ phase diagram of the $(1-x)\text{PbFe}_{1/2}\text{Nb}_{1/2}\text{O}_3 - x\text{PbTiO}_3$ system, plotted based on the dielectric (dots) and pyroelectric (triangles) data. The values of the pyroelectric effect measured in the quasi-static and dynamic modes are shown by white and black triangles, respectively. Rh (M), T, and C indicate the rhombohedral (monoclinic), tetragonal, and cubic phases, respectively.

5. PYROELECTRIC AND DIELECTRIC PROPERTIES OF THE $(1-x)$PBNB$_{2/3}$MG$_{1/3}$O$_3$ – xPBTIO$_3$ $(0.14 \leq x \leq 0.42)$ CERAMICS

For the system $(1-x)$PbNb$_{2/3}$Mg$_{1/3}$O$_3$ – xPbTiO$_3$ (PMN – PT) a phase diagram with such a succession of phases as cubic (C) $\rightarrow$ rhombohedral (Rh) $\rightarrow$ tetragonal (T) and morphotropic boundaries at $x \sim 0.15$ and 0.35 was considered as generally accepted for a long time [25]. Recently, it was revised because it had been established by experiments that inside the Rh – T phase transition there exits a monoclinic (M) phase [26 – 28] with the space group P_m (of the M_c-type in the notation of Vanderbilt and Cohen [29]) and the stability regions: $0.31 \leq x \leq 0.37$ at 20 K and $0.305 \leq x \leq 0.345$ at room temperature [28]. The results of studying some solid solutions (SS) of this system are reported in the literature. The more precise coordinates of existence of the morphotropic region corresponding to $0.28 \leq x \leq 0.42$ were obtained in Reference [30] based on the systematic detailed study of the X-ray structure parameters of the PMN – PT system in a whole range of solubility of the components ($0 \leq x \leq 1.0$).

The objects for studying were SS of the system $(1-x)$PbNb$_{2/3}$Mg$_{1/3}$O$_3$ – xPbTiO$_3$ ($0.14 \leq x \leq 0.42$). The ceramic samples of SS in the chosen concentration range were prepared by the columbite technology preventing the appearance of the intermediate phase with pyrochlore structure which substantially deteriorates many electrophysical properties of the ceramics.

Figure 17 shows the temperature dependences $i_{dyn}(T)$ (a) and $\varepsilon(T)$ (b) for the samples from MR-1 with the concentration of $x = 0.14$ (1), $x = 0.18$ (2), and $x = 0.20$ (3). With the increase of x one can observe the increase of absolute $i_{dyn}(T)$ values in the whole range of variations in the temperature from T_{room} to T_{max}. At $x = 0.2$, a maximum magnitude of $i_{dyn}(T)$ is reached for the sample of the PMN – PT ceramics under study in this MR.

Figure 18 presents the temperature dependences of $i_{dyn}(T)$ (a) and $\varepsilon(T)$ (b) for the samples from the MR-2 with the concentration of $x = 0.28$ (1), $x = 0.29$ (2), $x = 0.31$ (3), $x = 0.32$ (4), $x = 0.34$ (5), $x = 0.38$ (6) and $x = 0.42$ (7). Here, the maximum $i_{dyn}(T)$ values correspond to $x = 0.32$. The further increase of x leads to a drastic decrease of $i_{dyn}(T)$. It should be noted that, in the concentration range $0.27 \leq x \leq 0.31$, on the dependences $i_{dyn}(T)$ there appear the additional low-temperature maxima at T_{1m}, and depolarization of the samples takes place after passing the second maximum by $i_{dyn}(T)$ at T_{2m} (see Figure 18a, curves 1, 2 and 3). Below the temperature of the phase transition (PT) from ferroelectric (FE) into paraelectric (PE) state ($T_{2m\varepsilon}$), on the $\varepsilon(T)$ dependences one can also observe the anomalies in the form of smooth step-like rises in the region of T_{1m} (see Figure 18b, curves 1, 2 and 3). Figure 19 shows the x – T phase diagram for the PMN – PT system plotted using the data on the concentration dependences of $T_{1m}(x)$, $T_{2m}(x)$ and $T_{2\varepsilon}(x)$. Here, curves 1, 2 and 3 represent a transition from FE into PE state, the initial boundary of the temperature range of complete depolarization and a region of transition between the phases with rhombohedral and tetragonal distortions of the crystal lattice, respectively. Comparison of Figure 19 with the x –T phase diagram obtained from the results of pyroelectric, dielectric and X-ray studies reported in Reference [26, 28] shows their good agreement. Also, the results shown in Figure 19 enable one to determine a temperature range of the expected relaxor state in PMN – PT situated in the space between the T_{2m} and $T_{2m\varepsilon}$ curves.

The maximum magnitude of pyroelectric activity for all the studied compositions of the PMN – PT system determined for $x = 0.32$ may be due to the optimal concentration ratio of the Rh, M and T phases at this point. Such materials already find a wide application in multilayer capacitors with high dielectric permittivity, piezoelectric elements and electrostrictive actuators with large deformation, and extremely low hysteresis. The results obtained in the present work show that these materials are also potentially attractive for being used in uncooled detectors of infrared radiation.

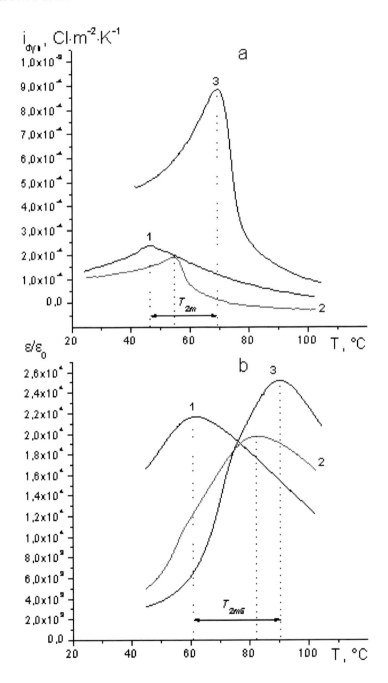

Figure 17. Dependences $i_{dyn}(T)$ (a) and $\varepsilon(T)$ (b) for the $(1-x)$PMN $-x$PT samples with values of $x = 0.14$, $x = 0.18$, and $x = 0.20$ (curves 1, 2 and 3, respectively) (MR-1).

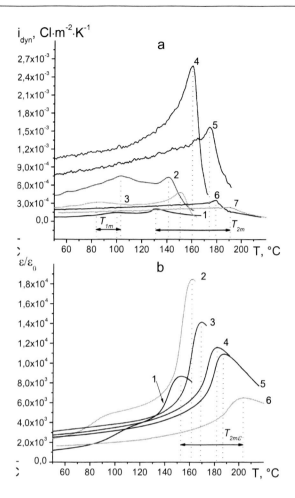

Fiure. 18. Dependences $i_{dyn}(T)$ (a) and $\varepsilon(T)$ (b) for the $(1 - x)$PMN – xPT samples with values of $x =$ 0.28, $x = 0.29$, $x = 0.31$, $x = 0,32$, $x = 0.34$, $x = 0.38$, and $x = 0.42$ (curves 1, 2, 3, 4, 5, 6, 7, respectively) (MR-2).

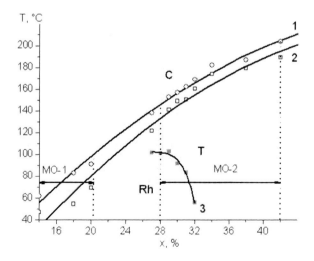

Figure 19. The $x - T$ phase diagram of the PMN – PT system obtained from the data on (1) $T_{2m\varepsilon}(x)$, (2) $T_{2m}(x)$, and (3) $T_{1m}(x)$.

6. Pyroelectric and Dielectric Properties of the $(1-x)$PMN $- x$PT $(0.14 \leq x \leq 0.42)$ Solid Solutions under the Action of a Bias Electric Field

A giant piezoelectric response of the single crystals of solid-solutions of ferroelectrics – relaxors based on the systems $(1-x)$PbMg$_{1/3}$Nb$_{2/3}$O$_3$ $- x$PbTiO$_3$ (PMN – PT) and $(1-x)$PbZn$_{1/3}$Nb$_{2/3}$O$_3$ $- x$PbTiO$_3$ has been under investigation for more than 10 years [31 – 34]. Considerably higher piezoelectric properties of the single crystals compared to those of the ceramics may be caused by the fact that in the crystals the electric field E may be applied in a certain crystallographic direction [33]. At the same time, similar high (more than 1500 – 2000 pC/N) values of the coefficients of the direct piezoelectric effect for the polycrystalline materials could not be obtained up to now [31]. In 2006, the giant piezoelectric response of single crystals-relaxors was successfully explained. It has been established that the maximum values of the piezoelectric coefficients can be observed at some finite electric field strength rather than at the zero one. This may be due to the presence of a critical point on the $E-T$ phase diagram and a critical behavior of the system near this point [34]. This assumption enables one to explain the results obtained more than 15 years ago which evidenced a drastic increase of the piezoelectric response of the PMN – PT ceramics in a direct electric field [35]. Similar results for a large group of ferroelectric ceramics were obtained later in Reference [36]. It was assumed that the effect of a giant electric field that is induced pyroelectric activity observed in the ceramic PMN-PT samples and some other relaxors [38, 39] is a critical phenomenon, too.

For the composition 0.82PMN $-$ 0.18PT, Figure 20 presents the temperature dependences of the relative dielectric permittivity $\varepsilon/\varepsilon_0$ at the 1 kHz frequency and the pyroelectric coefficient i_{dyn} measured in the regime of field heating (FH $-$ a, b) and field cooling (FC $-$ c, d), respectively, at different values of the bias electric field E. Similar results were obtained for the ceramics with $0.14 \leq x \leq 0.42$. Figure 21 shows the dependences $i_{dyn\ max}(E)$, $\varepsilon_{max}(E)$ and $T_{max,\varepsilon}(E)$ for compositions with $x = 0.26, 0.32$ and 0.34 obtained from the data similar to those shown in Figure 20.

Figure 21 (a, b and c) illustrates the field dependences of heights of the dielectric permittivity maxima $\varepsilon_{max}(E)$, their temperatures in the maxima $T_{max,\varepsilon}(E)$, and, also, heights of the pyroelectric coefficient maxima $i_{dyn\ max}(E)$, respectively. Figure 21 also presents the notation for the curves with respect to the PbTiO$_3$ content (x) and the direction of variation in temperature of the samples under study: FC for cooling and FH for heating.

It is well known that PMN – PT compositions exhibit typical relaxor properties in the range of $x < 0.2$ while with increasing titanium content in the range of $x > 0.2$ the relaxor behavior gradually weakens [31, 32, 35, 40].

Thus, Figure 21 presents typical ferroelectric-relaxor composition ($x = 0.18$), weak relaxor ($x = 0.26$) on the side of lower boundary of the morphotropic region (MR) for PMN – PT, and the very weak relaxors ($x = 0.32$ and $x = 0.34$) inside the MR. Let us consider them in accordance with the increase of the PbTiO$_3$ concentration.

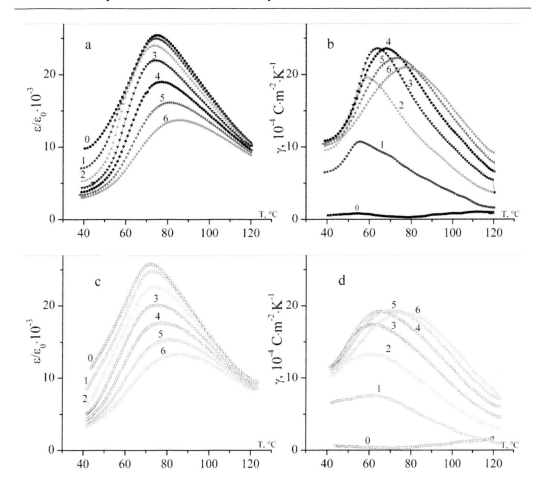

Figure 20. Dependences of the relative dielectric permittivity $\varepsilon(T)$ at 1 kHz and the pyroelectric coefficient $i_{dyn}(T)$ measured in the regime of heating in the electric field (FH – a, b) and cooling in the electric field (FC – c, d), respectively, at different values of the bias electric field E for 0.82PMN – 0.18PT ceramics. The values of E (in kV/cm) are given near the corresponding $\varepsilon(T)$ and $i_{dyn}(T)$ curves.

For $x = 0.18$, a typical feature is a considerable decrease of $\varepsilon_{max}(E)$ with increasing E, presence of temperature hysteresis at FH and FC, and reaching the points of maximum and minimum on the dependences $i_{dyn\ max}(E)$ and $T_{max,\varepsilon}(E)$ for FH at $E \sim 3$ kV/cm. Variation of the pyroelectric coefficient with E resembles a critical behavior and the data obtained suggest that at $E \sim 3$ kV/cm for PMN – PT composition with $x = 0.18$ there is a critical point, which is similar to the critical point on the liquid-vapor PT line [34].

For the composition with $x = 0.26$ one can also observe a considerable but hysteresis-free decrease of $\varepsilon_{max}(E)$; $i_{dyn\ max}(E)$ reaches the maximum values at the frequency of 1 kV/cm and, then, it does not change with increasing E, like to $T_{max,\varepsilon}(T)$.

The maximal value of $i_{dyn\ max}(E)$ for the PMN – PT ceramics ($35 - 37 \times 10^{-4}$ C·K^{-1}·m^{-2}) was observed for a composition with $x = 0.32$ in the range of the electric field strength from 3 up to 6 kV/cm. For this composition one can also observe a hysteresis-free dependence of $\varepsilon_{max}(E)$ at FH and FC.

As it is shown in Figure 21 for $x = 0.34$, the further increase of x up to 0.42 leads to a drastic decrease of $i_{dyn\ max}(E)$. The maximum values of $i_{dyn\ max}(E)$ for the PMN – PT ceramics

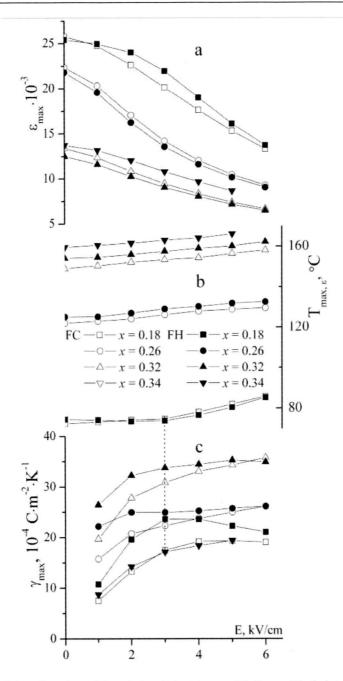

Figure 21. The heights of maxima of the relative dielectric permittivity $\varepsilon_{max}(E)$, their temperatures T_{max}, and the pyroelectric coefficient $i_{dyn\,max}(E)$, versus the bias electric field values E determined in the field cooling (FC) and field heating (FH) modes.

under study achievable in the MR at $x = 0.32$ may be supposedly due to the electric field that is induced additional contributions into the pyroelectric effect of the piezoelectric effect in microregions with the crystallographic orientations [011] and [111] of the polarization vector [41].

6. IRREVERSIBLE WIDENING OF THE TEMPERATURE RANGE OF THE ORTHORHOMBIC ANTIFERROELECTRIC PHASE IN PBZR$_{1-x}$TI$_x$O$_3$ (0.02 ≤ x ≤ 0.05) CERAMICS

Previously [42] we have ascertained the feasibility of reversible change in the temperature hysteresis value and diffusion of dielectric anomaly corresponding to the transition between antiferroelectric (AFE) and ferroelectric (FE) phases in PbZr$_{1-x}$Ti$_x$O$_3$ (PZT) ceramics with $0.03 \leq x \leq 0.05$ by varying the heating and cooling temperatures during thermal cycling. Here we report the results of studying the influence of long-term (more than 1 h) thermal exposure (thermostating) at $T > 450$ °C on the AFE $\leftrightarrow$ FE phase transition (PT) temperature.

The specimens were prepared by two independent methods. In the first method, we used the conventional ceramic technology based on two-step solid-phase synthesis at $T_1 = T_2 = 870$°C and $\tau_1 = \tau_2 = 7$ h with intermediate grinding, granulating of the synthesized powders, and sintering for 3 h at $T_{sint} = 1220 - 1240$ °C (depending on the composition). In the second method, we used the same stages of the conventional ceramic technology as in the first one but supplemented with application of $0.4Pb - (0.6 - x)BrO_3 - xGeO_2$ glass system rich in lead oxide in order to provide the required specimen strength during sintering. The X-ray analysis showed the obtained ceramics to have a perovskite structure with pseudomonoclinic distortion of the prototype unit cell at room temperature. The symmetry was truly orthorhombic with the unit cell parameters $a = 5.78$ Å, $b = 11.730$ Å, and $c = 8.203$ Å; no impurities were revealed.

The temperature dependences of the pyroelectric currents in the dynamic ($i_{dyn}(T)$) and quasi-static ($i_{st}(T)$) measurement modes were recorded simultaneously for each specimen. At the same time, the temperature dependences of the relative permittivity $\varepsilon(T) = \varepsilon^T_{33}/\varepsilon_0$ (where ε^T_{33} is the permittivity of the polarized sample, ε_0 is the permittivity of vacuum) were recorded by an E7 20 immittance meter. The information was recorded and processed and the programmer (temperature regulator) was controlled using personal computer equipped with data collection system and the L-CARD software.

Figure 22a shows the temperature dependences $\varepsilon(T)$, $i_{st}(T)$, and $i_{dyn}(T)$ that are typical for polarized PZT specimens with $x = 0.02$ in heating-cooling cycles. Curves 1 and 2 illustrate the behavior of $\varepsilon(T)$ upon heating and cooling and curves 3 and 4 present, respectively, the dependences $i_{dyn}(T)$ and $i_{st}(T)$ upon heating. Curves 5 and 6 are, respectively, the same dependences upon cooling (in relative units). In order to simplify the analysis, the abnormal changes in $\varepsilon(T)$, $i_{st}(T)$, and $i_{dyn}(T)$ on the temperature scale are shown dashed lines. Here, T_{1mh} and T_{1mc} are the temperatures of the AFE $\leftrightarrow$ FE PT upon heating and cooling and T_{2mh} and T_{2mc} are the PT temperatures to the paraelectric state and back to the ferroelectric phase, respectively, according to the $\varepsilon(T)$ data. Figure 22b shows the dependences similar to those in Figure 22a for the PZT specimen with $x = 0.02$, repolarized after annealing.

The simultaneous examination of the figures makes it possible to reveal their main common features and distinctions. Upon heating $\varepsilon(T)$ experiences a step increase near T_{1mh}, which is accompanied by the appearance of maxima in the dependences $i_{st}(T)$ and $i_{dyn}(T)$, thus indicating the occurrence of residual polarization in the specimen. The subsequent heating up to the Curie temperature (T_{2mh}) demonstrates a peak in $\varepsilon(T)$ and maxima in $i_{st}(T)$ and $i_{dyn}(T)$, which drop to zero at $T > T_{2mh}$, thus indicating the attenuation of the residual ferroelectric

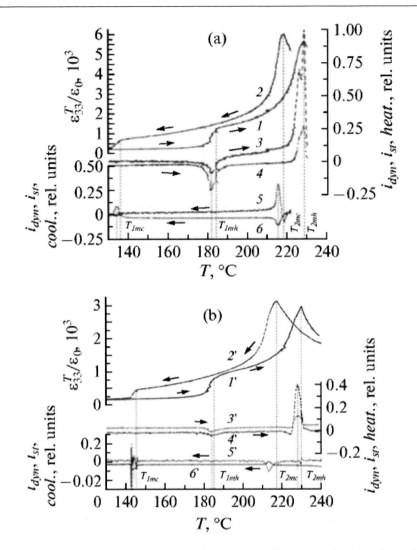

Figure 22. Temperature dependences $\varepsilon(T)$ upon (1) heating and (2) cooling; (3) $i_{st}(T)$, (4) $i_{dyn}(T)$ upon heating, (5) $i_{st}(T)$, and (6) $i_{dyn}(T)$ upon cooling (a) before and (b) after annealing; T_{1mh} and T_{1mc} are the AFE ↔ FE PT temperatures upon heating and cooling, respectively, and T_{2mh} and T_{2mc} are the PT temperatures in the paraelectric phase upon heating and cooling, according to the $\varepsilon(T)$ data for $x = 0.02$.

polarization. Upon cooling, the residual polarization is reinduced by the internal bias field and vanishes with the transition to antiferroelectric state through T_{1mc}.

The major distinction between Figures 22a and 22b is the increase in T_{1mc} by 10 °C upon 2h-annealing at 450 °C for the repolarized samples. The further experiments on hardening at temperatures up to 650 °C led to an additional increase in T_{1mc} by 2 – 4 °C, which was detected in the $\varepsilon(T)$ dependences. The results of the similar study for the ceramics with $x = 0.03$, $x = 0.04$, and $x = 0.05$ are shown in Fiure. 23.

Figure 23 shows the dependences of T_{2mh} (curve 1), T_{1mh} (curve 2), and T_{1mc} (curve 3) on the concentration x for the specimens before the annealing. Curve 4 (T_{1mc}) demonstrates the increase in the temperature range of existence of the antiferroelectric phase in the ceramic specimens after annealing. It follows from Figure 23 that the changes in the position of T_{1mc} on the temperature scale increase with increasing concentration x upon annealing. It should be

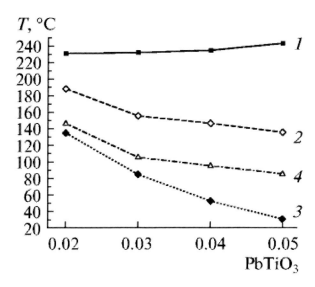

Figure 23. Dependences of (1) T_{2mh}, (2) T_{1mh}, and (3) T_{1mc} before annealing, and (4) T_{1mc} after annealing on the PbTiO$_3$ content.

mentioned that the repeated (more than 10 cycles) thermal cycling from T_{room} to T_{2mh} + 50 °C did not change the position of T_{1mc} on the temperature scale that corresponded to the specimens before and after annealing.

Starting from 1952, many works have been devoted to investigation of the transition mechanism from AFE to FE state in PZT solid solutions with various additives [43]. In order to interpret the observed effect (i.e., stabilization of the antiferroelectric state in a PZT binary system due to prolonged thermostating at $T > 450$ °C), we will consider its possible reasons. According to [44], based on the structural investigations by elastic neutron scattering, derivatography, and measurement of the temperature dependence of the conductivity of PZT specimens, the AFE ↔ FE phase transitions are mainly caused by polar clusters formed by PbTiO$_3$ unit cells and statically distributed within the PbZrO$_3$ matrix.

This suggestion was examined in Reference [45] by increasing the concentration of point defects upon quenching from 350 to 750 °C. The AFE ↔ FE PT temperature was determined optically, and its shift by 4 °C to lower temperatures upon quenching from 350 °C was ascertained. The rise of quenching temperature to 650 °C increased this shift by 10 °C. As was believed in [4], the rise in the concentration of point defects leads to the increase in the temperature range of existence of the FE phase due to the increased amount of ferroelectric clusters stabilized by the dipole moment fields of those defects.

The mesoscopic heterogeneities in systems of solid solutions based on Pb$_{0.85}$(Li$_{1/2}$La$_{1/2}$)$_{0.15}$(Zr$_{1-y}$Ti$_y$)O$_3$ were investigated in Reference [46] for polarized and subsequently annealed specimens by recording Debye powder patterns in filtered CrKα radiation. It was found that during annealing the disturbances of crystallographic long-range order have a mesoscopic scale and cause the formation of ion groups in the form of interlayers consisting of elements of ZrO$_2$ and TiO$_2$ binary oxides, coherently impregnated into the crystalline matrix structure.

The investigations performed by us on two independently prepared series of specimens revealed the following. Each of the specimens studied underwent the same thermodynamic process before annealing: (i) cooling from T_{sint} to T_{room}, and (ii) double heating to 700 °C followed by cooling to T_{room} with firing on silver electrodes. After such a thermal treatment, the specimens from the independent series gained the temperatures T_{1mh} in correspondence with Figure 2 (curve 3). According to the $\varepsilon(T)$ data, T_{1mc} is independent of polarization or complete depolarization of the specimen heated to temperatures below 400 °C. Thermostating at $T > 450$ °C leads to an irreversible shift of T_{1mc} according to Figure 2 (curve 4). The subsequent quenching leads only to further increase in T_{1mc}. This thermal effect apparently decreases the number of $PbTiO_3$ clusters [44] due to the generation of ion groups in the form of interlayers composed of elements of ZrO_2 and TiO_2 binary oxides, coherently impregnated into the crystalline matrix structure [46]. A special X-ray diffraction study needs to be performed to verify this suggestion.

It should be noted that any method for reducing the temperature hysteresis of AFE $\leftrightarrow$ FE PT at a purposeful selection of this transition on the temperature scale by varying the concentration x is promising for designing highly sensitive pyro- and piezoceramic transducers and converters of heat and mechanical energy to electric power.

ACKNOWLEDGMENTS

The work was carried out under partial financial support of Russian Foundation for Basic Research (project codes 07-02-12165 OFI and 09-02-92672 IND_a)

REFERENCES

[1]   Chynoweth, A.G. *Physical Review*, 1956, vol. 102, 705 – 714.
[2]   Chynoweth, A.G. *Physical Review*, 1960, vol. 117, 1235 – 1243.
[3]   Remeika, J. P. *J. Amer. Chem. Soc.*, 1954, vol. 76, 940-941.
[4]   Lieberman, D. S.; et.al. *J. Appl. Phys.*, 1955, vol. 26, 473 – 485.
[5]   Burfoot, J. C.; et.al., Brit. *J. Appl. Phys.*, 1969. vol. 2, 1168.
[6]   Jaffe, B.; Cook, W. R.; Jaffe, G. *Piezoelectric Ceramics*; Academic Press: London, New York, 1971; pp 1 – 288
[7]   Eremkin, V. V.; Smotrakov, V. G.; Fesenko, E. G. *Fiz. Tverd. Tela* (St. Petersburg), 1986, vol. 31, 156 – 161.
[8]   Noheda, B., *Curr. Opin. Solid State Mater. Sci.*, 2002, vol. 6, 27 –34.
[9]   Zakharov, Yu. N.; Borodin, V. Z.; Kuznetsov, V. G.; Smotrakov, V. G.; Eremkin, V. V.; Raevskaya, S. I.; Naskalova, O. V.; Sitalo, E. I.; Petin, G. P. *In Proc. Int. Symp. "Order, Disorder, and Properties of Oxides"* (ODPO-2005), Sochi, Russia, 2005. vol. 2, pp. 120-125.
[10] Pavelko, A. A.; Lutokhin, A. G.; Raevskaya, S. I.; Zakharov, Yu. N.; Malitskaya, M. A.; Raevskii, I. P.; Zakharchenko, I. N.; Sitalo, E. I.; Korchagina, N.A.; Kuznetsov, V.G. *Bull. Russ. Acad. Sci.: Physics*, 2010, vol. 74, 1104 – 1106.
[11] Reznitchenko, L. A.; Shilkina, L. A.; Razumovskaya, O. N.; Yaroslavtseva, E.A.; Dudkina, S. I.; Demchenko, O. A.; Yurasov, Yu. I.; Yesis, A. A.; Andryushina I. N. *Fiz. Tverd. Tela* (St. Petersburg), 2008, vol. 50, 1469 – 1475.

[12] Didkovskaya, O. S.; Rudenko, T. P.; Klimov, V. V.; Venevtsev,Yu. N. *Elektron. Tekh., Ser. 14: Mater.*, 1969, vol. l, 6 – 13.

[13] Rayevsky, I. P.; Novikov, M. S.; Petrukhina, L. A.; Gubaidulina, O. A.; Kuimov, A.Ye. *Ferroelectrics*, 1992, vol. 131, 327 – 329.

[14] Shonov, V. Yu.; Rayevsky, I. P.; Bokov, A. A. *Technical Physics*. 1996, vol. 41, 166 – 168.

[15] Zakharov, Yu. N.; Rayevsky, I. P.; Eknadiosians, E. I.; Piskaya, A. N.; Pustovaya, L. E.; Borodin, V. Z. *Ferroelectrics*, 2000, vol. 247, 37 – 46.

[16] Raevski, I. P.; Prokopalo, O. I.; Panich, A. E.; Pavlov, A. N.; Bondarenko E. I.; *Electric Conductivity and Posistor Effect in Perovskite Oxides*, SKNTs VSh Press: Rostov-on-Don, 2002; pp 1 – 280.

[17] Raevskii, I. P.; Kirillov, S. T.; Malitskaya, M. A.; Filippenko, V. P.; Zaitsev, S. M.; Kolomin, L. G. *Inorg. Mater.* 1988; vol. 24, 217 – 220.

[18] Rayevsky, I. P.; Bokov, A. A.; Bogatin, A. S.; Emelyanov, S. M.; Malitskaya, M. A.; Prokopalo, O. I. *Ferroelectrics*, 1992, vol. 126, 191 – 196.

[19] Ehses, K. H.; Schmid, H. Z., Z. *Kristallogr.*, 1983, vol. 162, 64 – 66.

[20] Dul'kin, E. A.; Rayevsky, I. P.; Emel'yanov, S. M. *Phys. Solid State*, 1997, vol. 39, 316 – 317.

[21] Sitalo, E. I.; Zakharov, Yu. N.; Lutokhin, A. G.; Raevskaya, S. I.; Raevski, I. P.; Panchelyuga, M. S.; Titov, V. V.; Pustovaya, L. E.; Zakharchenko, I. N.; Kozakov, A. T.; Pavelko, A. A.; *Ferroelectrics*. 2009, vol. 389, 107 – 113.

[22] Smolenskii, G. A.; Bokov, V. A.; Isupov, V. A.; et al., *Physics of Ferroelectric Phenomena*, Nauka: Leningrad, 1985; pp 1 – 396.

[23] Singh, S. R.; Pandey, D.; Yoon, S.; et al., *Appl. Phys. Lett.*, 2008, vol. 93, 182910-1 – 182910-3.

[24] Raevskaya, S. I.; Zakharov, Yu. N.; Lutokhin, A. G.; Emelyanov, A. S.; Raevski, I. P.; Panchelyuga, M. S; Titov, V. V.; Prosandeev, S. A., *Appl. Phys. Lett.* 2008. vol. 93, 042903-1 – 042903-3.

[25] Shrout, T. P.; Chang, Z. P.; Kim, N., Markgraf, S. *Ferroelectrics. Lett. Sec.* 1990, vol. 12, 63 – 72.

[26] Ye, Z.-G.; Noheda, B.; Dong, M.; Cox, D.; Shirane, G. *Phys. Rev. B.* 2001, vol. 64, 184114-1 – 184114-5.

[27] Kiat, J.-M.; Uesu, Y.; Dkhil, B.; Matsuda, M.; Malibert, C.; Calvarin, G. *Phys. Rev. B.* 2002. vol. 65. 064106.

[28] Noheda, B.; Cox, D. E.; Shirane, G.; Gao, J.; Ye, Z.-G. *Phys. Rev. B.* 2002, vol. 66, 054104-1 – 054104-10.

[29] Vanderbilt, D.; Cohen, M. H. *Phys. Rev. B.* 2001, vol. 63, 094108.

[30] [Reznichenko, L. A.; Shilkina, L .A.; Razumovskaya, O. N.; Yaroslavtseva, E. A.; Dudkina, S. I.; Verbenko, I. A.; Demchenko, O. A.; Andryushina, I. N.; Yurasov, Yu. I.; Esis, A. A. *Inorg. Mater.* 2008, vol. 45, 65 – 79.

[31] Park, S.- E.; Hackenberger, W. *Curr. Opin. Solid Mater. Sci.* 2002, vol. 6, 11 – 18.

[32] Emelyanov, A. S.; Raevskaya, S. I.; Savenko, F. I.; Topolov, V. Yu.; Raevski, I. P.; Turik, A. V.; Kholkin, A. L. *Solid State Commun.* 2007. vol. 143, 188 – 192.

[33] Fu, H.; Cohen, R. E. *Nature*, 2000, vol. 403, 281 – 283.

[34] Kutnjak, Z.; Petzelt, J.; Blinc, R., *Nature*, 2006, vol. 441, 956 – 959.

[35] Zhao, J.; Zhang, Q. M.; Kim, N.; Shrout, T. *Jap. J. Appl. Phys.*, 1995, vol. 34, 5658 – 5663.

[36] Turik, S. A.; Reznitchenko, L. A.; Rybjanets, A. N.; Dudkina, S. I.; Turik, A. V., Yesis, A. A., *J. Appl. Phys.*, 2005, vol. 97,.064102-1 – 064102-4.

[37] Raevskaya, S. I.; Emelyanov, A. S.; Savenko, F. I.; Panchelyuga, M. S.; Raevski, I. P.; Prosandeev, S. A.; Colla, E. V.; Chen, H.; Lu, S. G.; Blinc, R.; Kutnjak, Z.; Gemeiner, P.; Dkhil, B.; Kamzina, L. S. *Phys.Rev. B*. 2007, vol. 76, 11580R-1 – 11580R-4.

[38] Giniewicz, J. R.; Bhalla, A. S.;Cross, L. E. *Ferroelectrics*, 1991, vol. 118, 157 – 164.

[39] Smirnova, E. P.; Sotnikov, A. V. *Phys. Solid State*, 2006, vol. 48, 102 – 105.

[40] Zakharov, Yu. N.; Raevskaya, S. I.; Lutokhin, A. G.; Titov, V. V.; Raevski, I. P.; Smotrakov, V. G.; Eremkin, V. V.; Emelyanov, A. S.; Pavelko, A. A. *Ferroelectrics*. 2010, vol. 399. 20 – 26.

[41] Davis, M. ; Damjanovic, D. ; Setter, N. J. Appl. Phys. 2004, vol. 96, 2811 – 2815.

[42] Zakharov, Yu. N.; Raevskaya, S. I.; Borodin, V. Z.; Kuznetsov, V. G.; Raevski, I. P. *Phys. Solid State*, 2006, vol. 48, 1077 – 1078.

[43] Shirane, G.;Suzuki, K.; Takeda, A. *J. Phys. Soc. Jpn.*, 1952, vol. 7, 12.

[44] Morozov, E. M.; Smirnov, V. P.; Klimov, V. V.; Solov'ev, S. N. *Sov. Phys. Crystallogr.*, 1978, vol. 23, 64.

[45] Klimenchenko, E. N.; Baskov, N. V.; Kuznetsova, E. M.; et al., *In Proc. 7th. Int. Symp."Order, Disorder, and Properties of Oxides"* (ODPO-2004), Sochi, Russia, 2004. pp 131-132.

[46] Samoilenko, Z. A.; Pashchenko, V. R.; Ishchuk, V. M. Zh. *Tekh. Fiz.*, 1998, vol. 68, 43-47.

In: Ferroelectrics and Superconductors
Editor: Ivan A. Parinov

ISBN: 978-1-61324-518-7
©2012 Nova Science Publishers, Inc.

Chapter 4

DESIGNING OF MULTIFERROIC MATERIALS BASED ON PEROVSKITE AND SPINEL-LIKE COMPOUNDS: REACTIVITY AND REGIONS OF STRUCTURE STABILITY; PHASE FORMATION AND STEPWISE OPTIMIZATION OF TECHNOLOGY; RELAXATION DYNAMICS, UHF ABSORPTION AND SECONDARY PERIODICITY OF FERROMAGNETIC PROPERTIES

L. A. Reznichenko[*,1], *O. N. Razumovskaya*[1], *L. A. Shilkina*[1], *I. A. Verbenko*[1], *K. P. Andryushin*[1], *A. A. Pavelko*[1], *A. V. Pavlenko*[1], *V. A. Alyoshin*[1], *S. P. Kubrin*[1], *A. I. Miller*[1], *S. I. Dudkina*[1], *P. Teslenko*[2], *G. Konstantinov*[3], *M. V. Talanov*[4], *A. A. Amirov*[5], *A. B. Batdalov*[5], *V. M. Talanov*[6], *N. P. Shabelskaya*[6] *and V. V. Ivanov*[6]*

[1]Research Institute of Physics, Southern Federal University, Russia
[2]Physical Department, Southern Federal University, Russia
[3]Expert-Criminalistic Laboratory, Custom Department, Russia
[4]Department of High Technologies, Southern Federal University, Russia
[5]Institute of Physics, Dagestan Scientific Centre of Russian Academy of Sciences, Russia
[6]Chemical Engineering Department, South-Russian State Technical University, Russia

ABSTRACT

Recently, there has been renewed scientific interest in a vast class of substances, namely multiferroic (MF) combining ferroelectric and magnetic properties, which

---

*E-mail: ilich001@yandex.ru.

defined the organization of the work on atomic designing of such materials and mastering the technology of their production with practically important properties. A technological aspect for MF is of great importance because most of them will be inevitable based on bismuth ferrite (with high Curie and Néel temperatures) whose thermal instability is known. The objects for studying are as follows:

(i) $BiFeO_3$ modified with a wide spectrum of rare-earth elements (REE) La, Pr, Nd, Sm, Eu, Gd, Tb, Dy, Ho, Er, Tm, Yb and Lu incorporated as oxides stoichiometrically by the formula $(Bi_{1-x}A_x)FeO_3$ (A = REE, $0.00 \leq x \leq 0.20$, $\Delta x = 0.05$) at the stage of synthesis; (ii) binary systems of $BiFeO_3$ solid solutions with $NaNbO_3$, $KNbO_3$, $LiNbO_3$, $BaTiO_3$, $SrTiO_3$, $CaTiO_3$, $PbTiO_3$, $PbZrO_3$, $BiMnO_3$, $LaMnO_3$ and $CaMnO_3$; (iii) impurity-free and modified compounds $AFe_{2/3}W_{1/3}O_3$ (A = Ba, Sr, Pb) and $PbFe_{1/2}Nb_{1/2}O_3$ and solid solutions on their basis; (iv) spinel-like compounds AFe_2O_4 (A = Ni, Zn, Cu, Co) and solid solutions on their basis.

A work on optimization of processes of synthesis and recrystallization sintering was carried out, which enabled us to formulate the necessary conditions of achievement of a fine-grained state of the presynthesized powders facilitating realization of the sufficient rate of heterogeneous reaction, activity of synthesized powders during the sintering and, as a result, to obtain impurity-free high-density samples with an equilibrium microstructure. The phase formation in the systems was studied, the x–T diagrams (x is a concentration of the components) were shown, and the dielectric, thermal, magnetic and dissipative properties of the objects were studied in a wide range of temperatures (10 – 1000 K) and frequencies (10^6–10^9 Hz) of ac electric field. The objects, characterized by a presence of the spontaneous magnetization, the low- and high-temperature relaxations and the maximum UHF absorption, were identified. The secondary periodicity of properties of $BiFeO_3$ with REE was established. It is shown that the observed effects are related to the origination of symmetry phase transitions, the crystallochemical specifics of REE and the change of type of solid solutions. The objects having a high potential in practical applications were chosen.

1. INTRODUCTION

In recent years, great interest has been shown in the materials referred to as multiferroics in modern terminology (in view of their wide multifunctional potential for modern technology) [1]. An example of these materials are $AFe_{2/3}W_{1/3}O_3$ (A = Ba(BWF), Sr(SWF), Pb(PWF)) [2, 3], bismuth ferrite ($BiFeO_3$) [4], $PbFe_{1/2}Nb_{1/2}O_3$ (PFN). Bismuth ferrite, $BiFeO_3$, is a suitable object for designing the magnetoelectric materials owing to the high temperatures of its ferroelectric (FE) (T_C = 1083 K) and magnetic (T_N = 643 K) ordering [3, 5]. However, wide practical use of the materials on its basis is hindered by low level of magnetoelectric interaction caused, on the one hand, by the special features of its magnetic and crystal structure and, on the other hand, by a large difference between the temperatures of antiferromagnetic and FE phase transitions. One of the ways enabling one to optimize the properties of such objects is introduction of the rare-earth elements (REE) into their composition. This is connected with the unique magnetic properties of REE. Despite the fact that their intrinsic ferromagnetic ordering occurs only at very low temperatures, the REE magnetic nature (f-ferromagnetism) manifests itself in the enhanced exchange interaction between the other ferromagnetic ions resulting in the increased Nèel temperature. Incorporation of sufficiently rigid, highly ionized REE ions in place of easily deformable ones (e. g., Bi) inevitably leads to a reduction in stability of the FE state and, as a result, to a lowering of the Curie temperature. In addition, a partial replacement of Bi ions results in the increased compositional disorder in the system that may favor the appearance of new weakly ferromagnetic phases [6].

In this connection, a today problem is study and optimization of synthesis and sintering processes of the bismuth ferrite and its solid solutions with the REE ferrites having a composition of the formula $(Bi_{1-x}A_x)FeO_3$, where A = La, Pr, Nd, Sm, Eu, Yd, Tb, Dy, Ho, Ta, Lu and x = 0.05, 0.10, 0.15, 0.20. With this aim, it is developed a technology their preparation by using the method most available, simple and widely used in laboratory and industrial conditions, such as synthesis due to the reaction in a mixture of solid substances (solid-state synthesis) and sintering by applying the conventional ceramic technology (CCT). In adition, despite the great volume of data in the literature on this compound and its related materials, however, there are no comprehensive studies on the crystal and grain structure, Mössbauer effect, dielectric and magnetic properties of such samples. In this paper, we report the results of such investigations. Another multiferroic material, lead ferroniobate, belongs to the set of ferroelectromagnets with perovskite structure, the magnetic properties of which are related to the presence of Fe^{3+} ions in octahedral positions [3]. The transitions to the antiferromagnetic and polar phases in this compound occur, respectively, at the Nèel temperature ($T_N \sim 140 -$ 170 K) and the Curie temperature ($T_C \sim 368 - 385$ K), which depend on the thermodynamic history of the samples [7, 8]. To date, PFN in various solid states (ceramics, single crystals, polycrystals) has been studied fairly well [7–9], but a comprehensive analysis of the dielectric, magneto-electric, structure, and absorption properties of PFN over wide temperature and frequency ranges of identical samples obtained under different process conditions has great practical and theoretical interest. In this paper, we report the results of such analysis.

The group of these compounds is $AFe_{2/3}W_{1/3}O_3$. To date, PWF ceramics have been studied fairly well [10, 11], but there are much fewer data on BWF and SWF, possibly due to their structure instability and irreproducible characteristics [2, 12–14]. A comprehensive study of these compounds (in particular, their dielectric, magnetoelectric, structural, and absorption properties over wide temperature and frequency ranges) for identical samples obtained under different conditions is therefore a high priority task. In this paper, we present the results of such comprehensive investigations.

The intensive search new materials possessing both ferroelectric and magnetic properties in recent years has led to an avalanche of papers on creation of various solid solutions on the base of classic multiferroic ceramics [5]. At present time the $Bi_{1-x}A_xFeO_3$-type (A = La, Sr, Ca, K), $Bi(Fe_{1-x}B_x)O_3$-type (B = Mn, Cr, V) solid solutions and solutions with simultaneous replacement of A and B positions in perovskite-type structures (for example, $(1 - x)BiFeO_3 - xABO_3$, where A = Pb, Ba, Cd, Sr, B-Ti, Zr and others) are carefully studied. As it is expected, the relatively small dopes of secondary components to $BiFeO_3$ (up to $x \sim 0.2$) usually lower the temperatures of ferroelectric phase transitions. The identification of veritable phase diagrams of solid solutions' systems (for example "concentration – temperature", "concentration – pressure") is usually hindered due to possibility of instable intermediate phase formation by using different sample processing methods.

Among various solid solutions on the base of $BiFeO_3$ the $(1 - x)BiFeO_3 - xBiMnO_3$ system has special interest, because Mn^{3+} as well as Fe^{3+} ions are magnetoactive. The organization of $Mn^{3+}- Fe^{3+}$-type magnetic interactions together with $Fe^{3+}- Fe^{3+}$ and $Mn^{3+} -$

Mn^{3+} interactions in solid solutions can lead to new manifestations of order-disorder in compounds with component concentrations near to $x = 0.33$, 0.5, 0.67 and similar. For the complete understanding of various types of states (magnetic, electric, deformation) in multiferroics the careful research (x, T), (x, P), (x, E), (x, H)-type phase diagrams is required.

The study some compositions of $BiFe_{1-x}Mn_xO_3$ synthesized at different conditions (solid state reactions [15 – 17]) under pressure [18 – 19] in the form of thin films have shown that at different values of x at room temperature the perovskite phases of different symmetry are observed, namely: rhombohedral, orthorhombic and monoclinic. In [18] the phase diagram (x, T) was constructed based on the results of structure study of samples processed under pressure $(P \sim 3$ GPa).

In current paper the results of optimization $(1 - x)BiFeO_3 - xBiMnO_3$ solid solutions' compounds with phases stabilized at room temperature processing are present. Structure parameters of all the compounds in temperature range 293 K $\leq T \leq 1023$ K are also present, that will allow one to construct a stable phase diagram (x, T).

In addition, new mechanochemical process is successfully used for the synthesis of multiferroics during the recent time [20, 21]. In this process, solid-phase reactions of oxides are activated through the high mechanical energy and temperature arising during grinding. Such a process decreases high energy consumption for calcination stage, which makes the process simpler than that involving usual thermal synthesis. This is important for the synthesis of lead-containing compounds because the synthesis proceeds at a lower temperature and is carried out in containers isolated from the environment. The method also allows one to obtain the powders of materials in the nano-dispersed state, which promotes improvement of the microscopic characteristics of the final materials. In particular, they exhibit better sintering ability than the powders obtained by usual solid-phase synthesis or chemical methods involving solutions.

Mechanical activation has very big advantages for the synthesis of this compound since there is no need in heat treatment and expensive chemical products (in wet methods). The some measures should be taken by regarding the introduction of impurities during milling process. This preparation route although practically very simple but is a complex process which depends on many factors, for instance on physical and chemical parameters such as the dynamic conditions, temperature, material nature of grinding jars and others. For most of the compounds prepared by this technique various processing conditions such as ball to powder weight ratio, milling type and milling speed have significant effect on final product formed.

The number of works on the application of this method to the synthesis of piezoelectric materials is rather large. This allows us to systematize them for the purpose of establishing some regularities and formulating some recommendations concerning more optimal applications.

In this paper also some results of searching multiferroics on the basis of another structural type, namely crystals with spinel structure are present. The objects of our experimental and theoretical investigations are multicomponent oxide systems, including the chromites and ferrites of transition elements. Such selection is caused by the fact that some simple chromic spinels ACr_2O_4 (A = Mg, Zn, Cd, Hg, Fe, Co, Cu, Ni, Mn) discovered earlier demonstrate unusual magnetic and electric properties at low temperatures, appearing as a result of phase transitions [22 – 30]. Therefore the studies more complex chromites and ferrites with spinel structure are actual. The synthesis and physico-chemical research of three-component system is the first task of the searching new multiferroics with spinel-like structures. Some experimental results on phase states of two three-component systems and theoretical calculations of structure peculiarities of low-symmetrical spinel phases are also given in this paper.

2. PROCESSING AND INVESTIGATION METHODS

The formation reaction of BiFeO$_3$ has been studied by the authors previously using the differential thermal analysis (DTA). It has been established that the mixture Bi$_2$O$_3$ + Fe$_2$O$_3$ is characterized by the following five endoeffects, namely: (i) a polymorphic transformation of Bi$_2$O$_3$ at 1013 K, (ii) the melting of eutectics in the Bi$_2$O$_3$ – Fe$_2$O$_3$ system at 1063 K, (iii) a phase transition of BiFeO$_3$ at 1123 K, (iv) the incongruent melting of BiFeO$_3$ at 1193 K and 1223 K, that is indirectly indicative of the fact that the formation of the compound took place at the temperatures below 1123 K (Figure 1). This was substantiated by the structural studies from which it follows that the formation reaction of the compound begins at 873 K [31]. Parameters of the rhombohedrally disordered BiFeO$_3$ cell are a = 3.965 Å and α = 89°24' which agrees well with the reported data.

The (Bi$_{1-x}$A$_x$)FeO$_3$ solid solutions, where A = La, Pr, Nd, Sm, Eu, Gd, Tb, Dy, Ho, Er, Tm, Yb, Lu, and x = 0.05, 0.10, 0.15, 0.20 were also synthesized from the high-purity oxides Bi$_2$O$_3$, Fe$_2$O$_3$ and A$_2$O$_3$ at T_1 = 1023 K for 10 hours and T_2 = 1023 – 1073 K for 5 hours in dependence on composition. With increase of the REE content, the second calcination temperature T_2 was raised by ~ 10 K per every 4 mol % REE. This temperature was limited by the appearance of traces of a liquid phase. Two-stage synthesis provides the more

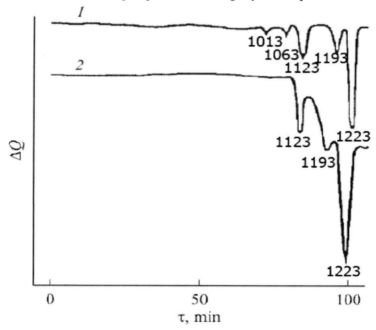

Figure 1. DTA curves: (1) the synthesized powders of composition Bi$_2$O$_3$ + Fe$_2$O$_3$, (2) BiFeO$_3$ compound.

complete proceeding the reaction and obtaining substances without any impurity of foreign phases. A unique feature of synthesis of the compounds of solid solutions under study is the increased strength of the sintered bodies. This effect is due to the proximity of the synthesis temperature to the incongruent melting one and, as a consequence, to the increased mutual diffusion of the components. It this case, for the repeated synthesis and, especially, for the following sintering it is necessary to use the sufficiently fine-grain powders the preparation of which is a fairly complicated technological problem because of the difficulties in crushing and grinding the sintered bodies formed after the calcination.

Typically, for the preparation of the ceramics of complex oxides it is used the powder sintering of the prescribed composition. As basis, we have accepted a standard technological process of manufacturing the piezoelectric ceramics in laboratory and industrial conditions, i.e. the conventional ceramic technology described in Reference [32]. The aim of the ceramic technology is, firstly, to obtain a monolithic, solid, poreless product (ceramics) and, secondly, to fabricate a product having the prescribed chemical composition. In combination with the optimal microstructure, this guarantees the achievement of the desirable physicochemical parameters in final sample.

The powders possessing the properties being necessary for pressing were obtained by incorporation of a plasticizer followed by their granulation. The latter is the process of preparation of the secondary grains with large density and mechanical strength. With this aim, the operation of molding the powder mixture is preceded by a choice of the plasticizer, its amount and a way its incorporation, a powder wetness, a method of granulation, a grain size and also a choice of the method and the pressure of pressing. As a plasticizer, in granulating the powders we have chosen the aqueous solution of polyvinyl alcohol (PVA) uniformly burning-down in temperature range $473 - 773$ K, that is to be important for keeping safety of the products. To increase a density of the small shaped products, in their processing double granulation was performed. With this goal, the moistened powder was briquetted, the briquette was crushed and rubbed first through $0.7 - 0.8$ mesh sieve and then through $0.3 - 0.4$ mesh sieve.

In order to mold the powder mixtures it was used the double-sided pressing at which the non-uniformity of the pressure distribution over the powder product volume is less pronounced than at the one-sided pressing. In order to realize the two-sided pressing in the press providing one-sided pressure we used the special metallic forks which were mounted between the matrix and the lower die. For the work it is practical to briquette the powder products of several typical sizes in dependence on their purposes.

The following stage of the work was sintering the prepared mixtures with aim to obtain a safe high-density ceramic product without any macroscopic defects in which it is possible to realize the desirable properties by means of the following technological operations.

Initially, the experimental specimens of 12 mm in diameter and of $2.5 - 3.0$ mm in height were sintered to choose the optimal temperature ensuring to obtain the ceramics having the highest density achievable at sintering by CCT ($90 - 95\%$ of theoretical). The pressed products were placed on the alund substrate, covered with alund crucible and sintered in a high-temperature chamber furnace. Then, the necessary technological conditions in order to obtain the measured specimens were chosen from the observation of density.

The compounds of $(1 - x)BiFeO_3 - xBiMnO_3$ with $0.00 \leq x \leq 0.50$, and step $\Delta x = 0.05$ were synthesized by the bulk state reaction method in two steps with transitional grinding.

Bi$_2$O$_3$ and Fe$_2$O$_3$, Mn$_2$O$_3$ oxides were chosen as the initial ones. Calcination at T_1 = 1063 K, t = 10 hours, and T_2 = 1073 K, t = 10 hours was chosen as the optimal time-temperature mode. In all of the compositions the admixture phases Bi$_2$O$_3$ и Bi$_2$(Fe,Mn)$_4$O$_9$ are present. The concentration of those admixtures in samples with $x > 0.3$ changes from 10 up to 25%, that witnesses on the incomplete synthesis and limited solubility of BiMnO$_3$ in BiFeO$_3$. Sintering the synthesized samples was made at T_{Sin} = 1143 K, 2h.

BWF and SWF samples were synthesized by solid phase reactions from high-purity BaCO$_3$ and SrCO$_3$ carbonates and WO$_3$ and Fe$_2$O$_3$ oxides (of pure and analytical grades) by using two-stage firing and intermediate grinding at temperatures T_1 = 1273 K and T_2 = 1473 K and exposure times τ_1 = 4 h and τ_2 = 2 h. The sintering conditions for BWF and SWF ceramic billets were varied: we used temperatures T_{sint1} = 1613 K and T_{sint2} =1623 K at $\tau_1 = \tau_2$ = 2.5 h for BWF and T_{sint1} = 1623 K and T_{sint2} = 1673 K at τ_{sint1} = 1 h and τ_{sint2} = 2.5 h for SWF. The PWF samples were obtained in a similar fashion from PbO, WO$_3$, and Fe$_2$O$_3$ oxides of pure and analytical grades at $T_1 = T_2$ = 1123 K, $\tau_1 = \tau_2$ = 4 h, T_{sint} = 1373 K, τ_{sint} = 2.5 h. Thermal treatment was performed in a Nabertherm LH 15/14 chamber furnace with a metal heater.

The samples were synthesized by solid phase reactions from high-purity PbO, Fe$_2$O$_3$, and Nb$_2$O$_5$ oxides (of pure and analytical grade) by using two-stage firing and intermediate grinding at temperatures $T_1 = T_2$ = 1123 K and exposure times $\tau_1 = \tau_2$ = 4 h. The ceramic billets were sintered at the temperature T_{sint} = 1373 K for 2 and 2.5 h (Nabertherm LH 15/14 chamber furnace with a metal heater).

Phase composition and synthesis completeness were checked by X-ray diffraction analysis. A precise study was performed on fine grained ceramic cakes by powder X-ray diffraction on a ULTIMA IV Rigaku diffractometer (Cu$_{K\alpha}$ radiation) with a graphite monochromator on the detector. The linear (a), angular (α), and volume (V, perovskite cell volume) parameters were calculated by the standard technique. The experimental (ρ_{exp}) sample density was found by hydrostatic weighing in octane. The X-ray density ($\rho_{X\text{-ray}}$) was calculated from the formula ρ_{Xray} = 1.66 M/V, where M is the unit weight in grams and V is the perovskite cell volume in E$_3$. The relative density ρ_{rel} was calculated from the formula ρ_{rel} = $(\rho_{exp}/\rho_{X\text{-ray}})\cdot100\%$. The grain structure of the samples was analyzed with a Hitachi TM-1000 scanning electron microscope (resolution of up to 30 nm, magnification of up to 104).

The Mössbauer spectra were measured on specially designed (Institute of Physics, Southern Federal University, IP SFU) MC1104Em spectrometer with [57]Co gamma-ray source in Cr matrix. The model interpretation of the spectra was performed using the UnivemMS program. Chemical isomer shifts of the spectra were estimated with respect to the metallic α-Fe.

The dielectric characteristics (relative permittivity $\varepsilon/\varepsilon_0$ and loss tangent) were measured in the temperature range 10 – 325 K at frequencies from 25 Hz up to 1.2 MHz by using precise WayneKerr 6500 B impedance analyzer. The samples were cooled in CCS_150 helium refrigerator cryostat of closed type (Cryogenics) down to ultra-low temperatures. The temperature was regulated by using a LakeShore 331temperature controller, which allowed us to set a specified temperature with an error of ± 0.01 K. During measurements, the samples were placed in the cryostat vacuum chamber evacuated by Boc Edwadrs turbo-molecular pump. In the temperature range 300 – 973 K at frequencies from 25 Hz up to 1 MHz, $\varepsilon/\varepsilon_0$ and loss tangent were measured by using precise Agilent E4980A LCR meter.

The magnetoelectric measurements in the temperature range 300 – 633 K and frequency range 25 Hz – 1 MHz were performed on specially designed automatic bench (IP SFU), composed of precise Agilent 4980A LCR meter, inductance coils forming dc magnetic field 0.6 T, and specially developed Kalipso v.2.0.0.27 software package that allowed us to automatically record the magnetoelectric spectra. The magnetoelectric interaction coefficient β_H was calculated from the formula $\beta_H = \{[\varepsilon(H)-\varepsilon(0)]/\varepsilon(0)\}\cdot 100\%$, where $\varepsilon(H)$ and $\varepsilon(0)$ are the relative permittivity in dc magnetic field and in its absence, respectively. The absorption spectra $L(f)$ and the voltage standing wave ratio (VSWR) were measured with system composed of sweep generator, broadband microstrip line (operating in the traveling-wave mode), and VSWR R2-61 scanning meter in frequency range 5.6 – 8.3 GHz.

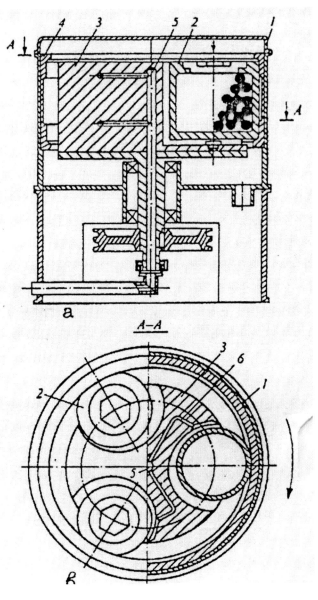

Figure 2. Planetary mill AGO-2.

We also used mechanical activation for preparation of several solid solutions. The devices most often used by researchers for synthesis purposes are the laboratory mills: planetary mill of Fritsch Pulverisette (Germany), shaker mill of Model 8000, Spex Industries Edison, NJ (USA) and high-energy planetary mill AGO-2, developed and patented by Institute of Solid State Chemistry and Mechanochemistry SB RAS (Novosibirsk) [33, 34]. For Fritsch Pulverisette mill, the rotation frequency of the cylinders with balls is 200 – 250 r. p. m., so the power input to the substance does not exceed 1 W/g. Spex 8000 has higher power and the frequency of rotation may be settled, so that the power up to 10 W/g may be achieved at the rotation frequency 900 r.p.m. Hovever, AGO-2 planetary mill exhibits better characteristics, energy consumption by the present mill changes from 10 to 60 W/g (Figure 2).

High rate of grinding and activation in planetary mills is provided by centrifugal forces arising at the rotation of each jar around its own axis and their common axis. These centrifugal forces exceed the gravity force by a factor of several tens, thus providing the energy of these mills by 2 – 3 orders of magnitude higher than that of common ball mills. Powder dispersion achieved in planetary mill after 2 min treatment is the same as that achieved in usual ball mill after the treatment for 10 – 12 h.

In laboratory planetary mills the jars operate due to frictional clutch with the cylindrical housing of the mill. The scheme and design of the apparatus are shown in Figure 2. The apparatus consists of housing 1, jars 2, carrier 3, guiding rim 4, channels 5 and 6 in the center and at the periphery of carrier for cooling water. The diameter of the jars is 70 mm, the working volume is 150 cm^3, ball mass of the substance under treatment is 10 g.

The device operates as follows. First, the water is admitted from water pipeline through the tube 5 and channels 6 into pockets for jars, carrier starts to rotate, the rings that are fixed on motor (no shown in the scheme) make jars rotate. The surface of jars and rings are separated from each other by water layer, similarly to hydrostatic bearings: the jars operate as a spindle while the ring acts as hydrostatic bearing. Under action of the centrifugal forces, the jars press to the guiding rim 4 and due to frictional clutch with its surface they rotate around their axis. Thus, the jars take part in two kinds of rotation: around their common and own axes. This technical solution excludes the necessity to use bearings in the most loaded part of the mill (the drive of jars) and increases operational lifetime of the apparatus. Intensive cooling of the jars by water decreases the contamination of the product with the material of grinding bodies as well as increases the degree of product activation since the annealing of defects is excluded.

3. EXPERIMENTAL RESULTS

3.1. Phase Stability, Crystal Structure and Spin Modulation in Solid Solutions of $Bi_{1-x}A_xFeO_3$

3.1.1. Discussion

Table 1 summarizes the results our choice of the optimal sintering temperature (T_{sint}).

Table 1. Experimental densities, ρ_{exp}, of the ceramics sintered at different temperatures of $BiFeO_3$ and its solid solution $(Bi_{1-x}A_x)FeO_3$, where A = La, Pr, Nd, Sm, Eu, Gd, Tb, Dy, Ho, Er, Tm, Yb, Lu

Composition of the ceramics	x value	ρ_{exp}, g/cm^3				
		1143 K	1163 K	1183 K	1203 K	1223 K
$BiFeO_3$	0	7.66	7.32	-	-	-
$(Bi_{1-x}La_x)FeO_3$	0.05	7.30	7.40	7.57	7.58	-
-"-	0.10	-	7.15	7.24	7.40	-
-"-	0.15	-	6.84	6.84	7.30	7.00
-"-	0.20	-	6.81	7.11	7.46	
$(Bi_{1-x}Pr_x)FeO_3$	0.05	-	6.84	7.14	7.42	7.17
-"-	0.10	-	6.88	7.05	7.28	7.27
-"-	0.15	-	6.55	6.78	6.90	7.25
-"-	0.20	-	6.40	6.56	6.94	7.62
$(Bi_{1-x}Nd_x)FeO_3$	0.05	7.26	7.67	7.66	7.53	-
-"-	0.10	-	7.08	7.18	7.32	-
-"-	0.15	-	7.15	7.21	7.26	7.06
-"-	0.20	-	6.99	6.65	7.24	7.14
$(Bi_{1-x}Sm_x)FeO_3$	0.05	-	7.56	7.57	7.60	-
-"-	0.10	-	7.46	7.50	7.52	-
1	2	3	4	5	6	7
-"-	0.15	-	6.79	7.09	7.40	
-"-	0.20	-	7.12	7.30	7.51	-
$(Bi_{1-x}Eu_x)FeO_3$	0.05	-	7.38	7.34	7.38	-
-"-	0.10	-	7.48	7.55	7.60	-
-"-	0.15	-	-	7.35	7.56	7.61
-"-	0.20	-	-	7.36	7.60	7.76
$(Bi_{1-x}Yd_x)FeO_3$	0.05	-	7.33	7.37	7.38	-
-"-	0.10	-	7.47	7.56	7.56	-
-"-	0.15	-	-	7.01	7.25	7.38
-"-	0.20	-	-	6.85	7.08	7.41
$(Bi_{1-x}Tb_x)FeO_3$	0.05	-	7.67	7.55	7.51	-
-"-	0.10	-	7.61	7.61	7.65	-
-"-	0.15	-	-	7.06	7.19	7.40
-"-	0.20	-	-	6.91	7.07	7.31
$(Bi_{1-x}Dy_x)FeO_3$	0.05	-	7.62	7.43	7.45	-
-"-	0.10	-	7.64	7.50	7.61	-
-"-	0.15	-	-	7.20	7.41	-
-"-	0.20	-	-	7.10	7.20	-
$(Bi_{1-x}Ho_x)FeO_3$	0.05	-	7.46	7.50	7.35	-
-"-	0.10	-	7.34	7.32	7,27	-
-"-	0.15	-	-	7.10	7.38	-
-"-	0.20	-	-	6.86	7.06	6.94
$(Bi_{1-x}Tm_x)FeO_3$	0.05	7.36	7.30	7.13	-	-
-"-	0.10	7.40	7.28	6.99	-	-
-"-	0.15	-	7.29	7.24	-	-
-"-	0.20	-	7.01	7.11	-	-
$(Bi_{1-x}Yb_x)FeO_3$	0.05	7.38	7.29	6.94	-	-
-"-	0.10	7.74	7.34	6.98	-	-

Composition of the ceramics	x value	ρ_{exp}, g/cm^3				
		1143 K	1163 K	1183 K	1203 K	1223 K
-"-	0.15	-	7.07	6.93	-	-
-"-	0.20	-	7.08	7.01	-	-
$(Bi_{1-x}Lu_x)FeO_3$	0.05	7.30	6.92	-	-	-
-"-	0.10	7.37	7.17	-	-	-
-"-	0.15	-	7.34	7.25	-	-
-"-	0.20	-	7.32	7.27	-	-

A special feature of sintering bismuth ferrite is low temperature of its incongruent melting (with decomposition) at 1193 K. It has been found that these sufficiently dense ceramics may be obtained at temperatures close to the melting temperature. This fact must be taken into consideration by selecting the temperatures because the undesirable appearance of traces of a liquid phase may lead to a breaking in stoichiometry of the substance (melting with decomposition). Therefore, a range of sintering temperatures is narrow 1123 – 1163 K. Experimental sinterings were performed with a step 10 – 20 K. Analysis of the results of selecting the optimal sintering temperatures has shown that for bismuth ferrite the liquid phase traces appear even at 1163 K and due to that because of the decrease in density at the $T_{sint.}$ rise to the above value of the optimal temperature is $T_{sint} = 1143$ K.

In the case of a partial replacement of bismuth by the REE (5 – 20 %) a thermal stability of bismuth ferrite increases and the appearance of a liquid phase is observed even at temperatures exceeding 1163 K. The range of sintering temperatures extends from 1143 – 1193 K, depending on the REE content. The rise in a sintering temperature above the upper limit results in the appearance of impurity phases. Thus, the bismuth ferrite ceramics and their solid solutions must be sintered at a maximum temperature preceding the appearance of the liquid or impurity phases. It should be noted here that in doing this we did not always succeed in obtaining the maximum density.

Table 2 presents the values of the optimal T_{sint}, ρ_{meas}, ρ_{x-ray} and ρ_{rel} for the sintered ceramic specimens of all the studied compositions and, also, the concentrations of the main impurity phase (Bi_2O_3) disclosed on the basis of the X-ray data.

Table 2. Characterization of the specimens of BiFeO₃ and (Bi₁₋ₓAₓ)FeO₃, where A = La, Pr, Nd, Sm, Eu, Gd, Tb, Dy, Ho, Er, Tm, Yb, Lu in density and phase composition

Composition of ceramic specimens	x value	T_{sint}, K	ρ_{exp} g/cm^3	ρ_{X-ray} g/cm^3	ρ_{rel}, %	Content of the impurity phase calculated from the relative intensity, %
$BiFeO_3$	0	1143	7.66	8.38	91.0	8
$(Bi_{1-x}La_x)FeO_3$	0.05	1183	7.57	8.28	92.0	6
-"-	0.10	1203	7.40	8.26	89.1	6
-"-	0.15	1203	7.30	8.17	89.4	6
-"-	0.20	1203	7.46	8.06	92.6	6
$(Bi_{1-x}Pr_x)FeO_3$	0.05	1203	7.42	8.32	89.2	6
-"-	0.10	1203	7.28	8.24	88.4	6
-"-	0.15	1223	7.25	8.19	88.5	6

Table 2. (Continued)

Composition of ceramic specimens	x value	T_{sint}, K	ρ_{exp} g/cm^3	$\rho_{X\text{-}ray}$ g/cm^3	ρ_{rel}, %	Content of the impurity phase calculated from the relative intensity, %
-"-	0.20	1223	7.62	8.10	94.1	6
$(Bi_{1-x}Nd_x)FeO_3$	0.05	1163	7.67	8.33	92.1	4
-"-	0.10	1203	7.32	8.28	88.4	4
-"-	0.15	1203	7.26	8.19	88.6	7
-"-	0.20	1203	7.24	8.19	88.4	5
$(Bi_{1-x}Sm_x)FeO_3$	0.05	1163	7.56	8.38	90.2	3
-"-	0.10	1183	7.50	8.28	90.6	10
-"-	0.15	1203	7.40	8.34	88.7	10
-"-	0.20	1203	7.51	8.20	91.6	10
$(Bi_{1-x}Eu_x)FeO_3$	0.05	1163	7.38	8.367	88.2	12
-"-	0.10	1183	7.55	8.309	90.86	6
-"-	0.15	1203	7.56	8.335	88.18	6
-"-	0.20	1203	7.60	8.421	90.25	6
$(Bi_{1-x}Gd_x)FeO_3$	0.05	1163	7.38	8.367	88.2	13
-"-	0.10	1183	7.56	8.319	90.87	4
-"-	0.15	1223	7.38	8.365	88.22	4
-"-	0.20	1223	7.41	8.472	87.46	4
$(Bi_{1-x}Tb_x)FeO_3$	0.05	1163	7.67	8.35	91.85	7
-"-	0.10	1163	7.61	8.32	91.50	8
-"-	0.15	1223	7.40	8.29	85.18	5
-"-	0.20	1223	7.31	8.46	82.39	6
$(Bi_{1-x}Dy_x)FeO_3$	0.05	1163	7.62	8.37	89.34	10
-"-	0.10	1163	7.64	8.30	91.99	8
-"-	0.15	1203	7.41	8.26	89.69	21
-"-	0.20	1203	7.20	8.19	87.91	25
$(Bi_{1-x}Ho_x)FeO_3$	0.05	1163	7.46	8.37	87.96	12
-"-	0.10	1163	7.34	8.31	88.28	9
-"-	0.15	1203	7.38	8.24	86.17	17
-"-	0.20	1203	7.06	8.19	83.70	24
$(Bi_{1-x}Er_x)FeO_3$	0.05	1143	6.86	8.19	83.95	9
-"-	0.10	1143	7.01	8.35	83.59	10
-"-	0.15	1163	6.93	8.29	83.96	17
-"-	0.20	1163	6.91	8.23	85.54	26
$(Bi_{1-x}Tm_x)FeO_3$	0.05	1143	7.36	8.35	86.98	6
-"-	0.10	1143	7.40	8.30	89.16	11
-"-	0.15	1163	7.30	8.25	87.76	15
-"-	0.20	1163	7.28	8.19	86.81	24
$(Bi_{1-x}Yb_x)FeO_3$	0.05	1143	7.38	8.34	87.37	16
-"-	0,10	1143	7.40	8.30	88.81	17
-"-	0.15	1163	6.93	8.24	84.06	54
-"-	0.20	1163	7.08	8.20	86.35	37
$(Bi_{1-x}Lu_x)FeO_3$	0.05	1143	7.30	8.35	85,07	50
-"-	0.10	1143	7.37	8.31	88.72	24
-"-	0.15	1163	7.34	8.25	87.92	22
-"-	0,20	1163	7.32	8.21	88.55	57

Table 3. Ionic radii of Bi and REE according to Belov and Bokii [35]

	r, Å	REE	r, Å	REE	r, Å	REE	r, Å	REE	r, Å
Bi^{3+}	1.20	La^{3+}	1.04	Sm^{3+}	0.97	Dy^{3+}	0.88	Tm^{3+}	0.85
		Pr^{3+}	1.00	Gd^{3+}	0.94	Ho^{3+}	0.86	Yb^{3+}	0.81
		Nd^{3+}	0.99	Tb^{3+}	0.89	Er^{3+}	0.85	Lu^{3+}	0.80

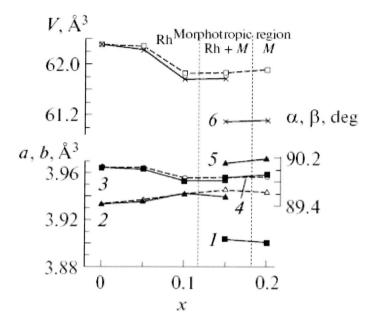

Figure 3. Dependences of the unit cell parameters and experimental and theoretical unit cell volumes of the (solid line) $Bi_{1-x}Nd_xFeO_3$ and (dotted line) $Bi_{1-x}La_xFeO_3$ solid solutions on x: (1) b of M-phase, (2) angle α, (3) a of Rh –phase, (4) ($a = c$) for M-phase, (5) angle β, (6) V_{theor}, and (7, 8) V_{exp} of the monoclinic and rhombohedral cells, respectively.

It follows from Table 2 that the most essential increase in thermal stability of $BiFeO_3$ takes place in solid solutions with the REE possessing the sufficiently large radii (La, Pr, Nd, Sm, Eu, Gd) (see Table 3) in the first half of the series (in this case, the optimal T_{sint} is equal to 1143 – 1163 K). For the REE having the small radiis (Tm, Yb, Lu) the optimal sintering temperatures were lesser than 1163 K even for the maximum content of incorporated REE.

When the REE radius decreases (in the whole series) a content of impurity phases increases, which is also indicative of the decreased stability of $(Bi_{1-x}A_x)O_3$. So, for the solid solutions with the relatively small ions (Yb, Lu) a content of impurities amounts equals to 50 %. The dependences of the structure parameters on the composition of $Bi_{1-x}A_xFeO_3$ ceramics at room temperature are shown in Figure 3. The rhombohedral (Rh) phase (typical of $BiFeO_3$) is retained over the concentration range in $Bi_{1-x}La_xFeO_3$ and up to $x = 0.15$ in $Bi_{1-x}Nd_xFeO_3$. A monoclinic (M) phase arises in ceramics containing Nd at $x > 0.10$, co-exists with the Rh phase in the range $0.10 < x \leq 0.15$ (morphotropic region), and independently crystallizes at $x > 0.15$. In the Rh phase, the structural parameters change in a similar fashion in both systems: the volume declines especially sharply at $0.05 \leq x \leq 0.10$, parameter a increases, and the rhombohedral angle α decreases with an increase in the A content. The constancy of the values of V_{exp} for Rh-phase and M-phase in the morphotropic region of the Nd-containing system is obviously due to the invar effect, characteristic for region of co-existence of the phases with different symmetry [36]. Study of the microstructure showed that at $x \leq 0.10$, BiFeO$_3$ and its solid solution Bi$_{1-x}$A$_x$FeO$_3$ are characterized by a multielement microstructure of the basic-connected matrix–pores–minority phase type, and the microstructure homogeneity increases with an increase in the (La, Nd) content x to more than 10 at. %. The Mössbauer spectrum of BiFeO$_3$ (Figure 4a) is a superposition of two Zeeman sextets and two paramagnetic doublets (their parameters are listed in the Table 4). The doublets correspond to a phase containing iron impurity, while the two Zeeman sextets are due to the presence of a spatial spin-modulated (SM) structure in BiFeO$_3$ [37–39].

Figure 4b shows as an example of Mössbauer spectrum for the Bi$_{0.8}$Nd$_{0.2}$FeO$_3$ ceramics. The spectral parameters (see Table 4) correspond to Fe$_{3+}$ ions in the oxygen octahedron. The presence of one Zeeman sextet in the Mössbauer spectrum of Bi$_{0.8}$Nd$_{0.2}$FeO$_3$ solid solution indicates the degradation of the SM structure in it. This is apparently related to the phase transition from the Rh-phase to the M-phase. The spectrum of Bi$_{0.8}$La$_{0.2}$FeO$_3$ retains two Zeeman sextets. Hence, in this case the SM structure is preserved at room temperature (Table 4). Analysis of the dielectric spectra of the solid solutions under study [40] revealed that they all lack a high temperature peak of relative permittivity $\varepsilon/\varepsilon_0$ (the Curie temperature), due to an increase in the ceramics conductivity. At the same time, a number of the dependences for Nd-containing samples demonstrate an additional low-frequency $\varepsilon/\varepsilon_0$ peak at 500 K (Figure 5). The character of the ε' anomaly and the corresponding ε'' value suggests that this anomaly is related to a change in the defect structure, due likely to the destabilization of the iron valence state in the samples. This process is enhanced by a rise in the polarizing effect of Nd compared with La. The magnetic susceptibility of the ceramics under study $\chi = 0.7 - 1.4$ relative units, is typical of antiferromagnets (Figure 5).

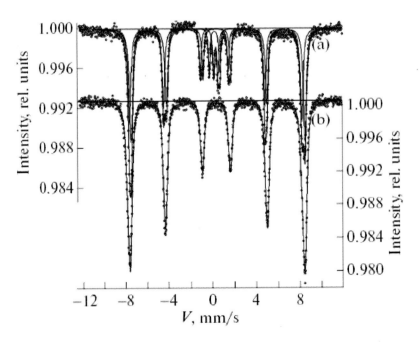

Figure 4. Mössbauer spectra of (a) BiFeO$_3$, and (b) Bi$_{0.8}$Nd$_{0.2}$FeO$_3$ ceramics, measured at room temperature in the rate range $\pm$ 12 mm/s.

Table 4. Parameters of the Mössbauer Spectrum of a BiFeO₃ ceramic sample

Component	δ, mm/s*	ΔE_Q, mm/s**	H, kOe***	S, %****
	BiFeO₃			
Sextet 1	0.3789 ± 0.0013	0.3353 ± 0.0026	495.81	45.31
Sextet 2	0.3817 ± 0.0017	−0.0964 ± 0.0033	491.53	43.08
Doublet 1	0.4419 ± 0.0026	0.5269 ± 0.0046		6.23
Doublet 2	0.1609 ± 0.0035	0.7806 ± 0.0071		5.38
	Bi₀.₈Nd₀.₂FeO₃			
Sextet 1	0.3889 ± 0.0007	0.0533 ± 0.0013	496.94 ± 0.05	−
	Bi₀.₈La₀.₂FeO₃			
Sextet 1	0.3623 ± 0.0034	−0.0269 ± 0.0028	496.97 ± 0.07	80.51
Sextet 2	0.5055 ± 0.033	0.1057 ± 0.0072	496.23 ± 0.18	15.95
Doublet	0.125 ± 0.0073	0.5596 ± 0.0126		3.54

Notes: * δ is the chemical isomer shift.
** ΔE_Q is the quadrupole splitting.
*** H is the magnetic field on Fe⁵⁷ nuclei.
**** S is the spectral-component area.

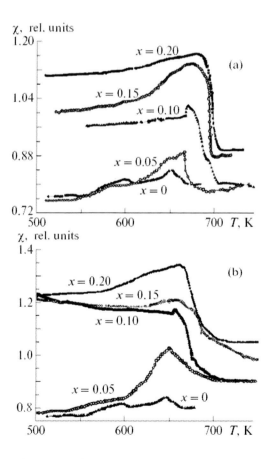

Figure 5. Temperature dependences of differential magnetic susceptibility of the Bi₁₋ₓNdₓFeO₃ and Bi₁₋ₓLaₓFeO₃ systems (a, b); the numbers on the curves correspond to the A content in the solid solution.

At the same time, χ and the magnetoelectric effect [40] grow with an increase in modifier concentration, thus indicating a stronger magnetic exchange interaction between Fe^{3+} ions. The dependences of the magnetoelectric effect on the A content contain two plateaus: the first ($0.05 \leq x \leq 0.10$) is apparently caused by a change in the dominant mechanism of solid solution formation (from interstitial to substitutional [40]), while the second ($0.15 \leq x \leq 0.20$) could be due to the invar effect characteristic of regions of the coexistence of different phases or phase states (Rh and M, in our case).

3.1.2. Conclusions

(1) A phase transition from the Rh-phase to the M-phase occurs in the $Bi_{1-x}Nd_xFeO_3$ system into range $0.10 \leq x \leq 0.20$ at room temperature, whereas the system with La exhibits no such change.

(2) According to the Mössbauer data, the spatially modulated spin structure is retained over the range of La concentrations but degrades in the morphotropic region (at the transition from the rhombohedral phase to the monoclinic phase) in the Nd-containing system. An increase in the content of rare-earth elements enhances the magnetic properties and magnetoelectric interactions in all of the solid solutions under study.

(3) Correlations were revealed between dielectric and magnetic properties and phase composition of the studied solid solutions. The character of the dielectric spectra of the solid solutions is determined by large extent of the phenomena induced by the conductivity resulting from change in the Fe valence state.

3.2. Microstructure, Dielectric and Mössbauer Spectroscopy of $PbNb_{1/2}Fe_{1/2}O_3$

3.2.1. Discussion

X-ray diffraction analysis (see Table 5) showed that the obtained ceramics were single-phase and had a rhombohedral structure with the lattice parameters $a = 4.014$ Å and $\alpha = 89.94°$ at room temperature. These values are in good agreement with the literature data ($a = 4.014$ Å, $\alpha = 89.92°$) [41]. The samples were characterized by a large ρ_{rel} value (> 95%) and a well developed fine-grain microstructure without pores (Figure 6).

Table 5. X-ray diffraction data for PFN ceramics

No	Composition	Unit-cell type	Parameters, Å	$\rho_{X\text{-ray}}$ g/cm^3	ρ_{exp} g/cm^3	ρ_{rel} %
1	$PbNb_{1/2}Fe_{1/2}O_3$ $T_{sint} = 1373$ K $\tau_{sint} = 2.5$ h	Rhombohedral	$a = 4.014$ $\alpha = 89.94$	8.47	8.07	95.28

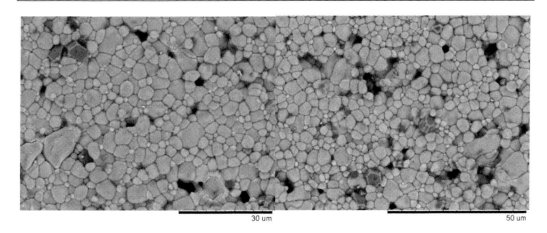

Figure 6. Microstructure fragments of PFN ceramics.

These data indicate the high quality of the synthesized ceramics and reliability of the obtained results. The room temperature Mössbauer spectrum of PFN (Figure 7a) was a paramagnetic doublet whose parameters (Table 6) corresponded to Fe^{3+} ions in an octahedral environment.

The quadrupole splitting in the PFN spectra could be due to the asymmetric environment of the intraoctahedral Fe^{3+} ions caused by local inhomogeneities of degree of the composition ordering of Fe^{3+} and Nb^{5+} ions. This is confirmed by the Mössbauer spectra of highly ordered perovskite $A(Fe_{1/2}Sb_{1/2})O_3$ (where A = Ca, Sr, Ba, Pb) containing no only a doublet but also a singlet that can be assigned to compositionally ordered regions, whereas the doublet is produced by regions with violated long-range order in the ion environment [6]. When a magnetic state is formed in the material, the Mössbauer spectrum is transformed into a sextet. As a result, we can determine the transition temperature to the magnetic state. In order to accomplish this, it is sufficient to measure the intensity one of the Mössbauer lines.

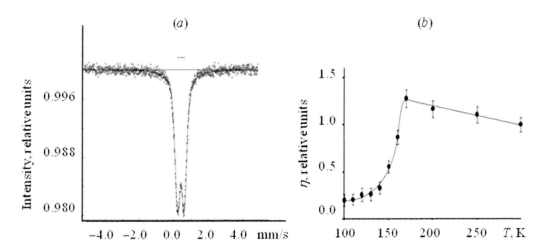

Figure 7. (a) Room temperature Mössbauer spectrum, and (b) temperature dependence of the relative amplitude of PFN Mössbauer lines.

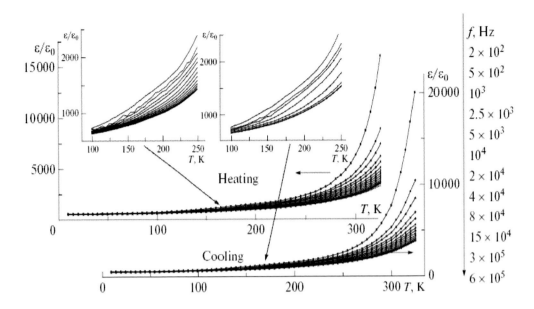

Figure 8. Temperature dependences of $\varepsilon/\varepsilon_0$ for PFN in the range 10 – 325 K at frequencies from 25 Hz to 1 MHz at heating and cooling.

Table 6. Mössbauer parameters for PFN ceramics

Component	δ, mm/s	ΔE_Q, mm/s	G, mm/s
Doublet	0.4133 ± 0.0008	0.3951 ± 0.004	0.3764 ± 0.005

Notes: δ is the chemical isomer shift, ΔE_Q is the quadrupole splitting, and G is the spectral line width.

The temperature dependence of the relative amplitude $\eta(T)$ of PFN Mössbauer lines is shown in Figure 7b. The sharp decrease in η in the range 150 – 170 K is related to the formation of an antiferromagnetic state. The antiferromagnetic transition temperature in the PFN sample under consideration was thus about 160 K. The temperature dependences of $\varepsilon/\varepsilon_0$ and tan δ in the range from cryogenic to room temperatures 10 – 325 K are shown in Figures 8 and 9. In the range 150 – 170 K (i.e., near the antiferromagnetic transition) both dependences exhibit anomalies both upon heating and cooling that are most pronounced at low frequencies. Figures 10 and 11 show the temperature dependences of $\varepsilon/\varepsilon_0$ and tan δ at high temperatures.

The temperature dependences of β_H in the heating and cooling modes are present in Figure 9 (insets 1, 2). It can be seen that T_C = 369 K. A jump is observed below T_C in the temperature dependences of β_H due to the Curie temperature T_C shifting to lower temperatures ($\Delta T_C \sim 4$ K) under an external magnetic field. This explains the temperature behavior of β_H near the phase transition. The microwave absorption spectra, measured in the frequency range of 5.6 – 8.3 GHz, reveal the loss in this range to be about 10 – 13 dB and have mainly a relaxation nature. An analysis of the experimental data for the PFN samples prepared at different exposure times (2 and 2.5 h at T_{sint} = 1373 K) show they are virtually identical. It indicates the high stability of the properties of the ceramic samples under study.

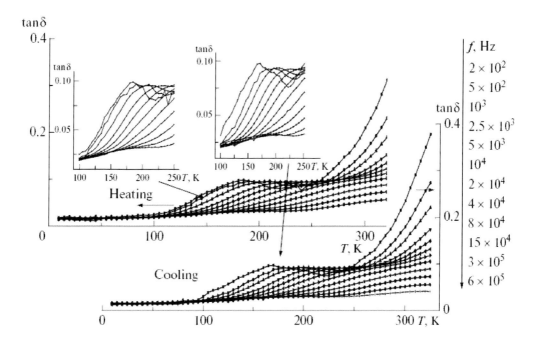

Figure 9. Temperature dependences of tan δ for PFN in the range 10 – 325 K at frequencies from 25 Hz up to 1 MHz at heating and cooling.

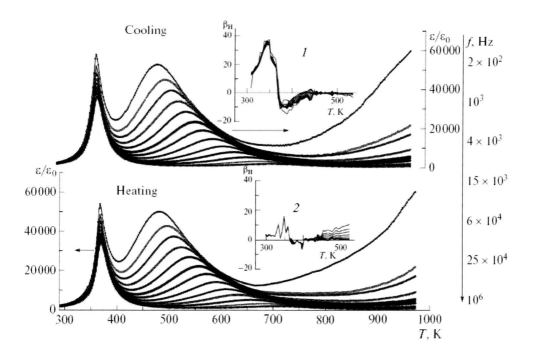

Figure 10. Temperature dependences of $\varepsilon/\varepsilon_0$ for PFN in the range 300 – 973 K at frequencies from 25 Hz up to 1 MHz at heating and cooling. Insets *1* and *2* show the temperature dependences of β_H.

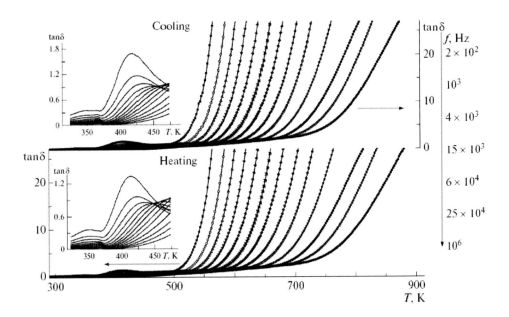

Figure 11. Temperature dependences of tan δ for PFN in the range 300 – 973 K at frequencies from 25 Hz up to 1MHz at heating and cooling.

3.2.2. Conclusions

The X-ray diffraction and microstructure analyses of the synthesized PFN ceramic samples showed their high quality, the results obtained are quite reliable. The Mössbauer study of the PFN ceramics allowed us to determine the antiferromagnetic temperature transition ($T_N \sim 160$ K) (near which the temperature dependences of $\varepsilon/\varepsilon_0$ and tan δ in the heating and cooling modes exhibit anomalies) related to the magnetoelectric interaction. Based on the temperature and frequency dependences of $\varepsilon/\varepsilon_0$ and tan δ the polar transition temperature was found to be $T_C \sim 369$ K. The revealed jump of β_H near T_C was due to the transition temperature shift to lower temperatures in an external magnetic field. Comparative analysis of the experimental data for the samples with different thermodynamic history demonstrated the high stability of the ceramic properties. The obtained results can be used to design multifunctional materials and devices based on them.

3.3. Effect of Alkaline-Earth Elements Nature on the Structure and Properties of AW$_{1/3}$Fe$_{2/3}$O$_3$ Compounds

3.3.1. Discussion

X-ray diffraction analysis (see Table 7) showed that the obtained SWF and PWF ceramic samples were single-phase ones and had, respectively, tetragonal ($a = 3.940$ Å, $c = 3.954$ Å) and cubic ($a = 3.980$ Å) structures at room temperature, whereas the BWF ceramics were two-phase ones and had cubic ($a = 8.131$ Å) and hexagonal (< 10%, $a = 5.764$ Å, $c = 14.131$ Å) phases co-existing in them at room temperature.

Table 7. X-ray diffraction data

No	Composition	Unit cell type	Parameters, Å	ρ_{X-ray} g/cm^3	ρ_{exp} g/cm^3	ρ_{rel} %
1	BaW$_{1/3}$Fe$_{2/3}$O$_3$ T_{sint} =1613K T_{sint} =2.5 h	Cubic + Hexagonal (< 10%)	a = 8.131 a = 5.764 c = 14.131	6.94	6.46	93.08
2	SrW$_{1/3}$Fe$_{2/3}$O$_3$ T_{sint} = 1673K T_{sint} = 2.5 h	Tetragonal	a = 3.940 c = 3.954	6.33	5.80	91.62
3	PbW$_{1/3}$Fe$_{2/3}$O$_3$ T_{sint} = 1373K T_{sint} = 2.5 h	Cubic	a = 3.980	9.31	8.44	90.66

The samples of all the compounds were characterized by large value of ρ_{rel} (> 90 %), absence of impurities, and well-developed fine-grain microstructure with plate-like (BWF) or isotropic (SWF and PWF) grains (Figure 12).

Figure 12. (Continued).

Figure 12. (Continued).

Figure 12. Microstructure fragments of the (a, b) BWF, (c, d) SWF, and (e, f) PWF ceramics.

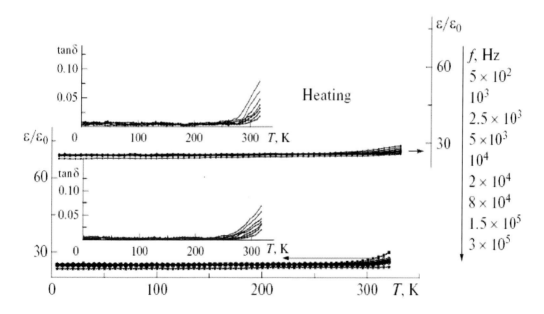

Figure 13. Dependences of $\varepsilon/\varepsilon_0$ and tan δ for BWF in the temperature range 10 – 325 K at frequencies from 25 Hz up to 1 MHz at heating and cooling

The temperature dependences of $\varepsilon/\varepsilon_0$ and tan δ for these compounds, measured in the range from cryogenic up to room temperatures 10 – 325 K in the heating and cooling modes, are shown in Figures 13 – 15.

Their dielectric properties differ considerably: the $\varepsilon/\varepsilon_0$ ratio is small (< 50) for BWF and does not increase, whereas SWF and PWF exhibit an increase in $\varepsilon/\varepsilon_0$ with an increase in T. The $\varepsilon/\varepsilon_0$ ratios vary in the ranges 100 – 30000 for SWF and 800 – 3500 for PWF. For SWF, $\varepsilon/\varepsilon_0$ increases sharply at $T \geq 200$ K (most significantly at low frequencies), while for PWF a strongly relaxing maximum is formed near the same temperature ($T \sim 180$ K) (in addition, becomes inverted in this range), and a frequency "stratification" of the temperature dependences of $\varepsilon/\varepsilon_0$ occurs in the range 250 – 350 K, which is transformed into a strongly relaxing fan-like maximum at high temperatures 400 – 600 K (see Figure 16).

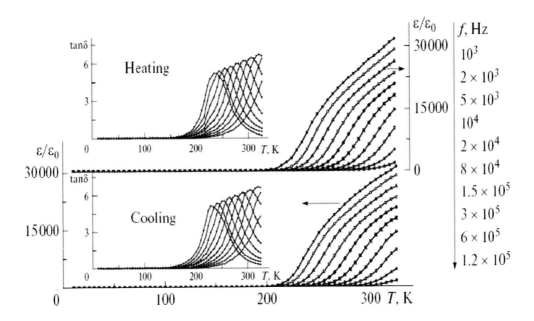

Figure 14. Dependences of $\varepsilon/\varepsilon_0$ and tan δ for SWF in the temperature range 10 – 325 K K at frequencies from 25 Hz up to 1 MHz at heating and cooling.

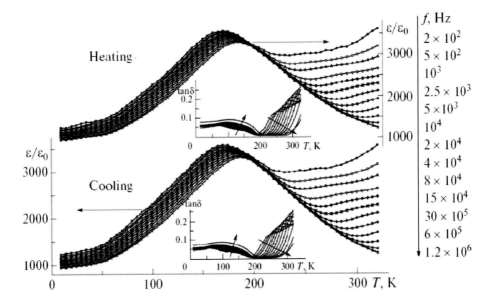

Figure 15. Dependences of $\varepsilon/\varepsilon_0$ and tan δ for PWF in the temperature range 10 – 325 K at frequencies from 25 Hz up to 1 MHz at heating and cooling.

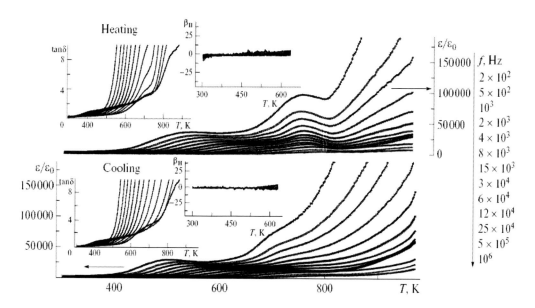

Figure 16. Dependences of $\varepsilon/\varepsilon_0$, tan δ, and β_H for PWF in the temperature range 300 – 973 K at frequencies from 25 Hz up to 1 MHz at heating and cooling.

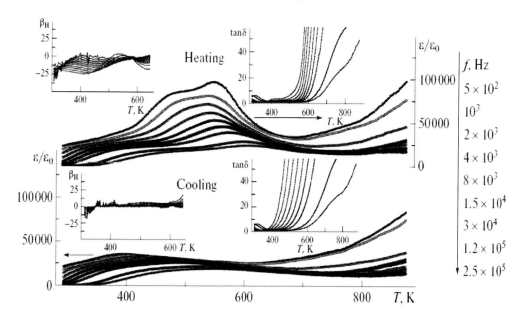

Figure 17. Dependences of $\varepsilon/\varepsilon_0$, tan δ, and β_H for SWF in the temperature range 300 – 973 K at frequencies from 25 Hz up to 1 MHz at heating and cooling.

In this range, anomalies are also observed for SWF (a two humped relaxing maximum arises, the shape of which partially changes upon cooling) (see Figure 17) and BWF (a weakly pronounced relaxing maximum, see Figure 18). Similar anomalies were observed earlier by studying the high-temperature dielectric properties of $PbFe_{0.5}Nb_{0.5}O_3$ [42] and $Bi_{0.8}Nd_{0.2}FeO_3$ [40]. An analysis and comparison of the results obtained here and in References [40, 42] suggest an explanation of the vacancy mechanism, which is related to the

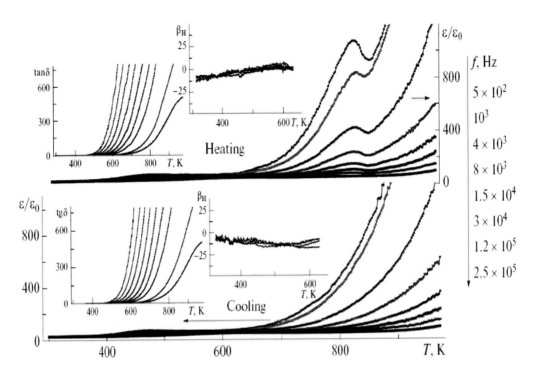

Figure 18. Dependences of $\varepsilon/\varepsilon_0$, tan δ, and β_H for BWF in the temperature range 300 – 973 K at frequencies from 25 Hz up to 1 MHz at heating and cooling.

change in the valence state of Fe^{3+} [43]. The additional maxima in the range 750 – 850 K observed for some PFW and BWF samples were most likely caused by defect annealing. The β_H measurements and microwave absorption spectroscopy in the frequency range 5.6 – 8.3 GHz revealed the following: all the compounds under study had low β_H values indicative of weak interaction between the ferroelectric and magnetic subsystems, and low energy losses, approximately from –2 down to –3 dB for BWF and SWF and from –8 down to –12 dB for PFW. An analysis of the experimental data for the BWF and SWF samples prepared under different sintering conditions (temperatures and durations) revealed the good reproducibility of their properties, i.e. the high process efficiency of these compounds.

3.3.2. Conclusions

(1) X-ray diffraction and microstructural analyses of the obtained ceramic samples showed that they all had a fairly high relative density ρ_{rel} (> 90%). PFW and SWF are single-phase specimens (cubic and tetragonal structures, respectively) and BWF is two-phase specimen presenting mixture of cubic and hexagonal (< 10%) phases. All of the samples had a well-developed fine-grain microstructure (layered in BWF and isotropic in PFW and SWF).

(2) Analysis of the temperature and frequency dependences of $\varepsilon/\varepsilon_0$ and tan δ for these compounds revealed radical difference in their dielectric properties: BWF was characterized by a small $\varepsilon/\varepsilon_0$ value (< 50) that barely changes up to 400 K, whereas SWF and SWF exhibit non-monotonous behavior of $\varepsilon/\varepsilon_0$ with one and two maxima at ~ 200 K and 400 – 600 K, due likely to changes in the defect structure of the samples (induced in particular by a change in

the iron valence state). The additional maxima in the range 750 – 850 K observed for some samples are most likely due to defect annealing.

(3) Study of the magnetoelectric and absorption properties of the ceramic compounds revealed small β_H values (indicating weak interaction between the ferroelectric and magnetic sub-systems) and low microwave losses.

(4) Comparative analysis of the experimental results for BFW and SWF samples prepared with different sintering temperatures and durations showed they were virtually identical (i.e. the process is highly efficient).

3.4. $x - T$ diagrams of $(1 - x)\text{BiFeO}_3 - x\text{Bi}B\text{O}_3$ Solid Solutions

The results of X-ray diffraction profiles' study of ceramic samples for all the studied compounds of $(1 - x)\text{BiFeO}_3 - x\text{BiMnO}_3$ system at room temperature are shown in Table 8. It was established that the compounds with $0.05 \leq x \leq 0.25$ are characterized by the $R3c$ orthorhombic phase, and those with $0.30 \leq x \leq 0.50$ by the $Pbnm$ phase. In Table 8 the cell parameters and cell volumes, as well as R_p values are shown.

The presence in $x \geq 0.3$ samples of $Pbnm$ phase, that was previously observed in pure BiFeO_3 at $1100 \text{ K} \leq T \leq 1200 \text{ K}$ [44, 45] witnesses that the temperature values of phase transitions of ferroelectric $R3c$ phase into $Pbnm$ one in these compounds occur below room temperature. It is still unclear if $Pbnm$ phase is paraelectric or antiferroelectric. The $Pbnm$ phase may be centrosymmetric (non-polar) at the scale of cell unit, but might be non-centrosymmetric $Pbam$ phase in PbZrO_3 [46].

Table 8. Structural parameters of $(1 - x)\text{BiFeO}_3 - x\text{BiMnO}_3$ compounds at room temperature

x	Space group	Perovskite cell unit parameters $a, b, c,$ Å			$\alpha, \beta,$ deg.		Perovskite cell unit volume, Å^3	$R_p,$ %
		a	b	c	α	β		
0.05	$R3c$	3.959	-	-	89.447	-	62.719	5.96
0.10	$R3c$	3.960	-	-	89.484	-	60.707	5.67
0.15	$R3c$	3.951	-	-	89.540	-	63.705	4.34
0.20	$R3c$	3.965	-	-	89.559	-	63.214	4.83
0.25	$R3c$	3.973	-	-	89.592	-	63.252	3.78
0.30	$Pbnm$	3.953	3.960	3.953	-	90.560	62.821	4.38
0.35	$Pbnm$	3.984	3.988	3.984	-	90.652	63.887	4.23
0.40	$Pbnm$	3.942	3.952	3.942	-	90.482	63.292	5.09
0.45	$Pbnm$	3.954	3.975	3.954	-	90.457	62.755	5.70
0.50	$Pbnm$	3.969	3.980	3.969	-	90.596	63.630	3.79

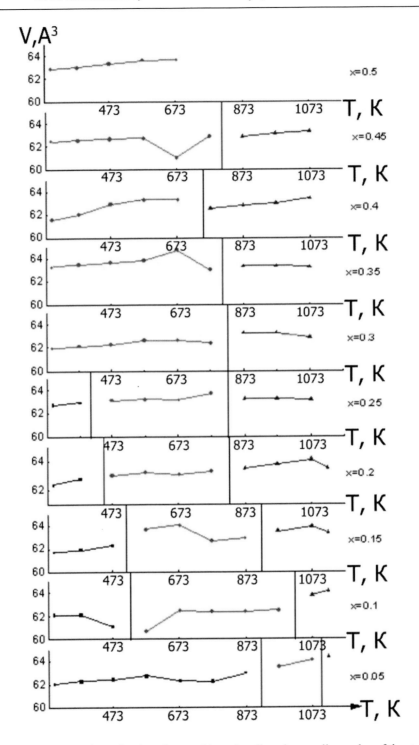

Figure 19 The temperature dependencies of perovskite sub-cells volumes all samples of the studied system.

In Figure 19 the temperature dependencies of perovskite sub-cells volumes for all samples of the studied system are shown, that allow one to determine the concentration-temperature phases' areas of existence i.e. state the phase diagram (x, T) (see Figure 20).

The temperature research of phase changes in the samples of the $(1 - x)\text{BiFeO}_3 - x\text{BiMnO}_3$ system have shown that in range $0.05 \leq x \leq 0.25$ of the samples the area of the *Pbnm* phase existence widens, the temperatures of phase transitions in paraelectric cubic phase decrease monotonously enough with the increase of x.

Figure 20 presents the phase diagram of $(1 - x)\text{BiFeO}_3 - x\text{Bi}B\text{O}_3$ $(0.00 \leq x \leq 0.50)$ solid solutions, obtained by methods of X-ray analysis. The phase diagram differs from the one presented in [15] by more precise representation of phase boundaries. Besides that, the orthorhombic phase that was interpreted by us as *Pbnm* is described in [15] only at the level of orthorhombic syngony with rather complicated superstructure cell. One can suppose that the diffraction reflexes of admixture phases were mistaken for the superstructure ones. After the analysis of volumes' temperature dependencies of perovslite cell units the following peculiarities can be sorted out. Firstly, at the transition from *R3c* phase to *Pbnm* one (in compounds with $0.05 \leq x \leq 0.25$) volume V_p decreases insignificantly. This can evidence on the ferroelectric transition from ferroelectric phase to the non-polar one. It is well-known, such transitions are accompanied by a jump-like cell volume reduction. Secondly, the transitions from *Pbnm* phase to *Pm3m* one (in compounds with $0.05 \leq x \leq 0.35$) are accompanied by the increase of V_p, that can evidence on deformation phase transitions connected with tilting of oxygen octahedra rotations in *Pbnm* phase.

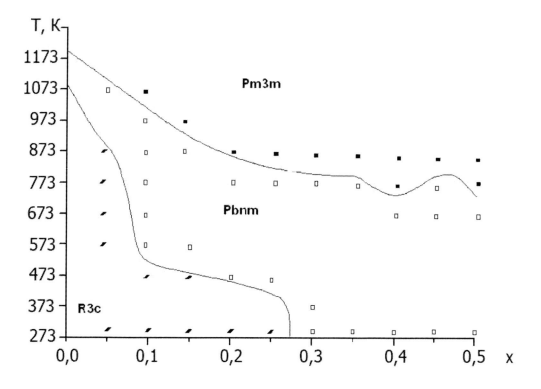

Figure 20. The phase diagram of $(1 - x)\text{BiFeO}_3 - x\text{Bi}B\text{O}_3$ $(0.00 \leq x \leq 0.50)$ solid solutions, obtained by the methods of X-ray analysis.

3.5. Phase States in Multiferroics with Spinel Structure

The systems of solid solutions both $CuCr_2O_4 - FeCr_2O_4 - NiCr_2O_4$ and $CuCr_2O_4-NiCr_2O_4 - NiFe_2O_4$ are characterized by complex phase equilibrium. The existence of multicritical points at $x - T$ diagrams of binary solutions $NiFe_{2-x}Cr_xO_4$, $Ni_{1-x}Cu_xCr_2O_4$ and $Fe_{1-x}Ni_xCr_2O_4$ is the important peculiarity of these systems [47 – 49].

In the system $Cu_xNi_yFe_{1-x-y}Cr_2O_4$ were determined the areas of existence of cubic $Fd3m$–phase (C), tetragonal $I4_1/amd$-phases with relations $c/a < 1$ (T_1) and $c/a > 1$ (T_2), $I\overline{4}2d$-phase with relation $c/a < 1$ (T_3) (see Figure 21). The rhombic $Fdd2$-phase or $Fddd$-phase (R) is fixed one together only with all phases pointed above (except T_1). The existence of the next three-phase points has been fixed at room temperature: (i) $x = 0.10$, $y = 0.79$ is the point of co-existence of C, T_3, and R–phases; (ii) $x = 0.05$, $y = 0.87$ is the the point of co-existence of C, T_2, and R–phases; (iii) $x = 0.10$; $y = 0.84$ is the point of co-existence of C, T_2, T_3, and R–phases.

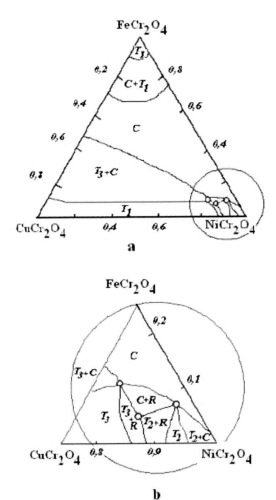

Figure 21. Phase states in the system $CuCr_2O_4 - FeCr_2O_4 - NiCr_2O_4$: (a) general picture of phase states, (b) the part of the phase states diagram near $NiCr_2O_4$ composition.

In the system $Ni_{1-x}Cu_xFe_{2y}Cr_{2(1-y)}O_4$, $(x + y) \leq 1$, it is stated the existence at room temperatute in $(T - x - y)$–diagram of three multiphase points: (i) $x = 0.14$, $y = 0.06$ is the point of co-existence of C, T_3 and R–phases; (ii) $x = 0.05$, $y = 0.05$ is the point of co-existence of C, T_2 and R–phases; (iii) $x = 0.10$, $y = 0.03$ is the point of co-existence of C, T_2 and R–phases (see Figure 22). At the diagram there are morphotropic areas presenting mixture of the phases $C + T_2$, $C + T_3$, $C + R$, $T_2 + R$, and $T_3 + R$ [50, 51]. Rhombic phase exists only together with other phases.

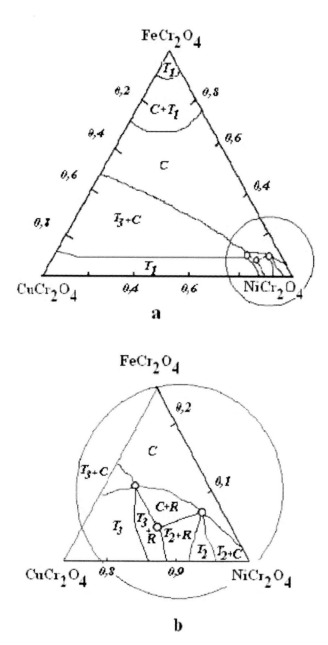

Figure 22. Phase states in the system $NiFe_2O_4 - NiCr_2O_4 - CuCr_2O_4$: (a) general picture of phase states, (b) the part of the phase states diagram near $NiCr_2O_4$ composition.

Ternary spinel systems considered were synthesized firstly. The existence of a great number of tetragonal and rhombic phases in them defines the interest to investigation no only their structural peculiarities, but physico-chemical properties, too, which may be used with practical purposes.

Structural mechanisms of phase transitions from $Fd3m$-phase to tetragonal $I\bar{4}2d$, $I4_1/amd$ phases and rhombic ($Fdd2$) phase are analyzed in detail in [52 – 57]. The transformation of occupied lattice complexes of $Fd3m$-phase of AB_2O_4 composition in the sequence of transformation $Fd3m \rightarrow I\bar{4}2d \rightarrow Fdd2 \rightarrow I4_1/amd$ may be given as

$$\begin{pmatrix} D \\ T \\ D4xxx \end{pmatrix}_{Fd3m} \rightarrow \left[2\begin{pmatrix} {}^vD \\ {}^vTF_clx \\ {}^vD4xyz \end{pmatrix}_{I\bar{4}2d} \rightarrow \begin{pmatrix} D \\ Txx \\ D4xyz \end{pmatrix}_{F\bar{4}d2_1} \right] \rightarrow \begin{pmatrix} D \\ Dxy \\ 2Dxy \end{pmatrix}_{Fdd2} \rightarrow$$

$$\rightarrow \left[\begin{pmatrix} D \\ T \\ D4xxz \end{pmatrix}_{F4_1/ddm} \rightarrow 2\begin{pmatrix} {}^vD \\ {}^vT \\ {}^vD4xz \end{pmatrix}_{I4_1/amd} \right].$$

By writing the lattice complexes of groups are used symbols recommended in [22]. In this case both $F\bar{4}d2_1$ and $F4_1/ddm$ are used as analogs of corresponding tetragonal groups $I\bar{4}2d$ and $I4_1/amd$ (with a divisible face-centered cell and identical to I-cells) to symmetric characteristics of occupied lattice complexes in F-cells [58].

The above-mentioned transformations from tetragonal $I\bar{4}2d$-phase to phombic $Fdd2$-phase and from rhombic $Fdd2$-phase to tetragonal $I4_1/amd$-phase are the second-order phase transitions. They are defined by continuous changes of displacement vectors for cations B with octahedral coordination and for oxygen anions.

$$B: \begin{bmatrix} (f,\bar{f},0) \\ (g,\bar{g},0) \end{bmatrix}_{I\bar{4}2d} \rightarrow \begin{bmatrix} (d_1,d_2d_3) \\ (e_1,e_2,e_3) \end{bmatrix}_{Fdd2} \rightarrow \begin{bmatrix} (0,0,0) \\ (c,c,2\bar{c}) \end{bmatrix}_{I4_1/amd}$$

In order to obtain $I\bar{4}2d$-phase with relation $c/a < 1$ from the initial $Fd3m$-phase the components of displacement vectors f and g should satisfy the following conditions: $f > 0$ и $g > 0$. In this case the condition of continuity of the transformation $I\bar{4}2d \rightarrow Fdd2$ for correlations of component vectors d and e of displacement allows us to write down the follows relations:

$$d_1 = 0;\ d_2 < 0;\ 0 > d_3 > \frac{\sqrt{3}}{2}d_2,\ \text{and}\ e_1 > 0;\ e_2 < \frac{2}{\sqrt{3}}e_1;\ -e_1 < e_3 < e_1 + \frac{\sqrt{3}}{2}e_2.$$

At transition from *Fdd2*-phase to tetragonal *I4₁/amd*-phase with relation $c/a > 1$ the component of displacement vector c for oxygen atoms must be positive.

Let us analyze the crystalline chemical construction of phases in the sequence of transformations: $Fd3m \to I\bar{4}2d \to Fdd2 \to I4_1/amd$. The structure of cubic phase may be presented as close packing of polyhedral layers parallel to planes of [001] type. In these layers isolated B_4O_4-hexahedra and AO_4-tetrahedra are alternated with each other as on a chess-

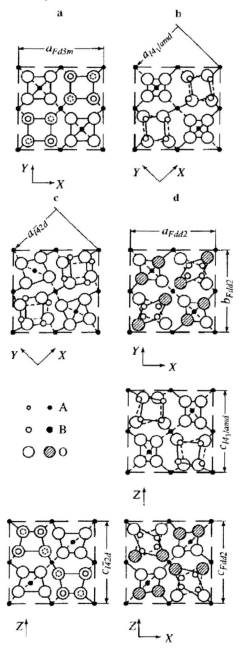

Figure 23. Idealizated images of projection of polyhedric layered *Fd3m*–phase (a), *I4₁/amd*-phase (b), *I$\bar{4}$2d*-phase (c), and *Fdd2*–phase (d) of AB_2O_4 composition.

board (see Figure 23, a). Then the transition to tetragonal $I\bar{4}2d$-phase with relation $c/a < 1$ must be accompanied by contra-rotation of two tetrahedra B_4 and O_4, composing B_4O_4-hexahedron, around the axis [001] of a cubic cell. This will lead to monoclinic pressing of BO_6-octahedra and to appearance of a set of three non-equivalent distances R_{B-O}: $2R'_{B-O}$ + $2R''_{B-O} + R'''_{B-O}$.

The similar rotation of atomic groups of oxygen in AO_4-tetrahedra will cause their tetragonal lengthening and lowering the positional symmetry for cations A. However, all four distances of R_{A-O} type on left side are the same (see Figure 23, b). Rhombic distortion $I\bar{4}2d$-phase at morphotropic transition $I\bar{4}2d \rightarrow Fdd2$ must lead to local pressing of tetragonally lengthened AO_4-tetrahedra, lowering interatomic distances R_{O-O} in the direction of [100] type of tetragonal phase and stretching the local pressed $BO_3O'_3$-octahedra in the directions, close to $[\bar{0}01]$. Change of the symmetry at transition will cause the formation of the sets of interior distances: $2R_{A-O} + R'_{A-O} + R''_{A-O}$ and six non-equivalent distances of R_{B-O} type (see Figure.23, c). The transition to tetragonal $I4_1/amd$-phase with relation $c/a > 1$ from rhombic $Fdd2$-phase must lead again to tetragonally distorted pressed AO_4-tetrahedra with the same interatomic distances R_{A-O}. Then, the distortion of oxygen O_4-tetrahedron in B_4O_4-hexahedron must lead to more high-symmetric monoclinically distorted BO_6-octahedron and formation of the following set of distances: $4R_{B-O} + 2R'_{B-O}$ (see Figure 23, d).

For the all new phases obtained there will be carried out the research of temperatute dependences of magnetic and electric properties. The further studies will be connected with synthesis, physico-chemical investigation of properties new ternary systems based on the structural type of spinel and with theoretical interpretation of phase transitions in the materials obtained.

REFERENCES

[1] Zvezdin, A.K.; Loginov, A.S.; Meshkov, G.A.; Pyatakov, A.P. *Izv. Russian Akad. Nauk, Ser. Fiz.* 2007, *vol. 71, No. 1*, 1604.

[2] Fesenko, E.G. *Perovskite Family and Ferroelectricity.* Atomizdat: Moscow, 1972, pp 1-248.

[3] Venevtsev, Yu. N.; Gagulin, V.V.; Lyubimov, V.N. *Ferroelectromagnets.* Nauka: Moscow, 1982, pp 1-224.

[4] Amirov, A. A.; Batdalov, A.B.; Kallaev, S.N.; et al. *Phys. Solid State.* 2009, *vol. 51, No 6*, 1189.

[5] Smolenski, G.A.; Chupis, I.E. *Uspehi Fizichesckih Nauk.* 1982, *vol. 137, No 3*, 415 - 448.

[6] Murashov, V.A.; Rakov, D.N.; Economov, N.A.; et al. *Fiz. Tverd. Tela.* 1990, *vol. 32, No 7*, 2156-2159.

[7] Malyshkina, O.V.; Baranova, E.V.; Gavrilova, N.D.; Lotonov, A.M. *Pis'ma Zh. Tekh. Fiz.* 2007, *vol. 33, No 18*, 70.

[8] Yang, Y.; Lui, J.M.; Huang, H.B.; et al. *Phys. Rev. B.* 2004, *vol. 70*, 132101.

[9] Landold, A.U.; Borstein, D. *Ferroelektrika ünd Verwandte Sübstanzen*, Springer: Berlin, 1981, pp 1-95.

[10] Mitoseriu, L.; Piaggio, P.; Nani, P.; Carnasciali, M.M. *Appl. Phys. Lett.* 2002, *vol. 81*, 5006.

[11] Ivanov, S.A.; Erricsson, S.-G.; Tellgren, R.; Rundlof, H., *Mat. Res. Bull.* 2004, *vol. 39*, 2317.

[12] Matzen, G.; Poix, P. J. *Solid State Chem.*, 1980, *vol. 33*, 341.

[13] Seveque, F.; Delamoge, P.; Poix, P.; Miche, A.L. *C.R. Seances Acad. Sci. Ser. C.* 1969, *vol. 269*, 1536.

[14] Lu, C.H.; Shinozaki, K.; Ishizawa, N.; et al. *J. Mater. Sci. Lett.*, 1988, *vol. 7*, 1078.

[15] Janculescu, A.; Gheorghik, F.P.; Postolache, P.; et al. *J. Alloys and Compounds.* 2010, *vol. 504*, 420-526.

[16] Higuchi, T.; Hattori, T.; Sakamoto, W.; et al. *Jap. J. Appl. Phys.* 2008, *vol. 47, No 9*, 7570-7573.

[17] Selbach, S.M.; Tybell, T.; Einarsurd, M.-A.; Graude T. *Phys. Rev. B.* 2009, *vol. 79*, 214113.

[18] Azuma, M.; Kanda, H.; Belik, A. A.; et al. *J. Magnetism and Magn. Mater.* 2007, *vol. 310*, 1177-1179.

[19] Belik, A.A.; Hayashi, N.; Azuma, M.; et al. *J. Solid State Chem.* 2007. *vol. 180, Is. 12*. 3401-3407.

[20] Avvakumov, E.G.; et al. *Mechanochemical Synthesis in Inorganic Chemistry.* Nauka, Siberian Branch: Novosibirsk. 1991, pp 1-210.

[21] Boldyrev, V. V.; Avvakumov, E.G.; Gusev, A. A.; et al. *Fundamental Basis of Mechanical Activation, Mechanosynthesis, and Mechanochemical Technologies.* Publishing House of the Siberian Branch of the Russian Academy of Sciences: Novosibirsk. 2009, pp 1-343.

[22] Kagomiya, I.; Sawa, H.; Siratori, K.; et al. *Ferroelectrics.* 2002, *vol. 268*, 327-332.

[23] Ueda, H.; Mitamura, H.; Goto, T.; Ueda, Y. *Phys. Rev.* 2006, *vol. B73*, 094415.

[24] Menyuk, N.; Dwight, K.; Wold, A. *J. Phys. (Paris).* 1964, *vol. 25*, 528-537.

[25] Tomiyasu, K.; Fukunaga, J.; Suzuki, H. *Physical Review.* 2004, *vol. B70*, 214434-1 – 214434-12.

[26] Yamasaki, Y.; Miyasaka, S.; Kaneko, Y.; et al.. *Phys. Rev. Letters.* 2006, *vol. 96*, 207204-1 – 207204-4.

[27] Arima, T.; Watanabe, Y.; Taniguchi, K.; et al. *J. Magnetism and Magnetic Materials.* 2007, *vol. 310*, 807–809.

[28] Klemme, S.; O'Neill, H.S.C.; Schnelle, W.; Gmelin, E. *American Mineralogist.* 2000, *vol. 85*, 1686–1693.

[29] Ueno, G.; Sato, S.; Kino, Y. *Acta Cryst. C.* 1999, *vol. 55*, 1963–1966.

[30] Ishibashi, H.; Yasumi, T. *J. Magnetism and Magnetic Materials.* 2007, *vol. 310*, 610–612.

[31] Razumovskaya, O.N.; Kuleshova, T.B.; Rudkovskaya, L.M. *Inorganic Materials.* 1983, *vol. 19, No 1*, 113-115.

[32] Zhuravlev, G.I. *Chemistry and Technology of Ferrites.* Khimiya: Leningrad, 1970, pp 1-191.

[33] Avvakumov, E.G.; Potkin, A.M.; Bertznyak, V.M. Patent RF. No.1584203A1, B O2 C17/08. *Planetary mill.* Publ. 18.06.87.

[34] Avvakumov, E.G, *Mechanical Activation Methods of Chemical Processes.* Nauka, Siberian Branch: Novosibirsk. 1986, pp 1-304.

[35] Bokii, G.B. *Crystallochemistry.* Nauka: Moscow. 1971, pp 1-400.

[36] Reznichenko, L.A.; Shilkina, L.A.; Razumovskaya, O.N.; et al. *Crystallogr. Rep.* 2006, *vol. 51, No 1,* 87.

[37] Zalessky, A.V.; Frolov, A.A.; Khimich, T.A.; Bush, A.A. *Phys. Solid State.* 2003, *vol. 45, No 1,* 141.

[38] Zalessky, A.V.; Frolov, A.A.; Khimich, T.A.; et al. *Europhys. Lett.* 2000, *vol. 50,* 547.

[39] Pokatilov, V.S.; Pokatilov, V.V.; Sigov, A.S. *Phys. Solid State,* 2009, *vol. 51, No 3,* 552.

[40] Verbenko, I.A.; Aleshin, V.A.; Kubrin, S.P.; et al. In *Proc. II Interdisciplinary Int. Symp. "Media with Structural and Magnetic Ordering" (Multiferroics-2009), Rostov-on-Don, Sochi.* 2009, 41.

[41] Battle, P.D.; Gibb, T.C.; Herold, A.J.; Hodges, J.P. *Mater. Chem,* 1995, *vol. 5,* 75.

[42] Gufan, Yu. M.; Pavlenko, A.V.; Reznichenko, A.A.; et al. *Izv. Russian Akad. Nauk, Ser. Fiz.* 2010, *vol. 74, No. 8.*

[43] Feng, L.; Ye, Z.G. *J. Solid State Chem.* 2002, *vol. 163,* 484.

[44] Palai, R.; Katiyar, R.S.; Schmid, H.; et al. *Phys. Rev.* 2008, *vol. B77,* 014110.

[45] Arnold, D.C.; Knight, K.S.; Morrinson, F.D.; Lightsfoot, P. *Phys. Rev. Lett.,* 2009, *vol. 102,* 027602.

[46] Ivanov, S.A.; Nordblad, P.; Tellgren, R.; et al. *Solid State Sciences.* 2008, *vol. 10,* 1875-1885.

[47] Antoshina, L.G.; Vehter, B.G.; Kaplan, M.D., et al. *Fizika Tviordogo Tela.* 1981, *vol. XXIII, No 7,* 1965-1969.

[48] Kose, K.; Iida, S. *J. Appl. Phys.* 1984, *vol. LV, No 6,* 2321-2323.

[49] Kino, Y.; Miyahara, S. J. *Phys. Soc. Japan.* 1966. *vol. XXI,* 2732.

[50] Ivanov, V.V.; Talanov, V.M.; Shabielskaya, N.P. *Cust Sources of Current: Proc. 6-th Int. Conf. 19-21 September 2000.* Nabla: Novocherkassk. 2000, *pp* 35-36.

[51] Ivanov, V.V.; Talanov, V.M.; Shabielskaya, N.P. *Izv. Vuzov. Severo-Kavkazskii Region. Tehnicheskie Nauki.* 2001, *No 1,* 91-95.

[52] Ivanov, V.V.; Talanov, V.M.; Shabielskaya, N.P. *Inorganic Materials.* 2001, *vol. 37. No 8.* 990-996.

[53] Ivanov, V.V.; Talanov, V.M.; Shabielskaya, N.P. *Izv. Vuzov. Severo-Kavkazskii Region. Tehnicheskie Nauki.* 2001, *No 4,* 104-105.

[54] Ivanov V.V.; Talanov V.M.; Shabielskaya N.P. *Izv. Vuzov. Severo-Kavkazskii Region. Tehnicheskie Nauki.* 2001, *No 4,* 105-106.

[55] Talanov V.M. *Izv. Akad. Nauk SSSR. Inorganic Materials.* 1989. *vol. 25, No 6,* 1001-1005.

[56] Talanov, V.M. *Izv. Akad. Nauk SSSR. Inorganic Materials.* 1989, *vol. 25, No 5,* 867-869.

[57] Ivanov, V.V.; Talanov, V.M. *Inorganic Materials.* 1995, *vol. 31, No 2,* 258-261.

[58] Fisher, W.; Burzlaff, H.; Hellner, E.; Donney, J.D.H. *US Dep. Commerce. Nat. Bur. Stand.:* Washington. 1975, *vol. 134.* pp 1-178.

In: Ferroelectrics and Superconductors
Editor: Ivan A. Parinov

ISBN: 978-1-61324-518-7
©2012 Nova Science Publishers, Inc.

Chapter 5

INVESTIGATIONS OF CONDUCTIVE AND MECHANICAL PROPERTIES OF SUPERCONDUCTIVE COMPOSITES BASED ON ASYMPTOTICAL METHODS AND PHASES WITH NEGATIVE STIFFNESS

Ivan A. Parinov[*,1] *and Shun Hsyung Chang*[≠,2]

[1]Vorovich Mechanics and Applied Mathematics Research Institute,
Southern Federal University, Russia
[2]Department of Microelectronics Engineering, National Kaohsiung Marine University,
Kaohsiung, Taiwan, Republic of China

ABSTRACT

The different approaches and methods to describe microstructure and macroscopic properties of various heterogeneous materials have been widely applied in sciences on composites. At the same time there is insufficient number of the studies for superconductors. In review form the chapter presents different methods of the composite studies. Due to sharp difference of conductivity of normal and superconducting components of superconductive composites, and also geometrical features of various phases, the asymptotic approaches to solve the problems become very effective and attractive. The results that studies of conductive and mechanical properties of superconductive composites are fulfilled by taking into account the influnce of inclusion coverage (for fibers and grains) on the composite conductivity, effects of non-ideal contacts of inclusions and matrix, structure irregularity and cluster formation. In these cases, for joining the solutions are used Pade two-point approximations and method of asymptotically-equivalent functions. Then, it should be expected that the materials being near some phase transitions will demonstrate a negative stiffness at microscale due to a set of causes. Ferroelectrics and superconductive ferroelastics in vicinity of structural phase transitions possess components with stiffness that achieve minimum values and

---

[*]E-mail: ppr@math.rsu.ru, 200/1, Stachki Ave, Rostov-on-Don, 344090, Russia.
[≠]E-mail: shchang@mail.nkmu.edu.tw No.142, Haijhuan Rd., Nanzih District, Kaohsiung City 81143, Taiwan, ROC.

aspire to zero at the critical temperature. Below this critical temperature initiates a domain or band structure. The composites with negative stiffness may be used in studies of single domains of the ferroelastics and ferroelectrics. Anomalous changes of composite properties could be predicted if inclusions have a negative module compared in value with matrix module. These studies are also present in the chapter.

1. INTRODUCTION

Definition of composite properties (in particular, for superconductors) in dependence on the microstructure and phase features is one of the main problems in mechanics of composite materials. Simplest models operate with properties of single components (or phases) and their concentrations. More complex computation methods take into account different micromechanical effects (e. g. interactions of adjacent components, non-ideal contacts between them, etc.). Multi-volume monographs [25, 93] have been devoted to basic investigations of numerous composite properties based on very common statements of the problems.

The averaged models in the composite studies have been initiated by Voigt [144], who proposed to define the effective elastic moduli by using simple arithmetic averaging stiffness of components, and by Reuss [120], who proposed to determine the effective elastic moduli by using averaging compliance of the components. Basing on the theorems on minimums of potential and additional energies Hill [53] stated that the Voigt and Reuss methods give respectively upper and lower estimations for the effective elastic moduli. These wide bounds for the estimations have been significantly narrowed by Hashin and Shtrikman [45, 48, 51] by using a variation method. Non-variation estimations could be obtained with help of Pade approximations [5]. The strong estimations which coupled one with other conductive and elastic properties of heterogeneous structures have been found in References [39, 97]. Numerous more precise estimations of structure-sensitive properties for various composites and methods their computations have been also present [101, 137]. Additionally with strict methods for estimation of effective composite properties, numerous approximation methods have been proposed. For example, in the virial expansion method [126], the effective moduli are calculated by using power series expansions on concentrations one of the components.

In this case, as rule, it is possible to construct only first approximation corresponding to the case of single inclusion into infinite matrix. The results obtained on the base of this method are applied to dilute fraction of inclusions. In this approach the medium properties as near inclusion, as far from the inclusion are assumed to be equal to the matrix characteristics. The three-phase model [65, 141] gets over this shortage by assuming that far from inclusion the medium properties are found by effective composite characteristics. This calculation scheme allows one to obtain a good approximation in the cases of dilute and mean concentrations of inclusions. In the case of maximal concentration of inclusions it could be found asymptotic solution in which as a small parameter is used the distance between adjacent inclusions. References [8, 62] present first terms of asymptotic expansions for effective conductivity of composites with strong difference of matrix properties from inclusion ones to being near-contacting one with others. In the case known effective properties at the dilute and high concentrations of inclusions a joining of solutions by using two-point Pade approximation [129, 131] and the method of asymptotically equivalent functions [3, 4] allows one to calculate effective properties for all spectrum of the volume concentrations. The exact solutions for composites with regular structure could be obtained by using representation of elastic fields as infinite series on periodic function of complex variables [41, 143].

The theory of effective moduli is based on principle of averaging corresponding properties of heterogeneous materials and composites. In this case the physical processes in these media are modeled by equations in partial derivations with rapidly changing factors. Asymptotic solutions for these equations could be stated by using two-scale expansions. Firstly, multi-scale method and averaging method have been developed for ordinary differential equations [15, 103]. These methods may be also applied to equations in partial derivations whose solution components change in different way on spatial coordinates [84, 85]. The mathematical basis of the averaging method in mechanics of composites is present in monographs [6, 10, 117, 124].

Important role in theory of composites plays the problem of ideal contact between components. Obviously, interfaces are the concentrators of mechanical stresses and significantly define strength properties of composites. Weaken couples between components lead to both diminishing effective properties and initiating defect growth (dislocations, microcracks, microvoids, etc.). The fracture processes of component interfaces and corresponding cohesion relations have been studied, in particular in References [31, 105, 142]. Micromechanical influence of boundary could be taken into account by introducing thin intermediate layer (or component coverage) [140]. In the case of lower stiffness of the coverage compared with inclusion (so-called, soft boundary), it is initiated the case of non-ideal contact. In contrast case of rigid contact, the coverage to be an additional carrying element.

The investigation methods of conductive and mechanical properties of composites include, in particular statistical mechanics of multi-particle systems, analysis of correlation functions, percolation theory, mathematical modeling, using the variation bounds and interaction between different properties. Review of those studies and corresponding results for superconductive composites will be present in first part of this chapter. The couples between conductive and elastic properties will be investigated in second part of the chapter. Due to sharp difference of conductivity normal and superconducting components of superconductive composites, and also geometrical features of various phases, the asymptotic approaches to solve the problems become very effective and attractive. The numerical results for conductive and mechanical properties of superconductive composites will be obtained in the third part of the chapter by taking into account influence of inclusion coverage (for fibers and grains) on the composite conductivity, effects of non-ideal contacts between inclusions and matrix, structure irregularity and cluster formation. Finally, it is known that ferroelectrics and superconducting ferroelastics in vicinity of structural phase transitions possess components with stiffness that achieve minimum values and aspire to zero at the critical temperature. Below this critical temperature initiates a domain or band structure. Therefore, the composites with negative stiffness could be used in studies of single domains of the ferroelastics and ferroelectrics. Anomalous changes of composite properties could be predicted when inclusions have a negative module compared in value with the matrix module. These studies will be discussed in the fourth part of the work.

2. Conductivity and Elasticity
of Superconductive Composites

The studies of percolation features of composites allow microstructure modeling to design superconductive systems with optimal transport properties. It is known, that the critical exponents for conductive and elastic properties in general case differ from their lattice values due a singular distribution of 'necks' coupling single cells near percolation threshold [34]. Let phase 1 is conductive or elastic medium and phase 2 corresponds to void space with infinity-small conductive and elastic properties. Then there is the percolation threshold (critical volume fraction) $c_{2p} = 1 - c_{1p}$ that, when $c_2 > c_{2p}$ the system ceases to support a conductivity or mechanical load that leads to discontinuity of the phase 1. In this case the critical exponents t and f for effective conductivity σ_0 and effective Young modulus E_0 can be defined as

$$\sigma_0 \sim (c_1 - c_{1p})^t \text{ at } c_1 \to c_{1p} + 0; \tag{1}$$

$$E_0 \sim (c_1 - c_{1p})^f \text{ at } c_1 \to c_{1p} + 0. \tag{2}$$

2.1. Two-Phase Composites with Phase-Interchange Properties

It should be noted that in difference from a lattice percolation, the conductivity threshold and the threshold of standard Bernoulli connectivity generally do not coincide with estimation of the percolation threshold in continuum. Consider the case of simultaneous percolation (bi-continuity) both phases in two-phase composite. In particular, this example is the case of two-dimensional (2D) two-phase model of the chessboard type (black squares corresponds to a superconductive phase and white squares define a normal-conductive phase) belonging to the material class with phase-reverse symmetry. Two-phase continuous system of dimension d is bi-continuous one at $c_{2p} < c_2 < 1 - c_{2p}$ for $d \geq 2$ under condition that the percolation threshold $c_{2p} < \frac{1}{2}$ [137]. Obviously, that the bi-continuity in 2D case may be achieved with more difficulties compared with the dimensions $d \geq 3$. However the introduction of criterion of the long-range connectivity allows decreasing c_{2p} lower than $\frac{1}{2}$ and introducing the bi-continuity. Because, conductivity could be occurred through common angle points of adjacent superconductive cells coupled only by the diagonal connection then the percolation threshold in this case is found as $c_{2p} = 1 - p_c^s \approx 0.4073$ (where p_c^s is the standard percolation threshold in square lattice by taking into account only nearest neighbors). At the same time, the Bernoulli connectivity threshold for superconductive phase is equal to $c_{2p} = p_c^s \approx 0.5927$. So, this system will be bi-continuous at $p_c^s < c_2 < 1 - p_c^s$. The result firstly has been obtained in Reference [125], where noted importance of conductivity through angle points.

Relations coupling effective properties of two-phase heterogeneous material with effective properties the same microstructure but with phase-interchange components are named by the phase-interchange relations. When these strict relations are applied to wide class of heterogeneous materials then they can ensure useful tests for analytical and numerical estimations of effective properties. In Reference [63] it has been treated heterogeneous material in 2D statement consisting of arbitrary lattice of the same parallel cylinders symmetrical both as in x-direction as in y-direction. It has been proved that the effective conductivity in the x-direction σ_0^x is coupled with the effective conductivity of the phase-interchange composite in the y-direction σ_0^y as

$$\sigma_0^x(\sigma_1,\sigma_2)\sigma_0^y(\sigma_2,\sigma_1) = \sigma_1\sigma_2 . \tag{3}$$

In this equation the first argument in σ_0^x (or σ_0^y) defines conductivity of matrix phase (or inclusion) and the second argument signs conductivity of inclusion (or matrix phase). In Reference [28] it has been shown that the effective conductivity of 2D two-phase composites with phase distributions, which to be statistically equivalent one to other, defined by mean geometrical value $(\sigma_1\sigma_2)^{1/2}$, where $\sigma_i (i = 1, 2)$ was the conductivity of i-th phase. Then these results have been generalized [87] and, in particular it has been shown that Equation (3) applied to any 2D two-phase composite (ordered or not) until x and y to be principal axes of effective conductivity tensor σ_0. In this case the geometrical symmetry is not required. If composite is macro-isotropic then

$$\sigma_0(\sigma_1,\sigma_2)\sigma_0(\sigma_2,\sigma_1) = \sigma_1\sigma_2 . \tag{4}$$

Further if two various composites have phase distributions which to be statistically equivalent one to other then $\sigma_0(\sigma_1,\sigma_2) = \sigma_0(\sigma_2,\sigma_1) = \sigma_0$ and Equation (4) leads to that the effective conductivity is given via mean geometrical value of the phase conductivities [28] that is $\sigma_0 = (\sigma_1\sigma_2)^{1/2}$.

Consider two different isotropic composites but with the same microstructure, namely: (i) the composite consisting of phase 1 with conductivity σ_1 and superconductive phase 2 ($\sigma_2 = \infty$), which has effective conductivity σ_∞, and (ii) the composite consisting of phase 1 with conductivity $\tilde{\sigma}_1$ and absolutely isolating phase 2 ($\tilde{\sigma}_2 = 0$), which has effective conductivity σ_0. At condition that the phase 2 in these composites no percolating the values σ_∞ and σ_0^{-1} remain finite. Then Equation (4) allows one to couple σ_∞ and σ_0 one with other. Consider the composite consisting of two phases with conductivities σ_1 and $\lambda\tilde{\sigma}_1$ (where λ is the dimensionless parameter). Denote via σ_λ и $\tilde{\sigma}_\lambda$ the effective conductivity above composite and corresponding composite with phase-interchange components, respectively. It is evident that

$$\sigma_\infty = \lim_{\lambda \to \infty} \sigma_\lambda \; ; \; \sigma_0 = \lim_{\lambda \to \infty} \widetilde{\sigma}_\lambda \tag{5}$$

In this case the effective conductivities σ_∞ и σ_0 satisfy to equality $\sigma_\infty \sigma_0 = \sigma_\lambda \widetilde{\sigma}_\lambda$, that is directly followed from Equation (4) in the limit defined by equalities (5). For continuous system near percolation threshold ($c_{2p} = 1 - c_{1p}$) by assuming usual power-law dependence and taking into account that $\widetilde{\sigma}_1 = \sigma_1$ we obtain:

$$\sigma_0/\sigma_1 = A_0 (c_1 - c_{1p})^t \text{ at } c_1 \to c_{1p} + 0 \, ; \tag{6}$$

$$\sigma_\infty/\sigma_1 = A_\infty (c_1 - c_{1p})^{-s} \text{ at } c_1 \to c_{1p} + 0 \, , \tag{7}$$

where $t > 0$ and $s > 0$ are the critical exponents, A_0 and A_∞ are the corresponding dimensionless amplitudes. Obviously in these cases, only one phase can percolate therefore the system is no bi-continuous. Moreover, the exponents t and s in Equations (5) satisfy to equality $t = s$, and the amplitudes A_0 and A_∞ satisfy to equality $A_0 A_\infty = 1$. These results are directly followed from relations (6) and (7), and also above discussion at $\widetilde{\sigma}_1 = \sigma_1$. The obtained results satisfy not only to the case of continuous percolation but to the percolation in discrete structure.

In the case of 2D polycrystal consisting of identical monocrystals, and effective conductivity tensor σ_0 with the property of rotational symmetry (to being isotropic tensor), the effective conductivity $\sigma_0 = (\sigma_a \sigma_b)^{1/2}$, where σ_a, σ_b are the principal values (or eigenvalues) of the conductivity tensor. Moreover, the effective conductivity does not depend on microstructure details. In 3D case that equality is absent and instead arise upper and lower bounds. Simplest bounds for macroisotropic polycrystals are the mean-arithmetic bound and mean-harmonic one:

$$\left[\frac{1}{3} \left(\frac{1}{\sigma_a} + \frac{1}{\sigma_b} + \frac{1}{\sigma_c} \right) \right]^{-1} \le \sigma_0 \le \frac{1}{3} (\sigma_a + \sigma_b + \sigma_c) \, . \tag{8}$$

Except very special cases of isotropic 2D two-phase media, there are no phase-interchange equalities for effective elastic moduli. In general case, as it has been shown [40], the phase-interchange relations for isotropic two-phase media in 2D and 3D cases have forms of inequalities.

2.2. Conductivity of Composites with Periodic Arrays of Superconductive Inclusions

Lord Rayleigh [81] was first who considered effective conductivity the periodic lattice of spheres based on simple cubic grid. In the next years this problems has been studied by numerous investigators which used the Rayleigh method. Specific results have been achieved in the case of arbitrary volume fraction of spheres for simple cubic (sc) structures [91], and also for body-centered cubic (bcc) and face-centered cubic (fcc) lattices [89]. Asymptotic formula in the Rayleigh form for these three cubic systems is found (for dimensionless effective conductivity $\lambda_0 = \sigma_0/\sigma_1$) as

$$\lambda_0 = 1 + \frac{3\beta c_2}{1 - \beta c_2 - a_1 \beta^2 c_2^{10/3}/(1 + 2\beta/7)},\tag{9}$$

where for sc-lattice $a_1 = 1.6772$, for bcc-lattice $a_1 = 0.073886$, for fcc-lattice $a_1 = 0.0060503$, and β at any dimension d is determined in the form:

$$\beta = \frac{\sigma_2 - \sigma_1}{\sigma_2 + (d-1)\sigma_1}.\tag{10}$$

Asymptotic solutions of higher orders have been obtained for sc-structure [91] and two other types of cubic lattices [89]. Then corresponding asymptotic formulae have been determined for 2D case of disc-shaped inclusions [113]. For discs disposed in knots of square lattice (square array) we have

$$\lambda_0 = 1 + \frac{2\beta c_2}{1 - \beta c_2 - 0.305827\,\beta^2 c_2^4},\tag{11}$$

and for discs disposed in knots of triangle lattice (hexagonal array) we have

$$\lambda_0 = 1 + \frac{2\beta c_2}{1 - \beta c_2 - 0.075422\,\beta^2 c_2^6}.\tag{12}$$

In References [8, 62] has been stated that for volume fraction near to maximum values of the inclusion package, the interaction between spheres concentrates in the regions neighboring to contact point. The local analysis of the region between two almost touching spheres together with the known disposition of nearest neighbors allows one to define principal contribution of the inclusion interactions for all system. For simple cubic array of identical superconductive spheres into matrix with conductivity σ_1 near percolation threshold of the particle phase $c_{2p} = \pi/6$, the dimensionless effective conductivity λ_0 is found [62] as

$$\lambda_0 = -\frac{\pi}{2}\ln(c_{2p} - c_2).\tag{13}$$

In Reference [62] it has been also obtained corresponding asymptotic solution for the dimensionless effective conductivity λ_0 of square array of identical superconductive circular cylinders into matrix with conductivity σ_1 near percolation threshold of the particle phase $c_{2p} = \pi/4$ as

$$\lambda_0 = \frac{\pi^{3/2}}{2}(c_{2p} - c_2)^{-1/2}.$$ (14)

2.3. Effective Medium Approximation for Definition of Effective Conductivity of Superconductive Composite

Effective medium approximations apply to estimate effective properties into broad range changing the volume fraction and also phase properties. However, these approximations can take into account only very simple microstructure information, for example an inclusion concentration and shape. Therefore, while the effective medium approximations can provide qualitative trends in behavior of effective properties for dispersed media they cannot make quantitative forecasts for general cases and hence it is important to understand the bounds of applicability these models in relation to specific microstructures.

The approximations of Maxwell type are one of typical approximations for effective medium. In the case of spherical inclusions in 3D approximation the approximate scheme has been assumed in Reference [86]. At arbitrary spatial dimension d and spherical non-interactive inclusions with conductivity σ_2 disposing into matrix with conductivity σ_1 the next relation called Maxwell approximation is carried out for effective conductivity σ_0 [137]:

$$\frac{\sigma_0 - \sigma_1}{\sigma_0 + (d-1)\sigma_1} = c_2 \left[\frac{\sigma_2 - \sigma_1}{\sigma_2 + (d-1)\sigma_1} \right].$$ (15)

Equation (15) coincides with well-known optimal Hashin – Shtrikman bounds that coincides with the proper lower bound if $\sigma_2 \geq \sigma_1$, and with the proper upper bound if $\sigma_2 \leq \sigma_1$. In the case of superconductive spheres in relation to matrix ($\sigma_2/\sigma_1 = \infty$) Equation (15) takes the form:

$$\lambda_0 = \frac{\sigma_0}{\sigma_1} = \frac{1 + (d-1)c_2}{1 - c_2}.$$ (16)

Thus, the composite becomes superconductive at trivial percolation threshold ($c_{2p} = 1$).

Then, firstly, self-consistent approximation for the case of spherical inclusions has been assumed in Reference [19] and then developed in References [76, 77]. In this case an effect of whole material outside inclusion is caused by homogeneous medium (matrix), whose effective conductivity is unknown value and should be calculated. With this aim it is required that the perturbations of homogeneous field introduced by inclusions are equal to zero in average. In the case of macroisotropic composite consisting of M various types of the spherical inclusions with volume fractions c_i and conductivities σ_i ($i = 1,\ldots, M$), this leads to the following equation:

$$\sum_{i=1}^{M} c_i \frac{\sigma_i - \sigma_0}{\sigma_i + (d-1)\sigma_0} = 0 . \tag{17}$$

This equation defines self-consistent approximation and as it is evident to be square-root equation for effective conductivity σ_0. For two-phase case ($M = 2$) its solution takes the form:

$$\sigma_0 = \frac{\alpha + \sqrt{\alpha^2 + 4(d-1)\sigma_1\sigma_2}}{2(d-1)} , \tag{18}$$

where $\alpha = \sigma_1(dc_1 - 1) + \sigma_2(dc_2 - 1)$ when both $\sigma_i > 0$. The restrictions of applicability of the self-consistent approach to actual media are connected with the next causes: (i) one does not include information on spatial distribution of inclusions or possible correlations between properties of adjacent inclusions, that creates difficulties in its application in the case of composites without phase-inverse symmetry, (ii) this approximation assumes existence of effective medium only outside the spherical inclusion considered, that it is evident impossible in the package case of the same spheres because between them always there are voids, (iii) violation of the self-consistent approximation occurs at its application to dispersed structures with strong-contrast phase conductivities. In the case of infinite contrast it is predicted no trivial but error percolation threshold. For example, in the case of two phases (18) gives the simple relation in the limit case of superconductive inclusions (phase 2):

$$\lambda_0 = \frac{\sigma_0}{\sigma_1} = \frac{1}{1 - dc_2} \text{ at } \sigma_2/\sigma_1 = \infty. \tag{19}$$

Thus, the phase 2 percolates (that is the composite becomes superconductive) at the critical volume fraction $c_{2p} = 1/d$ independently of microstructure details. However, it is evident that the percolation threshold c_{2p} should change into wide limits from one composite to other being very sensitive to microstructure.

The differential approximations based on effective medium are other popular approximations, whose scheme firstly has been introduced in [19]. In difference from self-consistent approach the differential approximations do not consider each phase symmetrically. However, the both approximate approaches use the same idea of incremented homogenization for two-phase composite in which indices 1 and 2 define matrix phase and inclusions, respectively. By assuming that the effective conductivity $\sigma_0(c_2)$ is known at one value of c_2, consider $\sigma_0(c_2)$ as intrinsic conductivity of composite and assume that $\sigma_0(c_2 + \Delta c_2)$ presents an effective conductivity of the composite after that the small fraction $\Delta c_2/(1 - c_2)$ of the composite with the intrinsic properties has been substituted by inclusions of phase 2. In mean, the fraction $\Delta c_2/(1 - c_2)$ of the composite with the intrinsic properties should be replaced by the material phase 2 with aim to change whole fraction of phase 2 up to value of $c_2 + \Delta c_2$. Then in the case of d-dimensional spherical inclusions for effective conductivity σ_0 we obtain

$$\sigma_0(c_2+\Delta c_2)-\sigma_0(c_2)=\sigma_0 c_2\left[\frac{\sigma_2-\sigma_0(c_2)}{\sigma_2+(d-1)\sigma_0(c_2)}\right]\frac{\Delta c_2}{1-c_2}d. \quad (20)$$

In the limit of $\Delta c_2 \to 0$ this relation transforms into differential equation:

$$(1-c_2)\frac{d\sigma_0}{dc_2}=\sigma_0\left[\frac{\sigma_2-\sigma_0}{\sigma_2+(d-1)\sigma_0}\right]d \quad (21)$$

with initial condition $\sigma_0(c_2=0)=\sigma_1$. The analytical solution of Equation (21) is determined as

$$\left(\frac{\sigma_2-\sigma_0}{\sigma_2-\sigma_1}\right)\left(\frac{\sigma_1}{\sigma_0}\right)^{1/d}=1-c_2. \quad (22)$$

The differential approximation defines preservation of continuity of the original matrix material in final composite [149].

In the case of superconductive phase 2 in relation to phase 1 equation (22) takes the form:

$$\lambda_0=\frac{\sigma_0}{\sigma_1}=\frac{1}{(1-c_2)^d} \text{ at } \sigma_2/\sigma_1=\infty. \quad (23)$$

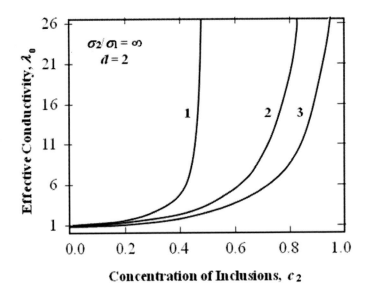

Figure 1. Comparison of three models of effective medium for effective conductivity (1 – self-consistent approximation, 2 – differential approximation, 3 – Maxwell approximation) in dependence on concentration of superconductive inclusions for 2D two-phase composite. The results correspond to formulae (16), (19), (23).

Figure 1 compares Maxwell, self-consistent and differential approximations in the case of existing superconductive phase into two-phase composite. In this case the numerical results are determined by formulae (16), (19) and (23), respectively. The behavior these three approximations in the case coincides qualitatively with the behavior of corresponding approximations for shear modulus in existence of absolutely rigid phase. Obviously, as the Maxwell approximation as the differential one badly fit for estimation of composites with inclusions consisting of great conglomerates.

2.4. Effective Medium Approximations for Definition of Effective Elastic Moduli

Maxwell approximation, self-consistent approach and differential approximations based on effective medium are also used to study elastic moduli. For example, for $M - 1$ different spheres ($M \geq 2$) with volume fractions c_i, bulk moduli K_i and shear moduli G_i, where $i = 2,...,$ M (value of the index $i = 1$ corresponds to matrix parameters), the Maxwell approximation for dimension d leads to the relations [137]:

$$\frac{K_0 - K_1}{K_0 + 2G_1(d-1)/d} = \sum_{i=1}^{M} c_i \frac{K_i - K_1}{K_i + 2G_1(d-1)/d}; \tag{24}$$

$$\frac{G_0 - G_1}{G_0 + H_1} = \sum_{i=1}^{M} c_i \frac{G_i - G_1}{G_i + H_1}; \quad H_1 = \frac{G_1[dK_1/2 + (d+1)(d-2)G_1/d]}{K_1 + 2G_1}. \tag{25}$$

When all inclusions are stiffer (softer) compared to matrix phase Equations (24) and (25) coincide with lower (upper) Hashin – Shtrikman – Walpole bounds [137]. At $d = 3$ these formulae are equivalent to Mori-Tanaka relations [105] based on other approach [145]. Comparison of relations for effective bulk and shear moduli obtained into framework of Maxwell approximations in the case of macroanisotropic composite, consisting of $M - 1$ various types one-directional isotropic ellipsoidal inclusions with the same shape, and test data has shown that these estimations could not be used at volume fraction of the inclusions greater than 20 – 30 % [13].

Self-consistent approximations for effective elastic moduli are totally similar to the relations obtained for effective conductivity. Firstly, the approximations have been obtained [21, 57] by using other approach than above one discussed for the effective conductivity. Additional approach for self-consistent approximations has been realized in [12, 42]. In the case of macroisotropic composites consisting of M various types of spherical inclusions with volume fractions c_i, bulk moduli K_i and shear moduli G_i, ($i = 1,..., M$), an effect of whole material without inclusion of i-th type creates homogeneous structure (matrix) with effective bulk modulus K_0, to being unknown. Then by using the condition of equality in mean to zero the perturbations of homogeneous field, the effective bulk modulus K_0 is selected in according with the condition of self-consistency thus the perturbations initiated by each type of the inclusions in average became zero. This leads to the next relation which is similar to Equation (17) for effective conductivity:

$$\sum_{i=1}^{M} c_i \frac{K_i - K_0}{K_i + 2G_0(d-1)/d} = 0. \tag{26}$$

This self-consistent estimation of the effective bulk modulus K_0 depends on unknown effective shear modulus G_0. By following to the same procedure that in the case of the bulk modulus, the self-consistent approximation for the effective shear modulus is found as

$$\sum_{i=1}^{M} c_i \frac{G_i - G_0}{G_i + H_0} = 0, \tag{27}$$

where

$$H_0 = \frac{G_0[dK_0/2 + (d+1)(d-2)G_0/d]}{K_0 + 2G_0}. \tag{28}$$

Bruggeman ideas [19] have been used to obtain solutions for effective elastic moduli K_0 and G_0 [17, 90] similar to above differential approximations for conductivity. Analogously to the case of conductivity for effective elastic moduli of two-phase composite (indices 1 and 2 correspond to matrix and inclusion) for dimension d we obtain the differential equation similar to (21):

$$(1-c_2)\frac{dK_0}{dc_2} = \left[K_0 + \frac{2(d-1)}{d} G_0 \right] \frac{K_2 - K_0}{K_2 + 2G_0(d-1)/d} \tag{29}$$

with initial condition $K_0(c_2 = 0) = K_1$. The corresponding equation for effective shear modulus G_0 is determined as

$$(1-c_2)\frac{dG_0}{dc_2} = (G_0 + H_0)\frac{G_2 - G_0}{G_2 + H_0} \tag{30}$$

with initial condition $G_0(c_2 = 0) = G_1$, where H_0 is found by relation (28). Differential approximations (29), (30) are realized because they present the special case of generalized differential scheme assumed in [107] and hence dispose between Hashin – Shtrikman bounds.

2.5. Cluster Expansions

Maxwell formula [86] for effective conductivity of dispersed spheres assumes absence their mutual interactions and to be sufficient for first-order presentation of solution on the sphere concentration c_2. These estimations may be modernized by taking into account couple, triplet, etc. mutual interactions. The problem is solved by using the method of cluster expansions. While in general case it is impossible to exactly calculate effective properties of arbitrary dispersed inclusions taking into account all orders of expansion on powers of c_2, the expansions for small density (non-interacting) inclusions present a good basis for verification of approximate theories and test data.

For a dispersion of the same mutual-interacting spheres in first approximation of the cluster expansions method, the effective conductivity σ_0 for any dimension d has the next form [137]:

$$\sigma_0 = \sigma_1 + d\sigma_1\beta_{21}c_2 + O(c_2^2),\tag{31}$$

where $\beta_{21} = \dfrac{\sigma_2 - \sigma_1}{\sigma_2 + (d-1)\sigma_1}$. In the case of $d = 3$ this result coincides with Maxwell approximation (15) for first-order expansion on powers of c_2. For superconductive spheres in relation to matrix ($\sigma_2/\sigma_1 = \infty$) Equation (31) becomes as

$$\lambda_0 = \frac{\sigma_0}{\sigma_1} = 1 + dc_2 + O(c_2^2).\tag{32}$$

By taking into account mutual-interacting inclusions [58], using the method [7], in which the integral describing mutual interaction of two solids was made absolutely convergent, it was estimated effective conductivity of dispersed spheres taking into account second-order expansion on powers of c_2. A broadening this renormalization technique in order to round conditionally convergent integrals and calculate terms of higher order has been carried out in Reference [59], in which estimated the values of σ_0 for all terms of the expansion.

In the case non-overlapping spheres the approach [58] allows one to obtain for superconductive inclusions (in the case of $d = 3$) the following estimation:

$$\lambda_0 = 1 + 3c_2 + 4.51c_2^2 + O(c_2^3)\tag{33}$$

By using other function of radial distribution relative to 'clearly divided' dispersed inclusions the next estimation has been also obtained [137]:

$$\lambda_0 = 1 + 3c_2 + 3c_2^2 + O(c_2^3).\tag{34}$$

It is not surprising that when spheres dispose sufficiently near one to other the effective conductivity is greater in comparison with the case of the clearly divided spheres. The approximate estimation for non-overlapping superconductive discs ($d = 2$), similar to (33) has been obtained in [114]:

$$\lambda_0 = 1 + 2c_2 + 2.74c_2^2 + O(c_2^3).\tag{35}$$

In 2D estimation factor 2.74 is replaced by 2.

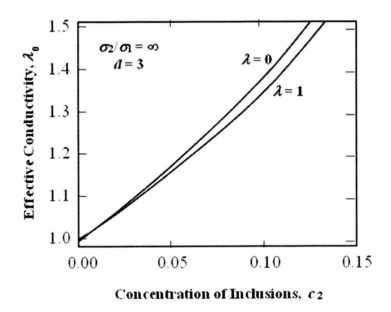

Figure 2. Comparison effective conductivity for systems with non-overlapping ($\lambda = 1$) and totally overlapping ($\lambda = 0$) superconductive spheres for 3D two-phase composite. The results correspond to formulae (33), (36).

The case of overlapping spheres has been studied in dependence on σ_2/σ_1 [133]. In the limit case of totally overlapping spheres for superconductive inclusions ($\sigma_2/\sigma_1 = \infty$) has been obtained the estimation:

$$\lambda_0 = 1 + 3c_2 + 7.56c_2^2 + O(c_2^3). \tag{36}$$

The comparative results for non-overlapping and totally overlapping spheres (Equations (33) and (36)) are present in Figure 2. Thus, in the case of account of the second-order expansion on sphere concentration, the clustering effect causes increasing the effective conductivity for superconductive inclusions at fixed value of c_2. Similar cluster expansions could be also applied to estimation of effective elastic moduli.

2.6. Strong Contrast Expansions

In Reference [18] has been proposed a procedure obtaining an expansion for perturbation caused by phase contrast with application of rational function for effective conductivity σ_0 in 3D case of two-phase isotropic medium. This procedure used the expansion of phase conductivities on powers of rational functions. This result has been expanded to isotropic medium of arbitrary dimension d [132]. By using third term of the expansion the effective conductivity defining exact estimation in the case when the inclusions do not form cluster structure is found in the form [132]:

Table 1. Three-point parameter ζ_2 in dependence on inclusion concentration c_2 for sc, bcc and fcc arrays of spheres and also for square and hexagonal arrays of ordered infinite-long circular cylinders

c_2	Three-point parameter ζ_2				
	sc	bcc	fcc	Square array	Hexagonal array
0.10	0.0003	0.0000	0.0000	3.398×10^{-5}	8.380×10^{-8}
0.20	0.0050	0.0007	0.0004	6.117×10^{-4}	6.034×10^{-6}
0.30	0.0220	0.0031	0.0021	3.540×10^{-3}	7.855×10^{-5}
0.40	0.0678	0.0107	0.0078	1.306×10^{-2}	5.149×10^{-4}
0.45	0.1104	0.0184	0.0136		
0.50	0.1738	0.0307	0.0232	3.833×10^{-2}	2.357×10^{-3}
0.60		0.0796	0.0619	9.965×10^{-2}	8.798×10^{-3}
0.66		0.1381	0.1095		
0.70			0.1596	0.2473	2.958×10^{-2}
0.71			0.1756		
0.76				0.4314	6.057×10^{-2}
0.78				0.5229	7.722×10^{-2}
0.80					9.888×10^{-2}
0.89					0.3409
0.90					0.4010
0.905					0.4364

$$\lambda_0 = \frac{1+(d-1)c_2\beta_{21}-(d-1)c_1\zeta_2\beta_{21}^2}{1-c_2\beta_{21}-(d-1)c_1\zeta_2\beta_{21}^2}, \tag{37}$$

where three-point parameter ζ_2 is determined for various structures in dependence on the inclusion concentration c_2 (see Table 1 [92]).

Numerical results for three-point approximation (37) in the case of different arrays of superconductive inclusions ($\sigma_2/\sigma_1 = \infty$) are present in Figure 3. It is evident that formula (37) ensures excellent coincidence of the results up to the maximal published data on the inclusion concentration (up to 95% of volume fraction). So, in this case, it is stated sensitivity of σ_0 to microstructure features in 3D case. The conclusion is not proper for Maxwell approximation (15) which is equivalent to Hashin – Shtrikman lower bound being no sensitive to specific lattice (see Figure 3).

In the case of self-consistent approach for high phase contrast similarly we obtain three-point approximation [137]:

$$c_2\frac{\sigma_0+(d-1)\sigma_1}{\sigma_0-\sigma_1}+c_1\frac{\sigma_0+(d-1)\sigma_2}{\sigma_0-\sigma_2}=2-d+(d-1)(c_2-\zeta_2)\beta_{21}, \tag{38}$$

and for superconductive inclusions ($\sigma_2/\sigma_1 = \infty$) Equation (38) gives the dependence:

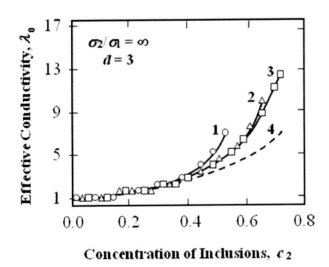

Figure 3. Effective conductivity of composites in dependence on concentration of superconductive spherical inclusions: the solid curves correspond to three-point approximations (37) for 1 – sc-lattice, 2 – bcc-lattice, 3 – fcc-lattice; circles, triangles and squares determine corresponding numerical results [89, 91]. Maxwell approximation (16) equivalent to Hashin - Shtrikman lower bound is shown by dashed line 4.

$$\lambda_0 = \frac{1+(d-1)(c_2-\zeta_2)}{c_1-(d-1)\zeta_2}, \qquad (39)$$

at condition $\zeta_2 \leq [d/(d-1)]c_1$. This condition is satisfied in 2D case for any values of ζ_2. Relation (39) states that phase 2 percolates (and the composite becomes superconductive) at $c_1 = (d-1)\zeta_2$. This percolation threshold in difference on the predicted threshold from (17) depends on microstructure due to three-point parameter ζ_2. For materials with symmetry and spherical cells Equation (39) predicts the same percolation threshold that formula (17), that is $c_2 = 1/d$.

2.7. Strong Bound Estimations for Conductive and Elastic Properties

Theoretical methods differing from variation approaches could be also applied to define the bounds of effective conductivity σ_0 and stiffness C_0 tensors. The analytical method for σ_0 [11] and stiffness C_0 [61] has stated analytical properties based on these effective characteristics in dependence on the properties of single phases. The Pade-approximations method [101] is closely connected with the analytical method because the conductivity tensor σ_0 presents Stieltjes function. Other theoretical approach named the translation method (or compensated compactness method) has been developed to state the bounds for tensors of σ_0 and C_0 [82, 83]. Broad set of experimental and theoretical methods and also computational approaches (in particular, by using finite-element software ANSYS and ACELAN) for definition elastic, conductive and other structure-sensitive properties of superconductive and piezoelectric materials and composites has been presented in the author's monographs [1, 110 – 112, 115, 116].

For macroisotropic 3D two-phase composites have been obtained the probable bounds defining effective conductivity and including information on volume fraction of the components [48]. One-point (trivial), two-point, three-point and four-point (which use two-point, three-point and four-point probable function, respectively) d-dimensional Hashin – Shtrikman bounds for σ_0 at $\sigma_2 \geq \sigma_1$ for these composites have been determined [137] and present below.

One-point bounds:

$$\left\langle \sigma^{-1} \right\rangle^{-1} \leq \sigma_0 \leq \left\langle \sigma \right\rangle. \tag{40}$$

Two-point bounds:

$$\sigma_L^{(2)} \leq \sigma_0 \leq \sigma_U^{(2)}, \tag{41}$$

where

$$\sigma_L^{(2)} = \left\langle \sigma \right\rangle - \frac{c_1 c_2 (\sigma_2 - \sigma_1)^2}{\left\langle \hat{\sigma} \right\rangle + (d-1)\sigma_1}; \; \sigma_U^{(2)} = \left\langle \sigma \right\rangle - \frac{c_1 c_2 (\sigma_2 - \sigma_1)^2}{\left\langle \hat{\sigma} \right\rangle + (d-1)\sigma_2}; \tag{42}$$

$\langle\langle (...) \rangle\rangle = (...)_1 c_2 + (...)_2 c_1$, and designation $\langle\langle (...) \rangle\rangle$ defines mean value of corresponding parameter.

Three-point bounds:

$$\sigma_L^{(3)} \leq \sigma_0 \leq \sigma_U^{(3)}, \tag{43}$$

where

$$\sigma_L^{(3)} = \left\langle \sigma \right\rangle - \frac{c_1 c_2 (\sigma_2 - \sigma_1)^2}{\left\langle \hat{\sigma} \right\rangle + (d-1)\left\langle \sigma^{-1} \right\rangle_\varsigma^{-1}}; \; \sigma_U^{(3)} = \left\langle \sigma \right\rangle - \frac{c_1 c_2 (\sigma_2 - \sigma_1)^2}{\left\langle \hat{\sigma} \right\rangle + (d-1)\left\langle \sigma \right\rangle_\varsigma}; \tag{44}$$

$$\left\langle (...) \right\rangle_\varsigma = (...)_1 \varsigma_1 + (...)_2 \varsigma_2; \; \varsigma_1 + \varsigma_2 = 1. \tag{45}$$

Four-point bounds:

$$\sigma_L^{(4)} \leq \sigma_0 \leq \sigma_U^{(4)}, \tag{46}$$

where

$$\sigma_L^{(4)} = \sigma_1 \left\{ \frac{1 + [(d-1)c_2 - \gamma_2/\zeta_2]\beta_{21} + (1-d)[c_1\zeta_2 + c_2\gamma_2/\zeta_2]\beta_{21}^2}{1 - [c_2 + \gamma_2/\zeta_2]\beta_{21} + [c_1(1-d)\zeta_2 + c_2\gamma_2/\zeta_2]\beta_{21}^2} \right\}; \qquad (47)$$

$$\sigma_U^{(4)} = \sigma_2 \left\{ \frac{1 + [(d-1)c_1 - \gamma_1/\zeta_1]\beta_{12} + (1-d)[c_2\zeta_1 + c_1\gamma_1/\zeta_1]\beta_{12}^2}{1 - [c_1 + \gamma_1/\zeta_1]\beta_{12} + [c_2(1-d)\zeta_1 + c_1\gamma_1/\zeta_1]\beta_{12}^2} \right\}; \qquad (48)$$

$$\gamma_1 - \gamma_2 = (d-2)(\zeta_2 - \zeta_1); \quad \beta_{ij} = \frac{\sigma_i - \sigma_j}{\sigma_i + (d-1)\sigma_j}; i \neq j. \qquad (49)$$

Three-point bounds (43) depend not only on the phase concentrations c_i, but on the three-point microstructure parameters ζ_i being multi-dimensional integrals consisting of three-point probable functions [132, 136]. Because $\zeta_i \in [0, 1]$ then bounds (43) improve Hashin – Shtrikman two-point bounds. At $\zeta_2 = 0$ bounds (43) coincide and equal to two-point lower bound (42). At $\zeta_2 = 1$ bounds (43) coincide and equal to two-point upper bound (42). At the same time four-point bounds (46) depend on c_i, ζ_i and four-point microstructure parameters η_i, stated by four-point probable functions [95]. Note, that in 2D case the four-point parameters transform in zero and hence the four-point bounds will be only defined by parameters c_i and ζ_i. The improved lower three-point bound is determined for the case $d = 3$ in the form [132]:

$$\frac{\sigma_L^{(3)}}{\sigma_1} = \frac{1 + (1 + 2c_2)\beta_{21} - 2(c_1\zeta_2 - c_2)\beta_{21}^2}{1 + c_1\beta_{21} - (2c_1\zeta_2 + c_2)\beta_{21}^2}. \qquad (50)$$

Usual upper and lower bounds (for example in the case of contrast or cluster phases) generally divergent one with other in the limit of infinite phase contrast. In the case of superconductive phase 2 relatively phase 1 ($\sigma_2/\sigma_1 = \infty$) upper bounds of finite order aspire to infinity because in the consideration take into account realizations in which phase 2 percolates while in reality one is not percolating. This is sequel of that the limited microstructure information introducing the bounds is to be insufficient for statement of the phase connectivity. Therefore, it is useful to obtain bounds no diverging into utmost case of infinite phase contrast.

The concept of "secure" sphere for creation of trial fields have been proposed in [64] to obtain limits for effective toughness of suspensions. The same idea has been proposed to define bounds of effective conductivity [139]. In this case, it is treated statistically isotropic system of the same spheres with radius R and conductivity σ_2 into matrix with conductivity σ_1. After circulation i-th sphere by greater "secure" sphere, the area of concentric shell between the actual sphere and the "secure" sphere contains phase 1 only. Then by considering trial fields into "secure" spheres and into matrix region without "secure" spheres, using principles of minimum of whole energy and additional energy for this composite structure there are stated lower and upper bounds for σ_0. For any dimension d in the case of superconductive inclusions of concentration c_2 the upper bound is found in the form [137]:

$$\lambda_0 \leq 1 + 2dc_2 \int_0^\infty \frac{x^d}{x^d - 1} RH_P(x)dx, \qquad (51)$$

where $2x = r/R$ is the dimensionless distance measured from center of each sphere, $H_P(x)dx$ is the probability that center of arbitrary particle into coordinate system connected with center of nearest particle disposes into range between x and $x + dx$. Lower bound in whole is defined by usual three-point lower bound. Note, that the factor before $H_P(x)$ in (51) contains pole at $x = 1$ (when the spheres touch). So, in difference of usual upper bounds, the upper bound (51) in the case of "secure" spheres remains finite for superconductive inclusions under condition of sufficiently rapid convergence of $H_P(x)$ to zero at $x \to 1$ in order to make the integral to be convergent. Upper bound (51) will give non-trivial result in the case absence of touching the spheres one to other which dispose into regular lattice with minimal distance between them equal to $2bR$.[1] In this case we obtain

$$2R H_P(x) = \delta(x - b). \qquad (52)$$

By using the results for H_P in the case of various arrays of inclusions, are determined non-trivial bounds for "secure" spheres under condition their touching [139]. The bounds of "secure" spheres types for σ_0 have been stated in [20] by using the method of complex variables. However, these bounds were obtained in terms of minimal distance between all adjacent inclusions (but no via H_P).

The strong d-dimensional Hashin – Shtrikman bounds for effective elastic moduli for two-phase isotropic composites are similar to corresponding bounds for effective conductivity and defined [137] as present below.

One-point bounds:

$$\left\langle K^{-1} \right\rangle^{-1} \leq K_0 \leq \left\langle K \right\rangle; \ \left\langle G^{-1} \right\rangle^{-1} \leq G_0 \leq \left\langle G \right\rangle. \qquad (53)$$

Two-point bounds:

$$K_L^{(2)} \leq K_0 \leq K_U^{(2)}, \qquad (54)$$

where

$$K_L^{(2)} = \left\langle K \right\rangle - \frac{c_1 c_2 (K_2 - K_1)^2}{\left\langle \hat{K} \right\rangle + 2[(d-1)/d]G_1} ; \ K_U^{(2)} = \left\langle K \right\rangle - \frac{c_1 c_2 (K_2 - K_1)^2}{\left\langle \hat{K} \right\rangle + 2[(d-1)/d]G_2} ; \qquad (55)$$

---

[1]Dimensionless parameter b is found below from formula (68).

$$G_L^{(2)} \le G_0 \le G_U^{(2)}, \tag{56}$$

where

$$G_L^{(2)} = \langle G \rangle - \frac{c_1 c_2 (G_2 - G_1)^2}{\langle \hat{G} \rangle + H_1}; \; G_U^{(2)} = \langle G \rangle - \frac{c_1 c_2 (G_2 - G_1)^2}{\langle \hat{G} \rangle + H_2}; \tag{57}$$

$$H_i = G_i \left[\frac{dK_i / 2 + (d+1)(d-2)G_i / d}{K_i + 2G_i} \right]. \tag{58}$$

Three-point bounds:

$$K_L^{(3)} \le K_0 \le K_U^{(3)}, \tag{59}$$

where

$$K_L^{(3)} = \langle K \rangle - \frac{c_1 c_2 (K_2 - K_1)^2}{\langle \hat{K} \rangle + 2[(d-1)/d]\langle G^{-1} \rangle_\zeta^{-1}}; \tag{60}$$

$$K_U^{(3)} = \langle K \rangle - \frac{c_1 c_2 (K_2 - K_1)^2}{\langle \hat{K} \rangle + 2[(d-1)/d]\langle G \rangle_\zeta}; \tag{61}$$

$$G_L^{(3)} \le G_0 \le G_U^{(3)}, \tag{61}$$

where

$$G_L^{(3)} = \langle G \rangle - \frac{c_1 c_2 (G_2 - G_1)^2}{\langle \hat{G} \rangle + \Xi}; \; G_U^{(3)} = \langle G \rangle - \frac{c_1 c_2 (G_2 - G_1)^2}{\langle \hat{G} \rangle + \Theta}. \tag{62}$$

In the case of three-point bounds for effective transverse shear modulus for fiber composites in 2D case ($d = 2$), the parameters Ξ and Θ satisfy to relations [137]:

$$\Xi = (2\langle 1/k \rangle_\zeta + \langle 1/G \rangle_\eta)^{-1}; \; \Theta = \frac{2\langle k \rangle_\zeta \langle G \rangle^2 + \langle k \rangle^2 \langle G \rangle_\eta}{\langle k + 2G \rangle^2}. \tag{63}$$

Here $k \equiv K$ for $d = 2$ and

$$\langle(...)\rangle_\eta = (...)_1\eta_1 + (...)_2\eta_2; \; \eta_1 + \eta_2 = 1. \tag{64}$$

In 3D case ($d = 3$), the parameters Ξ and Θ satisfy to relations [102]:

$$\Xi = \frac{\langle 128/K + 99/G \rangle_\zeta + 45\langle 1/G \rangle_\eta}{30\langle 1/G \rangle_\zeta \langle 6/K - 1/G \rangle_\zeta + 6\langle 1/G \rangle_\eta \langle 2/K + 21/G \rangle_\zeta}; \tag{65}$$

$$\Theta = \frac{3\langle G \rangle_\eta \langle 6K + 7G \rangle_\zeta - 5\langle G \rangle_\zeta^2}{6\langle 2k - G \rangle_\zeta + 30\langle G \rangle_\eta}. \tag{66}$$

While the three-point bounds for the bulk modulus depend on the component concentrations c_i and three-point microstructure parameters ζ_i [132, 136], the three-point bounds for the shear modulus depend additionally on other three-point microstructure parameters η_i, which at $d = 2$ and $d = 3$ are defined by multi-dimensional integrals including three-point probability functions obtained in [96] and [95], respectively.

Four-point bounds for elastic moduli of 3D two-phase composites have been constructed in Reference [102]. In this case the bounds for bulk modulus include three various microstructure parameters, and the bounds for shear modulus includes eight that parameters.

2.8. Improved Estimations of Range Bounds for Effective Conductivity of Superconductive Composites

More exact estimations of microstructure parameters allow improving three- and four-point bounds for effective conductivity σ_0. In particular, the results obtained for effective transverse conductivity for equilibrium arrays of ordered the same superconductive cylinders ($\sigma_2/\sigma_1 = \infty$) in 2D case and for superconductive spheres in 3D case are present in Figure 4 and Figure 5, respectively. Obviously, that in the case of superconductive inclusions all upper bounds aspire to infinity. Superconductive discs cannot form clusters for the treated range of volume fraction that is the conductive phase to be discontinuous. Really, for arbitrary equilibrium models of rigid spheres at any spatial dimension the contacts between particles occur only at maximal density of arbitrarily jammed particles. Figure 5 compare two- and three-point bounds for 3D model of superconductive spheres obtained in [67, 94] and also three-point approximation (37). Best lower bound provides a good estimation of σ_0, and three-point approximation to be absolutely exact.

The prediction of effective conductivity for systems with overlapping spheres is more interest because of non-trivial cluster formation. Figure 6 compares four-point bounds for the same overlapping discs (the percolation threshold $c_{2p} \approx 0.68$) [138], in the case of $\sigma_2/\sigma_1 = 1000$ with results of modeling [52]. For range $0 \le c_2 \le 0.4$ the lower bound provides a good estimation by assuming that proper cluster size Λ_2 of phase 2 is relatively small. The increasing c_2 from 0.4 up to 0.9 initiates crossover from lower bound to upper one corresponding to increasing Λ_2, which becomes macroscopically great at achievement of percolation threshold. The upper bound provides a good estimation for test data at $0.9 \leq c_2 \leq 1$ assuming that proper cluster size Λ_1 of phase 1 is relatively small. Figure 7 compares for mono-dispersed and poly-dispersed superconductive spheres three-point lower bound obtained in 3D case at $\sigma_2 \gg \sigma_1$ [97], which improves lower bound (44). In the poly-dispersed case is used three-point approximation (37). The effect of poly-dispersity leads to increasing σ_0.

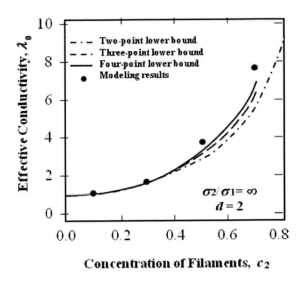

Figure 4. Lower bounds for effective transverse conductivity vs concentration of ordered the same superconductive cylinders. The results for point approximations are determined by formulae (41) - (49), and modeling data from [23, 66].

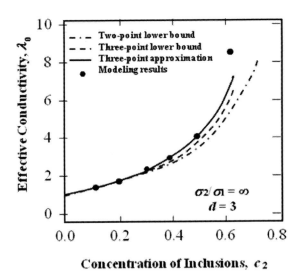

Figure 5. Bounds for effective conductivity in dependence on concentration of superconductive inclusions for arbitrary equilibrium arrays of spheres. Results for two- and three-point lower bounds are found by formulae (41), (42), (37), respectively; improved three-point approximation (50) obtained in [94] and modeling data in [67].

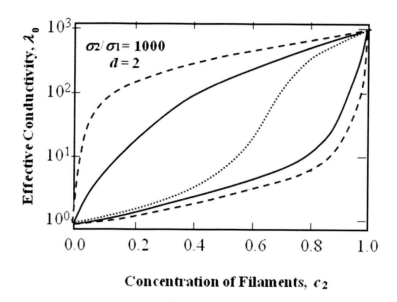

Figure 6. Bounds for effective conductivity in dependence on concentration of superconductive filaments for arbitrary arrays of ordered identical overlapping circular discs compare with modeling results (dots) [52]; four-point bounds (solid curves) are defined by formulae (46) - (49); two-point bounds (dashed curves) are found by formulae (41), (42).

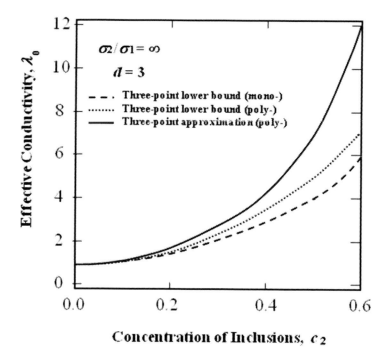

Figure 7. Three-point lower bounds (50) of effective conductivity for equilibrium systems of superconductive mono- and poly-dispersed spheres compared with exact three-point approximation (37) for poly-dispersed spheres.

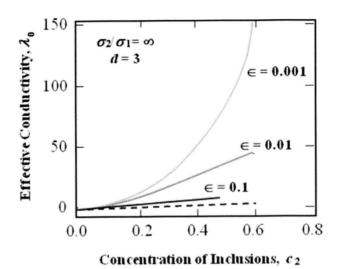

Figure 8. Upper bounds of effective conductivity (51) in model of 'secure' spheres for arbitrary equilibrium systems of superconductive inclusions for various values of dimensionless coverage thickness ϵ [139] (solid lines); dash line shows Hashin - Shtrikman lower bound.

As it is shown by formula (51) for upper bound of "secure" spheres at any concentration of superconductive inclusions c_2 the bound could remain finite in difference from usual bounds. These bounds have been estimated [139] for equilibrium sphere arrays in which each sphere with conductivity σ_2 and diameter D was covered by thin layer of matrix material with conductivity σ_1 so that diameter of each composite sphere equals to D_0. That coverage is required for preventing a mutual touching of internal spheres. It is interesting to study effects of diameter $D_0 > D$ and volume fraction $\tilde{c}_2 = (D_0/D)^3 c_2$, where c_2 is actual concentration of spheres. By estimating the function of probability density for nearest neighbors $H_p(x)$ at dimensionless distance $(D_0/D)x$ for volume fraction of inclusion $\tilde{c}_2$, we show the bounds of "secure" spheres for superconductive spheres in Figure 8 for several values of dimensionless thickness of the coverage $\epsilon = (D_0 - D)/D$. It is evident, that the bounds become sharper due to increasing ϵ. For cubic lattices of spheres in the case of dimension d by using estimation (52) for function $H_p(x)$ and upper bound of "secure" spheres (51) we obtain

$$\lambda_0 = \frac{\sigma_0}{\sigma_1} \leq 1 + d \frac{b^d}{(b^d - 1)} c_2 \text{ at } \sigma_2/\sigma_1 = \infty. \quad (67)$$

Here b is minimal distance between particles which depends on volume fraction as

$$b^d = c(d) c_2^{-1}, \quad (68)$$

where $c(d)$ is the constant stated by dimension d и and lattice type. For example, we obtain for d-dimensional simple cubic lattice

$$c(d) = \frac{\pi^{d/2}}{2^d \Gamma(1 + d/2)}.$$ (69)

The bound (67) is exact while of first order on c_2, and relatively sharp at moderately big volume values of superconductive inclusions as for $d = 2$ as for $d = 3$. Moreover, one is exact on all orders of c_2 as for $d = 1$ as for $d = \infty$.

3. COUPLES BETWEEN CONDUCTIVITY AND ELASTIC PROPERTIES

Statement of these couples is especially useful in the case of simpler test determination one property compared to other. Because effective properties of arbitrary media reflect definite morphological information on the medium then it could be expected that can pick out useful information on effective property given by using exact definition (experimental or theoretical) other effective property even when their corresponding constitutive equations are not coupled one with other. That couple relations provide estimation methods for possible range of values which could have other effective properties (in this case an achieved region is stated into multi-dimensional space of the properties). So, they have important applications for design of multi-functional composites.

It may be shown that the effective conductivity and effective elastic moduli are generally coupled one with other even while the first property is calculated from Laplace equation and the second ones are determined by vector equations of equilibrium. It has been shown for arbitrary d-dimensional isotropic two-phase medium [97] that if phase bulk moduli K_i are equal phase conductivities σ_i then effective bulk modulus K_0 is limited from above by effective conductivity σ_0. This result has been generalized on more common microstructure case with arbitrary topology demonstrating phases with non-negative Poisson's ratio [135]. At condition $K_2/K_1 \leq \sigma_2/\sigma_1$ for dimensionless effective bulk elastic modulus and effective conductivity has been stated the following dependence:

$$K_0/K_1 \leq \sigma_0/\sigma_1.$$ (70)

Moreover, it has been stated upper bound for effective shear modulus G_0 (into terms of σ_0 and effective Poisson's ratio v_0) and weaker bound determined only by effective conductivity σ_0:

$$\frac{G_0}{K_1} \leq \frac{\sigma_0}{\sigma_1} \frac{d[1 - v_0(d-1)]}{2(1 + v_0)}; \quad \frac{G_0}{K_1} \leq \frac{\sigma_0}{\sigma_1} \frac{d}{2}.$$ (71)

Above-mentioned bounds have interesting applications. For compressible phase 1 ($K_1 < \infty$) with finite conductivity ($\sigma_1 < \infty$) and non-compressible superconductive phase 2 ($K_1 = \infty$, $\sigma_1 = \infty$), it has been stated the next conclusion [97]: inequality (70) assumes that the composite could not be non-compressible and possess finite conductivity. However, this does not interdict that the composite will be compressible and superconductive. The corresponding

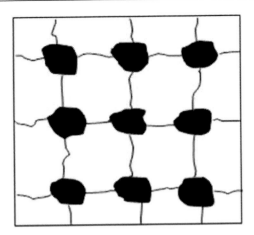

Figure 9. Scheme of compressible superconductive composite consisting of non-compressible superconductive phase (black color) and compressible normal phase (white color).

imaginary structure is present in Figure 9, where thin black superconducting couples of non-compressible phase connect big black blocks allowing the composite to be superconductive. Due to the black phase has finite shear modulus, the thin links will bend under influence of hydrostatic loading creating the compressible composite.

In order to obtain sharp bounds of effective properties for composites is used translational method [82, 83, 99]. The method is based on obtaining corresponding bounds for a comparative medium, whose local properties differ from original medium by tensor of constant translation being quasi-convex. This method has been applied to define sharp bounds of couples of the effective properties (σ_0, k_0) and (σ_0, G_0) in 2D case of two-phase isotropic composites with different structures in dependence on component concentrations c_i [37, 38]. These bounds cover definite regions into planes $\sigma_0 - k_0$ and $\sigma_0 - G_0$ realized by specific microstructures and so being optimal. In special example of absolutely rigid and superconductive phase 2 ($k_2/k_1 = \infty$, $G_2/G_1 = \infty$, $\sigma_2/\sigma_1 = \infty$) the bounds are determined by the next inequalities [137]:

$$\sigma_0 \geq \sigma_{1*}^{\infty}; \; k_{1*}^{\infty} \leq k_0 \leq k_{1*}^{\infty} + \max\left[\frac{k_1 + G_1}{2\sigma_1}; \frac{2k_2 G_2}{(k_2 + G_2)\sigma_2}\right](\sigma_0 - \sigma_{1*}^{\infty}); \qquad (72)$$

$$\sigma_0 \geq \sigma_{1*}^{\infty}; \; G_{1*}^{\infty} \leq G_0 \leq G_{4*}^{\infty} + \max\left[\frac{k_1 + 2G_1}{4\sigma_1}; \frac{k_2 G_2}{(k_2 + G_2)\sigma_2}\right](\sigma_0 - \sigma_{1*}^{\infty}), \qquad (73)$$

where

$$\sigma_{1*}^{\infty} = \frac{1 + c_2}{c_1}\sigma_1; \; k_{1*}^{\infty} = \frac{k_1 + G_1 c_2}{c_1};$$

$$G_{1*}^{\infty} = \frac{k_1 G_1 (1+c_2) + 2G_1^2}{(k_1 + 2G_1)c_1} \; ; \; G_{4*}^{\infty} = \frac{k_1 c_2 + 2G_1}{2c_1}. \tag{74}$$

In difference from Hashin – Shtrikman upper bounds for k_0 and G_0 upper bounds (72) and (73) do not divergent to infinity if σ_0 remains finite in this case of infinite contrast. Note that lower bounds for elastic moduli are independent on conductivity and coincide with corresponding Hashin – Shtrikman lower bounds. Moreover, upper bounds for k_0 and G_0 can depend on ratio of infinite moduli because very small quantity (that is the volume fraction of order $1/k_2$ and $1/\sigma_2$) of very rigid conductive phase can lead to finite effective properties.

In the case of superconductive absolutely rigid cylinders (phase 2) located in cells of triangle lattice (hexagonal array) disposed into matrix (phase 1) and satisfying to conditions: $k_2/k_1 = \infty$, $G_1/k_1 = G_2/k_2 = 0.4$ and $\sigma_2/\sigma_1 = \infty$, the exact results for effective conductivity [113] together with relations (72) and (73) introduce bounds for effective moduli. In the case of additional condition that the phase 1 defines inclination of upper bounds in (72), (73), that is $(k_1 + G_1)/(2\sigma_1) \geq 2k_1 G_2/[(k_2 + G_2)\sigma_2]$, Figure 10 shows that the cross-property bounds for effective elastic moduli very well coincide with corresponding model data[2] [30]. At the same time, in this case the standard variation upper bounds of the effective properties (for example, Hashin – Shtrikman upper bounds) divergent to infinity because they do not introduce information that the absolutely rigid phase is in fact to be discontinuous. On the contrary, the cross-property upper bounds use information that the phase with infinite contrast of properties is discontinuous due to using conductive properties

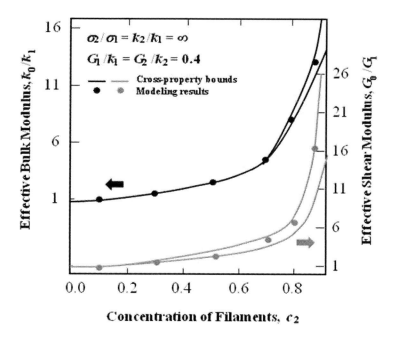

Figure 10. Comparison of cross-property bounds for effective bulk modulus (72) and shear modulus (73) for hexagonal arrays of superconductive cylinders with modeling results [30].

---

[2] Note, that only upper bounds for elastic moduli contain information on conductivity.

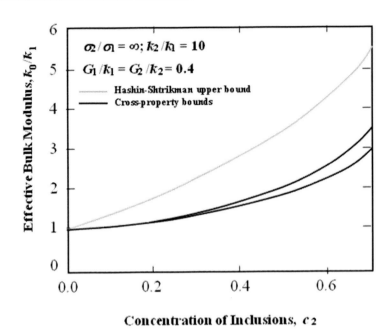

Figure 11. Cross-property bounds for effective bulk modulus for arbitrary arrays of superconductive non-overlapping cylindrical inclusions based on data for conductivity [66].

In the case of arbitrary array of superconductive cylinders ($\sigma_2/\sigma_1 = \infty$) for $k_2/k_1 = 10$, $G_1/k_1 = G_2/k_2 = 0.4$ the effective volume moduli are present in Figure 11. It is evident that the cross-property upper bound provides significant improvement in comparison with Hashin – Shtrikman upper bound for k_0.

In order to obtain the cross-property upper bounds of effective elastic and conductive properties of two-phase isotropic composites in 3D case the translational method has been used [14, 39]. In the case of absolutely rigid and superconductive phase 2 ($K_2/K_1 = \infty$, $G_2/G_1 = \infty$, $\sigma_2/\sigma_1 = \infty$) these bounds are determined by the following inequalities [137]:

$$\sigma_0 \geq \sigma_{1*}^{\infty};$$

$$K_{1*}^{\infty} \leq K_0 \leq K_{1*}^{\infty} + \max\left[\frac{3K_1 + 4G_1}{9\sigma_1}; \frac{6K_2 G_2}{(3K_2 + 4G_2)\sigma_2}; \frac{2G_2}{3\sigma_2}\right](\sigma_0 - \sigma_{1*}^{\infty}); \qquad (75)$$

where

$$\sigma_{1*}^{\infty} = \frac{1 + 2c_2}{c_1}\sigma_1; \quad K_{1*}^{\infty} = \frac{3K_1 + 4G_1 c_2}{3c_1}. \qquad (76)$$

In difference from Hashin – Shtrikman upper bounds stated for K_0 [14, 39], upper bounds (75) do not divergent to infinity if σ_0 remains finite in this case of infinite contrast. Note that the lower bound for K_0 does not depend on conductivity and coincides with corresponding Hashin – Shtrikman lower bound.

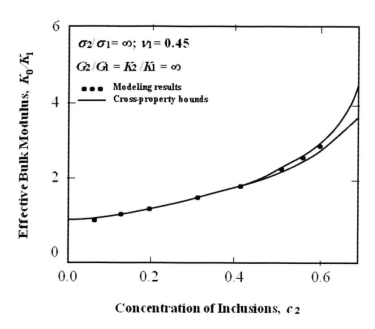

Figure 12. Comparison of cross-property bounds (75) for effective bulk modulus (fcc-array of superconductive spheres) with modeling results [108].

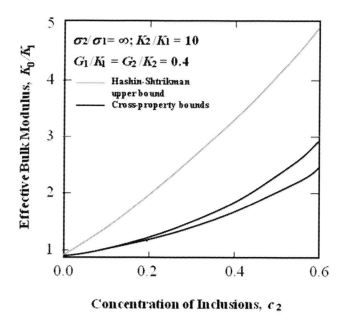

Figure 13. Cross-property bounds for effective bulk modulus for arbitrary arrays of superconductive non-overlapping spherical inclusions based on data for conductivity [67].

In the case of cubic array of superconductive spheres exact results [91] for effective conductivity of that array together with inequalities (75) could be applied to obtain elastic moduli bounds. For example, in the case of fcc-array of superconductive absolutely rigid spherical inclusions (phase 2) disposed into matrix (phase 1) and satisfying to conditions

$K_2/K_1 = \infty$, $G_2/G_1 = 0.4$ and $\nu_1 = 0.45$, the bounds are determined by inequalities (75). Moreover, by assuming that the phase 1 defines inclination of the upper bound in (75), we obtain dependence of effective bulk modulus on inclusion concentration (see Figure 12). In the figure are also present comparative results of Reference [108]. It should be noted, that only upper bound contains information on conductivity of the composite. The results show that at $c_2 \leq 0.5$ the cross-property upper bounds predict the bulk modulus of composite in practice exactly. In the case of higher concentration of inclusions coincidence with existing results is very well. Typical upper bounds (e. g. Hashin – Shtrikman upper bound) are divergent to infinity because they cannot take into account that the absolutely rigid phase is discontinuous in fact. On the contrary, the cross-property upper bound use important topological information on discontinuity of the phase with infinite contrast of properties by taking into account an information on conductivity of the composite.

In the case of equilibrium array of non-overlapping superconductive spheres in Reference [91] have been obtained results in dependence on volume fraction of inclusions [67] (see Figure 13). It is evident, that the results for the upper bound improve significantly corresponding results for Hashin – Shtrikman upper bound.

4. Asymptotical Methods for Definition of Properties of Superconductuve Composites

As it is known, high-temperature superconductive (HTSC) mono- and multi-filamentary tapes, wires, coated conductors and cables are filamentary composites in which superconductive brittle filaments are rounded by metallic (or from alloy) matrix. Due to infinite conductivity of superconductive component, in order to define effective properties it is convenient to use asymptotical approaches. In this case the introduced small parameters are defined not only by physical features of composite but proper relative sizes of filaments and matrix. Introduction, so-called 'slow' and 'fast' coordinates, defining solutions in macro-scale and micro-scale, respectively, allows one to effectively divide original boundary-value problem and take into account local oscillations that are present in final solution as corrections determined by order of small structural parameter. By introducing the periodicity cell characterizing the composite structure we obtain the required ingredient finalizing a formal statement of the problem. Similar asymptotical approach will be also developed below for granular superconductive composites.

4.1. Strong Bound Estimations for Conductive and Elastic Properties

Consider filamentary composites consisting of matrix Ω_1 and filaments Ω_2. It is obvious, that effective conductivity factor $\sigma_0 = \sigma_{33}$ in longitudinal direction x_3 could be determined by using mixture rule as

$$\sigma_{33} = (1 - c_2)\sigma_1 + c_2\sigma_2, \tag{77}$$

where c_2 is the volume fraction of filaments, σ_i are the conductivity factors of the matrix ($i = 1$) and filaments ($i = 2$).

Then treat the problem of calculation of the effective conductivity $\sigma_0 = \sigma_{11} = \sigma_{22}$ into transverse area $x_1 x_2$. In the stationary case the constitutive equation stated on the base of the conductivity equations and conservation equations for components $i = 1, 2$ has the form:

$$\sigma_i \nabla_{xx}^2 u_i = -f_i, \tag{78}$$

where $\nabla_{xx}^2 = \sum_k \dfrac{\partial^2}{\partial x_k^2}$ is the Laplace operator, σ_i are the conductivity factors of matrix ($i = 1$) and filaments ($i = 2$), u_i and f_i are the corresponding potentials and densities of volume sources, $k = 1, 2$. At the interface $\partial\Omega$ of the components there are considered the conditions of ideal contact (equality of potentials, and equality of fluxes), namely:

$$u_1 = u_2 , \; \left\{ \sigma_1 \frac{\partial u_1}{\partial \boldsymbol{n}} = \sigma_2 \frac{\partial u_2}{\partial \boldsymbol{n}} \right\}_{\partial\Omega}, \tag{79}$$

where $\partial/\partial\boldsymbol{n}$ are the derivatives on normal to $\partial\Omega$. Select in composite two spatial scales: (i) microscale determined by the distance l between centers of adjacent filaments and macroscale characterized by the specimen size L. Then, the small parameter $\varepsilon = l/L$ will define the order of the composite heterogeneity.

The solution of the boundary-value problem (78), (79) can be present as the expansion:

$$u_i = u_0(\boldsymbol{x}) + \varepsilon u_i^{(1)}(\boldsymbol{x}, \boldsymbol{y}) + \varepsilon^2 u_i^{(2)}(\boldsymbol{x}, \boldsymbol{y}) + \dots , \tag{80}$$

where $\boldsymbol{x} = \sum_k x_k \boldsymbol{e}_k$, $\boldsymbol{y} = \sum_k y_k \boldsymbol{e}_k$, $\boldsymbol{e}_k$ are the basis vectors of Cartesian coordinate system, $x_k = x_k$ and $y_k = \varepsilon^{-1} x_k$ are the 'slow' and 'fast' coordinates. The first term u_0 presents averaged part of the solution, changing in macroscale of whole specimen and no depending on 'fast' coordinates ($\partial u_0/\partial y_k$). The following terms $u_i^{(j)}$ ($j = 1, 2, 3, \dots$) introduce the corrections order of ε^j and describe local oscillations in microscale. Due to the medium periodicity $u_i^{(j)}$ satisfy also to the relative periodicity condition.

By taking into account the medium periodicity and using the perturbation method of boundary shape [43], it could be obtained first approximation for solution of the problem in polar coordinates (r, θ), introduced into periodicity cell ($r^2 = y_1^2 + y_2^2$; $\tan\theta = y_2/y_1$) [166] as

$$u_i^{(1)} = (C_{1i} r + C_{2i} r^{-1}) \frac{\partial u_0}{\partial \boldsymbol{n}} ; \; \frac{\partial}{\partial \boldsymbol{n}} = \frac{\partial}{\partial x_1} \cos\theta + \frac{\partial}{\partial x_2} \sin\theta , \tag{81}$$

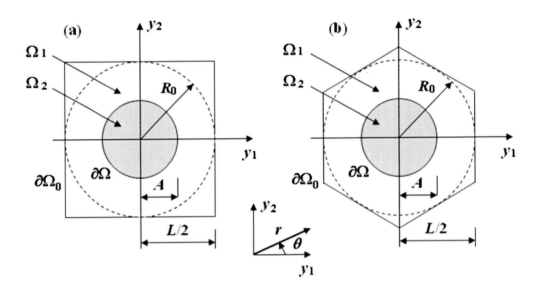

Figure 14. Periodicity cell in square (a) and hexagonal (b) lattices.

where

$$C_{11} = \frac{(\lambda_2 - 1)\chi}{D}; \quad C_{21} = -\frac{C_{11}A^2}{\chi}; \quad C_{12} = -\frac{(\lambda_2 - 1)(1-\chi)}{D}; \quad C_{22} = 0;$$

$$D = \lambda_2 + 1 - \chi(\lambda_2 - 1); \quad \chi = \left[\frac{A}{R(\theta)}\right]^2 = \frac{c_2}{c_{2\max}}\left[\frac{R_0}{R(\theta)}\right]^2; \quad R(\theta) = \frac{R_0}{\cos\delta}. \quad (82)$$

Here A is the filament radius, $R_0 = L/2$ is the radius of inscribed circle, $\lambda_2 = \sigma_2/\sigma_1$ is the dimensionless conductivity of filaments, $c_{2\max}$ is the geometrically-maximal volume fraction of filaments, $c_2 = \pi A^2 / S_0$, S_0 is square of the periodicity cell in 'fast' coordinates, $\delta = \theta - \theta_0 < 1$. In the case of square lattice (see Figure 14a) we obtain: $-\pi/4 \leq \delta \leq \pi/4$, $\theta_0 = \pi n/2$ ($n = 0, 1, 2, \ldots$), $c_{2\max} = \pi/4 = 0.7853$, $S_0 = L^2$ and for hexagonal lattice (see Figure 14b) it is carried out: $-\pi/6 \leq \delta \leq \pi/6$, $\theta_0 = \pi n/3$, $c_{2\max} = \pi\sqrt{3}/6 = 0.9068$, $S_0 = \sqrt{3} L^2/2$.

By assuming the same asymptotical order of conductivities for the components $\sigma_2/\sigma_1 = O(\varepsilon^0)$, taking into account recurrent sequence of boundary-value problems on the periodicity cell, including microscopic equations of conductivity and microscopic equations of ideal contacts, and applying the averaging operator on area of the periodicity cell, $S_0^{-1}\iint_{\Omega_0}(\cdot)dS$, where $dS = dy_1 dy_2$, we obtain the averaged equation [166] as

$$\sigma_1 \iint_{\Omega_1}[\nabla_{xx}^2 u_0 + \nabla_{xy}^2 u_1^{(1)}]dS + \sigma_2 \iint_{\Omega_2}[\nabla_{xx}^2 u_0 + \nabla_{xy}^2 u_2^{(1)}]dS = -\iint_{\Omega_1} f_1 dS - \iint_{\Omega_2} f_2 dS. \quad (83)$$

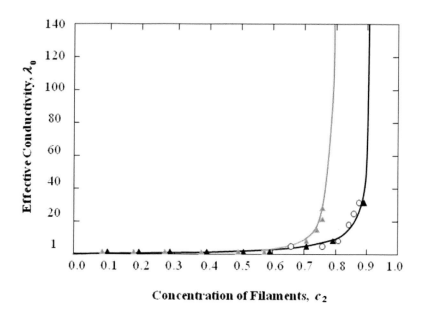

Figure 15. Effective conductivity of composites in dependence on concentration of superconductive filaments: grey and black curves correspond to solution (82) for square and hexagonal lattices; triangle show theoretical data [113] and circles present test data [113].

By substituting into (83) the relation for $u_i^{(1)}$ defined from (81) and (82), we obtain macroscopic conductivity equation with order of ε^0 in the form:

$$\sigma_0 \nabla_{xx}^2 u_0 = -f_0, \qquad (84)$$

where $f_0 = (1 - c_2)f_1 + c_2 f_2$ is the averaged density of volume sources. The effective conductivity σ_0 could be calculated after computation of integrals in Equation (83). By using numerical integration for dimensionless effective conductivity $\lambda_0 = \sigma_0/\sigma_1$ in the case of superconductive filaments ($\sigma_2 = \infty$), we obtain the dependences on concentration of superconductive components c_2 for square and hexagonal lattices (see Figure 15), which are compared with relative theoretical dependences calculated by using Rayleigh method [113]. Figure 15 presents also the test results [113] of conductivity in hexagonal lattice of superconductive cylinders. In the case of conductive cylinders with very high conductivity ($\sigma_2 \to \infty$) aspiring to contact ($c_2 \to c_{2\max}$), it has been also found the next asymptotical formula [113]:

$$\lambda_0 = M\pi \left\{ 2\frac{\ln \lambda_2}{\lambda_2} + \sqrt{2\left(\frac{c_{2\max}}{c_2}\right)^{1/2} - 1} \right\}^{-1}, \qquad (85)$$

where for square lattice $M = 1$ and for hexagonal lattice $M = \sqrt{3}$.

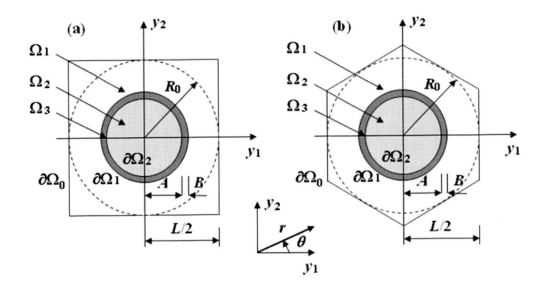

Figure 16. Periodicity cell in square (a) and hexagonal (b) lattices.

4.1.1. Effect of Filament Coverage on Composite Conductivity

In the case when filaments and matrix are divided ones with other by coverage layer Ω_3, the periodicity cell takes the form presented in Figure 16. The original conductivity equation coincides with Equation (78), in which $i = 1, 2, 3$. At the 'matrix-coverage' interfaces $\partial\Omega_1$ and 'inclusion-coverage' interfaces $\partial\Omega_2$ the conditions of ideal contact are selected to be similar conditions (79), namely:

$$\vec{u}_1 = u_3\,\vec{\partial}_{\Omega_1},\ \left\{\sigma_1\frac{\partial u_1}{\partial n} = \sigma_3\frac{\partial u_3}{\partial n}\right\}_{\partial\Omega_1},\ \vec{u}_2 = u_3\,\vec{\partial}_{\Omega_2},\ \left\{\sigma_2\frac{\partial u_2}{\partial n} = \sigma_3\frac{\partial u_3}{\partial n}\right\}_{\partial\Omega_2}. \quad (86)$$

By using the averaging method to boundary-value problem (78), (86) the approximate solution of the problem for the periodicity cell may be written in form (81), where factors C_{ki} are found as

$$C_{11} = -\chi(c_2 + c_3)[-\lambda_2 c_3 - \lambda_3(2c_2 + c_3) + \lambda_2\lambda_3(2c_2 + c_3) + \lambda_3^3 c_3]/D;$$

$$\begin{aligned}C_{12} = \{&\lambda_2[(\chi+1)c_2 + c_3]c_3 + \lambda_3[3(\chi-1)c_2 c_3 - 2(1-\chi)c_2^2 + \chi c_3^2] + \\ &\lambda_2\lambda_3[(1-3\chi)c_2 c_3 + 2(1-\chi)c_2^2 - \chi c_3^2] + \lambda_3^2[(1-\chi)c_2 - \chi c_3]c_3\}/D;\end{aligned} \quad (87)$$

$$\begin{aligned}C_{13} = \{&\lambda_2[(\chi-1)c_2 c_3 + 2c_2^2 + \chi c_3^2] + \lambda_3[(3\chi-1)c_2 c_3 + 2\chi c_2^2 + \chi c_3^2] + \\ &\lambda_2\lambda_3[(1-3\chi)c_2 c_3 + 2(1-\chi)c_2^2 - \chi c_3^2] + \lambda_3^2[(\chi+1)c_2 - \chi c_3]c_3\}/D;\end{aligned}$$

$$C_{21} = -C_{11}A^2/\chi;\ C_{22} = 0;\ C_{23} = 2A^2 c_2(c_2 + c_3)(\lambda_2 - \lambda_3)/D;$$

$$D = -\lambda_2[(1+\chi)c_2 + \chi c_3]c_3 - \lambda_3[(1+3\chi)c_2 c_3 + 2(1+\chi)c_2^2 + \chi c_3^2] +$$
$$\lambda_2\lambda_3[(3\chi-1)c_2 c_3 + 2(\chi-1)c_2^2 + \chi c_3^2] + \lambda_3^2[(\chi-1)c_2 + \chi c_3]c_3,$$

where $\lambda_3 = \sigma_3/\sigma_1$ is the dimensionless conductivity of the coverage layer conductivity; $c_3 = c_2 h(2 + h)$ is the coverage volume fraction, $h = B/A$ is the dimensionless thickness of the coverage.

The effective conductivity σ_0 is obtained after calculation of integrals in averaged equation [166]:

$$\sigma_1 \iint\limits_{\Omega_1}[\nabla_{xx}^2 u_0 + \nabla_{xy}^2 u_1^{(1)}]dS + \sigma_2 \iint\limits_{\Omega_2}[\nabla_{xx}^2 u_0 + \nabla_{xy}^2 u_2^{(1)}]dS + \sigma_3 \iint\limits_{\Omega_3}[\nabla_{xx}^2 u_0 + \nabla_{xy}^2 u_3^{(1)}]dS =$$
$$-\iint\limits_{\Omega_1} f_1 dS - \iint\limits_{\Omega_2} f_2 dS - \iint\limits_{\Omega_3} f_3 dS. \tag{88}$$

Then we find local increments $U_3^{(1)}$ and $U_3^{(2)}$ to the averaged electrical potential u_0 at both sides $\partial\Omega_1$ and $\partial\Omega_2$ of the coverage:

$$U_3^{(1)} = [u_3 - u_0]_{\partial\Omega_1} = \varepsilon u_3^{(1)}\Big|_{r=A+B} + O(\varepsilon^2) = a\left[C_{13}(1+h) + \frac{C_{23}}{A^2(1+h)}\right]\frac{\partial u_0}{\partial n} + O(\varepsilon);$$
$$\tag{89}$$
$$U_3^{(2)} = [u_3 - u_0]_{\partial\Omega_2} = \varepsilon u_3^{(1)}\Big|_{r=A} + O(\varepsilon^2) = a\left(C_{13} + \frac{C_{23}}{A^2}\right)\frac{\partial u_0}{\partial n} + O(\varepsilon),$$

where $a = \varepsilon A$ is the inclusion radius in 'slow' coordinates.

In composites with well-conductive filaments ($\lambda_2 \gg 1$) the most flux initiates at interfaces of the components. The fluxes $q_3^{(1)}$ and $q_3^{(2)}$ at the interfaces $\partial\Omega_1$ and $\partial\Omega_2$ determined along normal to the boundary are calculated as

$$q_3^{(1)} = -\sigma_3\left[\frac{\partial u_0}{\partial n} + \frac{\partial u_3^{(1)}}{\partial r}\right]_{r=A+B} + O(\varepsilon) = q_0 \frac{\lambda_3}{\lambda_0}\left[1 + C_{13} - \frac{C_{23}}{A^2(1+h)^2}\right] + O(\varepsilon); \tag{90}$$

$$q_3^{(2)} = -\sigma_3\left[\frac{\partial u_0}{\partial n} + \frac{\partial u_3^{(1)}}{\partial r}\right]_{r=A} + O(\varepsilon) = q_0 \frac{\lambda_3}{\lambda_0}\left[1 + C_{13} - \frac{C_{23}}{A^2}\right] + O(\varepsilon),$$

where $q_0 = -\sigma_0 \partial u_0/\partial n$ is the averaged flux via boundary $\partial\Omega$ in macroscale.

In the case of thin coverage ($h \to 0$) at $\lambda_3 < \lambda_2$ the contact conditions of matrix with inclusions satisfy to the model of soft (badly conductive) boundary defining the flux continuity ($q_3^{(1)} = q_3^{(2)}$) and the jump of electrical potential ($U_3^{(1)} > U_3^{(2)}$). On the contrary case, at $\lambda_3 > \lambda_2$ it is realized the model of rigid (well-conductive) boundary stating the flux discontinuity ($q_3^{(1)} > q_3^{(2)}$) and equality of the electrical potentials ($U_3^{(1)} = U_3^{(2)}$). In this case take place the next limit transitions:

at $\lambda_3 \to 0$ we have $U_3^{(1)} \to U_3|_{\lambda_2=0}$; $U_3^{(2)} \to U_3|_{\lambda_2=\infty}$; $q_3^{(1)} = q_3^{(2)} \to q_3|_{\lambda_2=0} = 0$; (91)

at $\lambda_3 \to \infty$ we have $U_3^{(1)} = U_3^{(2)} \to U_3|_{\lambda_2=\infty}$; $q_3^{(1)} \to q_3|_{\lambda_2=\infty}$; $q_3^{(2)} \to q_3|_{\lambda_2=0} = 0$. (92)

The results for dimensionless effective conductivity λ_0, obtained by using three-phase model [46] and compared with above asymptotical model lead to underestimated value of λ_0 at $\lambda_3 \to \infty$, because the three-phase model does not take into account spatial geometry of the filament package.

4.1.2. Effect of Non-Ideal Contact of Filaments and Matrix

Effect of the non-ideal contact between filaments and matrix could be modeled by assuming that the volume fraction of coverage $c_3 \to 0$ and its dimensionless conductivity $\lambda_3 \to 0$. In this case in dependence on the ratio λ_3/c_3 may be estimated different adhesion order for the composite components. With this aim assume that $h \to 0$, $c_3 = c_2 h(2 + h) \to 0$ and introduce dimensionless couple parameter a:

$\lambda_3 = (1 - a)c_3/a$, at $0 \le a \le 1$. (93)

The value of $a = 0$ corresponds to ideal contact, and $a = 1$ defines the case of total absence any contact between the components. Moreover, flux q_3 via the 'matrix-filament' interface $\partial\Omega$ is proportional to the jump of electrical potential $\Delta u_3 = [u_1 - u_2]_{\partial\Omega}$:

$q_3 = - (\sigma_{ef}/b)\Delta u_3$, (94)

where $b = \varepsilon B$ is the coverage thickness in 'slow' coordinates, σ_{ef} characterizes the interface conductivity.

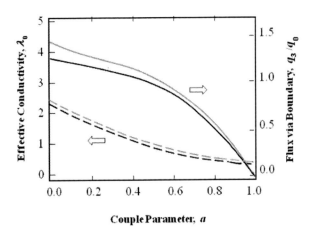

Figure 17. Influence of couple parameter on effective conductivity and flux via 'matrix-filament' interface: grey and black curves correspond to square and hexagonal lattice, respectively.

At the non-ideal contact ($a = 0$) $\Delta u_3 = 0$, and in total absence any contact ($a = 1$) $q_3 = 0$. The last case corresponds to the composite with absolutely non-conductive filaments (or voids). As a result, it has been estimated influence of the couple parameter a in the case of superconductive inclusions ($\lambda_2 = \infty$) on effective parameter λ_0 and flux q_3 (at $\theta = \theta_0$) at the 'matrix-filament' interface for the filament concentration $c_2 = 0.6$. The plots of these dependences are present in Figure 17.

4.1.3. Filamentary Composites of Irregular Structure

At accidental perturbation of periodical disposition of filaments into composite, present each filament center into circle with diameter d. By this the circles compose square lattice (Figure 18a) or hexagonal one (Figure 18b) with period l. Parameter $\beta = d/l$, $0 \leq \beta \leq \beta_{max}$ defines order of the structure irregularity. In the case when adjacent inclusions touch one to other the maximal value is calculated as $\beta_{max} = 1 - (c_2/c_{2\,max})^{1/2}$. It is assumed absence of the inclusion overlapping and no formation of cluster structure. If the filament conductivity is greater than the matrix conductivity ($\lambda_2 > 1$)[3], then there are introduced the corresponding bounds for dimensionless effective conductivity λ_0:

$$\lambda_L \leq \lambda_0 \leq \lambda_U. \tag{95}$$

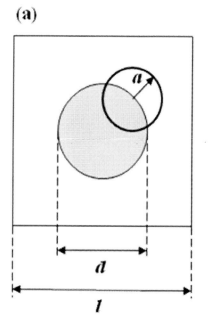
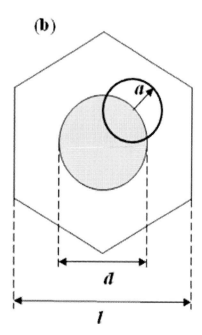

Figure 18. Periodicity cell for filamentary composites of irregular structure in square (a) and hexagonal (b) lattices.

---

[3] The solution for contrary case ($\lambda_2 < 1$) is obtained by using the formula $\lambda_0\big|_{\lambda_2=\lambda} = \dfrac{1}{\lambda_0\big|_{\lambda_2=1/\lambda}}$ [63].

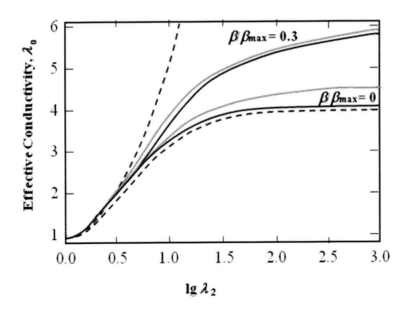

Figure 19. Effective conductivity of irregular composites: grey and black curves correspond to square and hexagonal lattice, respectively; Hashin - Shtrikman bounds are shown by dashed lines.

In the regular case (at $\beta = 0$) the effective conductivity has been defined in dependence on the filament concentration c_2 in the form [69]: $\lambda_0|_{\beta=0} = \Lambda_0(c_2)$. The strong lower bound λ_L is determined by the solution for corresponding regular structure in the form:

$$\lambda_L = \Lambda_0(c_2). \tag{96}$$

The upper estimation λ_U may be defined by substitution of accidently disposed filaments with radius a by periodic lattice of filaments with radius $a + d/2$ (so-called method of safe spheres [122]):

$$\lambda_U = \Lambda_0[(\sqrt{c_2} + \beta\sqrt{c_{2\max}})^2]. \tag{97}$$

Other known variation Hashin – Shtrikman estimations [45] give the next relations for extreme estimations:

$$\lambda_L = 1 + \frac{c_2}{(\lambda_2 - 1)^{-1} + (1 - c_2)/2}; \quad \lambda_U = \lambda_2 + \frac{1 - c_2}{(1 - \lambda_2)^{-1} + c_2/(2\lambda_2)}. \tag{98}$$

The numerical results presented in (Figure 19) for $c_2 = 0.6$ at very high conductivity of filaments ($\lambda_2 \to \infty$), significantly improve corresponding Hashin – Shtrikman estimations described by dashed lines.

4.1.4. Filamentary Composites of Irregular Structure

At accidental disposition of filaments, an increasing their concentration c_2 at proper critical value (percolation threshold) $c_2 = c_{2p}$ leads to formation of infinite cluster with fractal structure from contacting links of the inclusions. According to model [27, 29, 127] its conducting skeleton may be presented by a lattice with period (correlation radius) R_c:

$$R_c = \frac{1}{\left| c_2 - c_{2p} \right|^\nu} \text{ at } c_2 \to c_{2p}, \tag{99}$$

where l is the distance between the centers of contacting inclusions, ν is the index of the correlation radius. Value of R_c defines maximal size of finite clusters at $c_2 < c_{2p}$. Effective characteristics σ_0 the composite near the percolation threshold ($c_2 \to c_{2p}$) are defined by asymptotical dependences of type $\sigma_0 \sim \left| c_2 - c_{2p} \right|^t$, where t is the critical exponent of corresponding critical properties. The critical exponents depend on the spatial dimension. In 2D case we have $c_{2p} = 0.5$ and $\nu = 1.33$.

Below we obtain qualitative estimations for composite with superconductive filaments whose fraction aspires to the percolation threshold from below ($\sigma_2 = \infty$, $c_2 \to c_{2p} - 0$) [166]. Consider the region with proper size $R_c + l$, into of which to be cluster with maximal length R_c. Electrical resistivity of the region is equal

$$\rho_0 = \rho_1 \frac{l}{R_c + l}, \tag{100}$$

where $\rho_1 = 1/\sigma_1$ is the electrical resistivity of matrix and the effective conductivity is approximately determined as

$$\sigma_0 = \frac{1}{\rho_0} \approx \frac{\sigma_1}{(c_{2p} - c_2)^\nu}. \tag{101}$$

This formula allows one to state the singularity order no pretending on strict account of numerical coefficients.

At small concentration of filaments ($c_2 \to 0$) the dimensionless effective conductivity in 2D case is defined by Garnet formula [36] as

$$\lambda_0 = \frac{\sigma_0}{\sigma_1} = 1 - 2c_2 \left[\frac{(\sigma_1 + \sigma_2)}{(\sigma_1 - \sigma_2)} + c_2 \right]^{-1}. \tag{102}$$

For superconductive inclusions ($\sigma_2 = \infty$) we obtain

$$\lambda_0 = \frac{\sigma_0}{\sigma_1} = \frac{1 + c_2}{1 - c_2}. \tag{103}$$

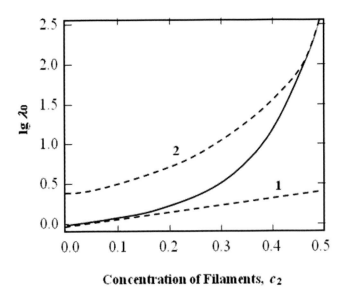

Figure 20. Effective conductivity of composite with superconductive filaments.

By joining relations (101) and (102), we define approximate analytical solution corresponding to any value from range $0 \leq c_2 \leq c_{2p}$. With this aim it is used the method of asymptotically equivalent functions [2, 128], allowing one to present the effective conductivity as

$$\lambda_0 = \frac{\sigma_0}{\sigma_1} = \frac{a_0 + a_1\xi + \xi^2(1+\xi)^\nu}{1 + b_1\xi + \xi^2 c_{2p}^\nu}, \qquad (104)$$

where $\xi = c_2/(c_{2p} - c_2)$. Asymptotic of relation (104) coincides with (101) at $c_2 \to c_{2p}$. Unknown factors a_0, a_1 and b_1 are calculated from the condition of coincidence of function expansion (104) into series on powers of c_2 at $c_2 \to 0$ and function expansion (103) into series on powers up to order $O(c_2^2)$, inclusive. As a result, we obtain $a_0 = 1$, $a_1 = 1 + c_{2p} + (1-c_{2p}^\nu)/(2c_{2p})$, $b_1 = 1 - c_{2p} + (1-c_{2p}^\nu)/(2c_{2p})$.

The numerical results for composite with superconductive filaments are present in Figure 20. Here, the solid line corresponds to formula (104), the dashed line 1 is determined by formula (103), and the dashed line 2 defined by formula (101).

4.2. Conductivity of Granular Composites

Consider granular composites consisting of matrix Ω_1 and spherical inclusions Ω_2. The periodicity cells (their 1/8 parts) for simple cubic (sc) and body-centered cubic (bcc) lattices are shown in Figure 21. In order to define effective conductivity $\sigma_0 = \sigma_{11} = \sigma_{22} = \sigma_{33}$ consider constitutive equation (78), where $i = 1, 2, 3$ and conditions of ideal contact (79). The solution of boundary-value problem (78), (79) is present again into form of expansion (80).

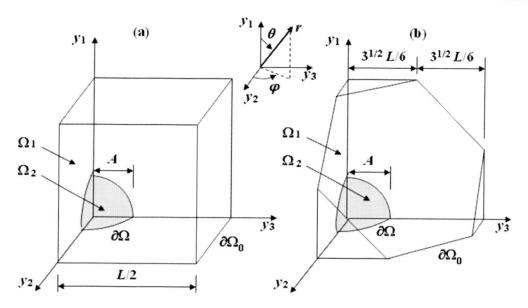

Figure 21. Periodicity cell of granular composites in sc-lattice (a) and bcc-lattice (b).

By taking into account the medium periodicity, using the method of perturbation of the boundary shape [43], we obtain first approximation for solution of the problem into spherical coordinates (r, θ, φ), introduced into periodicity cell ($r^2 = y_1^2 + y_2^2 + y_3^2$; $\tan\varphi = y_3/y_2$; $\cos\theta = y_1/r$), in the form [166]:

$$u_i^{(1)} = (C_{1i}r + C_{2i}r^{-2})\frac{\partial u_0}{\partial n}; \quad \frac{\partial}{\partial n} = \frac{\partial}{\partial x_1}\cos\theta + \frac{\partial}{\partial x_2}\sin\theta\cos\varphi + \frac{\partial}{\partial x_3}\sin\theta\sin\varphi, \quad (105)$$

where

$$C_{11} = \frac{(\lambda_2 - 1)\chi}{D}; \quad C_{21} = -\frac{C_{11}A^3}{\chi}; \quad C_{12} = -\frac{(\lambda_2 - 1)(1 - \chi)}{D}; \quad C_{22} = 0;$$

$$D = \lambda_2 + 2 - \chi(\lambda_2 - 1); \quad \chi = \left[\frac{A}{R(\theta,\varphi)}\right]^3 = \frac{c_2}{c_{2\max}}\left[\frac{R_0}{R(\theta,\varphi)}\right]^3. \quad (106)$$

Here A is the grain radius, $R_0 = L/2$ is the radius of inscribed sphere, $\lambda_2 = \sigma_2/\sigma_1$ is the dimensionless grain conductivity, $c_{2\max}$ is the geometrically maximal volume fraction of grains, $c_2 = 4\pi A^3/(3V_0)$, V_0 is the volume of the periodicity cell into 'fast' coordinates. For 1/16-part of the cell ($0 \le \theta \le \pi/2$, $0 \le \varphi \le \pi/4$) presented in Figure 21, relations for $R(\theta, \varphi)$ are shown in Table 2. At other values of angles θ and φ the function $R(\theta, \varphi)$ continues periodically. For sc-lattice we obtain (see Figure 21a) $c_{2\max} = \pi/6 = 0.5235$, $V_0 = L^3$ and for bcc-lattice (see Figure 21b) $c_{2\max} = \pi\sqrt{3}/8 = 0.6801$, $V_0 = 4\sqrt{3}L^3/9$.

Table 2. Function $R(\theta, \varphi)$

φ	θ		$R(\theta, \varphi)$
sc-lattice			
from 0 up to $\pi/4$	from 0 up to $\mathrm{atan}\left(\dfrac{1}{\cos\varphi}\right)$		$\dfrac{R_0}{\cos\theta}$
	from $\mathrm{atan}\left(\dfrac{1}{\cos\varphi}\right)$ up to $\pi/2$		$\dfrac{R_0}{\sin\theta\cos\varphi}$
bcc-lattice			
from 0 up to $\mathrm{atan}(1/2)$	from 0 up to $\mathrm{atan}\left[\dfrac{1}{2(\sin\varphi+\cos\varphi)}\right]$		$\dfrac{2R_0}{\sqrt{3}\cos\theta}$
	from $\mathrm{atan}\left[\dfrac{1}{2(\sin\varphi+\cos\varphi)}\right]$ up to $\mathrm{atan}\left[\dfrac{2}{\cos\varphi-2\sin\varphi}\right]$		$\dfrac{\sqrt{3}R_0}{\cos\theta+\sin\theta(\sin\varphi+\cos\varphi)}$
	from $\mathrm{atan}\left[\dfrac{2}{\cos\varphi-2\sin\varphi}\right]$ up to $\pi/2$		$\dfrac{2R_0}{\sqrt{3}\sin\theta\cos\varphi}$
from $\mathrm{atan}(1/2)$ up to $\pi/4$	from 0 up to $\mathrm{atan}\left[\dfrac{1}{2(\sin\varphi+\cos\varphi)}\right]$		$\dfrac{2R_0}{\sqrt{3}\cos\theta}$
	from $\mathrm{atan}\left[\dfrac{1}{2(\sin\varphi+\cos\varphi)}\right]$ up to $\pi/2$		$\dfrac{\sqrt{3}R_0}{\cos\theta+\sin\theta(\sin\varphi+\cos\varphi)}$

By applying the averaging operator on region of the periodicity cell $V_0^{-1}\iiint\limits_{\Omega_0}(\cdot)dV$, where $V = dy_1 dy_2 dy_3$, we obtain the averaged equation [166]:

$$\sigma_1\iiint\limits_{\Omega_1}[\nabla^2_{xx}u_0+\nabla^2_{xy}u_1^{(1)}]dV+\sigma_2\iiint\limits_{\Omega_2}[\nabla^2_{xx}u_0+\nabla^2_{xy}u_2^{(1)}]dV=-\iiint\limits_{\Omega_1}f_1dV-\iiint\limits_{\Omega_2}f_2dV. \quad (107)$$

By substituting into (107) the relation for $u_i^{(1)}$ from (105) and (106), we obtain macroscopic equation of conductivity for estimation of the effective conductivity σ_0.

The numerical results for dimensionless effective conductivity $\lambda_0 = \sigma_0/\sigma_1$ in the case of superconductive inclusions are shown in Figure 22. Moreover, there are present theoretical data obtained by using Rayleigh method [89, 91] and test results measuring electric conductivity of bcc-lattice consisting of superconductive spheres [89].

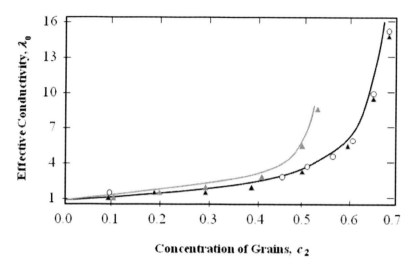

Figure 22. Effective conductivity of composites in dependence on concentration of superconductive grains: grey and black curves correspond to solution (106) for sc-lattice and bcc-lattices; triangles show theoretical data for sc-lattice [91] and bcc-lattice [89], and circles present test data for bcc-lattice [89].

Asymptotic of λ_0 in the case of strong interaction of adjacent superconductive grains is present in Figure 23. The solution (solid curves) good coincides with asymptotical formulae [8] (dashed curves) obtained for two very well conductive spheres aspiring to contact:

$$\lambda_0 = M_1 \ln \zeta - M_2 \text{ at } \lambda_2 = \infty, c_2 \to c_{2\max}; \tag{108}$$

$$\lambda_0 = M_1 \ln \lambda_2 - M_2 \text{ at } \lambda_2 \to \infty, c_2 = c_{2\max}, \tag{109}$$

where $\zeta = 1/[1-(c_2/c_{2\max})^{1/3}]$, $M_1 = \pi/2$, $M_2 = 0.7$ for sc-lattice and $M_1 = \pi\sqrt{3}/2$, $M_2 = 2.4$ for bcc-lattice.

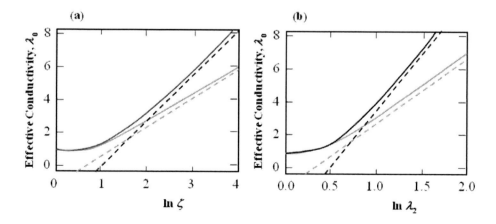

Figure 23. Effective conductivity of composites in the cases: (a) $\lambda_2 = \infty$, $c_2 \to c_{2\max}$; (b) $\lambda_2 \to \infty$, $c_2 = c_{2\max}$. Solid grey and black lines correspond to solutions defined for sc-lattice and bcc-lattices, dashed straight lines present corresponding asymptotical results [8].

4.2.1. Coverage Grain Effect on Composite Conductivity

In the case when simple cubic lattice of spherical grains and matrix are separated ones with other by coverage layer Ω_3, the periodicity cell (its 1/8 part) takes form shown in Figure 24. Original equation of conductivity coincides with Equation (78), in which $i = 1, 2, 3$. At 'matrix-coverage' interface $\partial\Omega_1$ and 'inclusion-coverage' interface $\partial\Omega_2$ conditions of ideal contact (86) are selected.

By using the averaging method for boundary-value problem (78), (86), approximate solution of the problem for periodicity cell may be written in form (81), where factors C_{ki} are determined as

$$C_{11} = -\chi(c_2 + c_3)[-\lambda_2 c_3 - \lambda_3(3c_2 + 2c_3) + \lambda_2\lambda_3(3c_2 + c_3) + 2\lambda_3^2 c_3]/D;$$

$$C_{12} = \{\lambda_2[(\chi+2)c_2 + \chi c_3]c_3 + \lambda_3[5(\chi-1)c_2 c_3 + 3(\chi+1)c_2^2 + 2\chi c_3^2] + \\ \lambda_2\lambda_3[3(1-\chi)c_2^2 - \chi c_3^2 + (1-4\chi)c_2 c_3] + \lambda_3^2[2(1-\chi)c_2 - 2\chi c_3]c_3\}/D;$$

$$C_{13} = \{\lambda_2[\chi c_2 c_3 + 3c_2^2 + \chi c_3^2] + \lambda_3[(5\chi-2)c_2 c_3 + 3\chi c_2^2 + 2\chi c_3^2] + \\ \lambda_2\lambda_3[(1-4\chi)c_2 c_3 + 3(1-\chi)c_2^2 - \chi c_3^2] + \lambda_3^2[2(1-\chi)c_2 - 2\chi c_3]c_3\}/D;$$

$$C_{21} = -C_{11}A^3/\chi; \quad C_{22} = 0; \quad C_{23} = 3A^3 c_2(c_2 + c_3)(\lambda_2 - \lambda_3)/D;$$

$$D = -\lambda_2[(2+\chi)c_2 + \chi c_3]c_3 - \lambda_3[(4+5\chi)c_2 c_3 + 3(2+\chi)c_2^2 + 2\chi c_3^2] + \\ \lambda_2\lambda_3[(4\chi-1)c_2 c_3 + 3(\chi-1)c_2^2 + \chi c_3^2] + \lambda_3^2[2(\chi-1)c_2 + 2\chi c_3]c_3,$$

(110)

where $\lambda_3 = \sigma_3/\sigma_1$ is the dimensionless conductivity of the coverage layer, $c_3 = c_2 h(3 + 3h + h^2)$ is the coverage volume fraction, $h = B/A$ is the dimensionless thickness of the coverage.

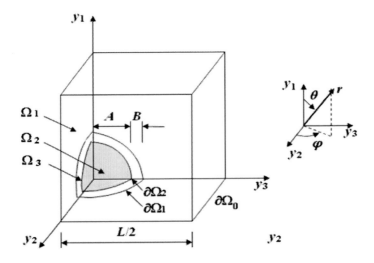

Figure 24. Periodicity cell of granular composites in sc-lattice.

The effective conductivity σ_0 is calculated computing integrals in the averaged equation:

$$\sigma_1 \iiint_{\Omega_1}[\nabla^2_{xx}u_0 + \nabla^2_{xy}u_1^{(1)}]dV + \sigma_2 \iiint_{\Omega_2}[\nabla^2_{xx}u_0 + \nabla^2_{xy}u_2^{(1)}]dV + \sigma_3 \iiint_{\Omega_3}[\nabla^2_{xx}u_0 + \nabla^2_{xy}u_3^{(1)}]dV =$$
$$-\iiint_{\Omega_1} f_1 dV - \iiint_{\Omega_2} f_2 dV - \iiint_{\Omega_3} f_3 dV. \qquad (111)$$

The local electrical potentials $U_3^{(1)}$ and $U_3^{(2)}$, and also the fluxes $q_3^{(1)}$ and $q_3^{(2)}$ at interfaces $\partial\Omega_1$ and $\partial\Omega_2$ are determined again by using formulae (89) and (90), in which A^2 is substituted on A^3.

For thin coverage ($h \to 0$) at $\lambda_3 < \lambda_2$ the conditions of matrix-inclusion contact satisfy to the model of badly conducting boundary ($q_3^{(1)} = q_3^{(2)}$, $U_3^{(1)} > U_3^{(2)}$). On the contrary, at $\lambda_3 > \lambda_2$ is realized the model of well conducting boundary ($q_3^{(1)} > q_3^{(2)}$, $U_3^{(1)} = U_3^{(2)}$). Again, the results for dimensionless effective conductivity λ_0, based on three-phase model [46], which does not take into account spatial package of grains in comparison with above asymptotical model, lead to underestimated value of λ_0 at $\lambda_3 \to \infty$.

4.2.2. Effect of Non-Ideal Contact of Grains and Matrix

Assume that the volume fraction of coverage $c_3 \to 0$ and its dimensionless conductivity $\lambda_3 \to 0$. Introduce dimensionless couple parameter a (93) and suppose that $h \to 0$, $c_3 = c_2 h(3 + 3h + h^2) \to 0$. Again, the value of $a = 0$ corresponds to ideal contact, and $a = 1$ defines total absence any contact between components. In this case the flux q_3 via 'matrix-grain' interface $\partial\Omega$ is proportional to the jump of electrical potential Δu_3 (94). At ideal contact ($a = 0$) $\Delta u_3 = 0$, and at total absence any contact between components ($a = 1$) $q_3 = 0$. The last case corresponds to composite with absolutely non-conducting grains (or voids). As a result, it may be estimated influence of the couple parameter a in the case of superconductive inclusions ($\lambda_2 = \infty$) on dimensionless effective conductivity λ_0 and flux q_3 (at $\theta = \theta_0$, $\varphi = \varphi_0$). The plots these dependences are shown in Figure 25.

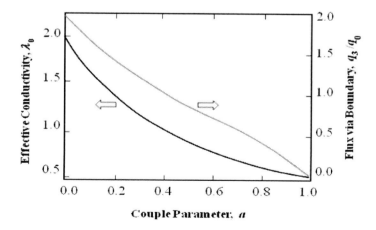

Figure 25. Effect of couple parameter on effective conductivity and flux via 'matrix-grain' interface.

4.2.3. Granular Composites of Irregular Structure

At accidental perturbation of periodical disposition of grains into composite, we present any grain centre into sphere of diameter d (in this case these spheres compose sc-lattice or bcc-lattice with period l). Parameter $\beta = d/l$, $0 \leq \beta \leq \beta_{max}$ defines order of the structure irregularity. In the case when adjacent inclusions touch one to other the maximal value is calculated as $\beta_{max} = 1 - (c_2/c_{2\max})^{1/2}$. It is assumed absence of the inclusion overlapping and no formation of cluster structure. If the grain conductivity is greater than the matrix conductivity ($\lambda_2 > 1$), then there are introduced the corresponding bounds for dimensionless effective conductivity λ_0 (95).

In the regular case (at $\beta = 0$) the effective conductivity has been defined in dependence on the grain concentration c_2 in the form [69]: $\lambda_0|_{\beta=0} = \Lambda_0(c_2)$. The strong lower bound λ_L is determined by the solution for corresponding regular structure in the form:

$$\lambda_L = \Lambda_0(c_2). \tag{112}$$

The upper bound λ_U may be defined by substitution of accidently disposed grains with radius a by periodic lattice of grains with radius $a + d/2$ by using again the method of safe spheres [122]:

$$\lambda_U = \Lambda_0[(c_2^{1/3} + \beta c_{2\max}^{1/3})^3]. \tag{113}$$

Other known variation Hashin – Shtrikman bounds [48] give the next relations for extreme estimations:

$$\lambda_L = 1 + \frac{c_2}{(\lambda_2 - 1)^{-1} + (1 - c_2)/3} ; \quad \lambda_U = \lambda_2 + \frac{1 - c_2}{(1 - \lambda_2)^{-1} + c_2/(3\lambda_2)}. \tag{114}$$

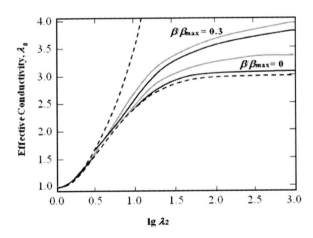

Figure 26. Effective conductivity of irregular composites: grey and black curves correspond to sc-lattice and bcc-lattices, dashed lines present Hashin – Shtrikman bounds.

More precise estimations (112) and (113), shown by solid curves in Figure 26 for $c_2 = 0.4$ at high conductivity of grains are significantly better than corresponding Hashin – Shtrikman bounds presented by dashed lines.

4.2.4. Cluster Conductivity of Granular Composites

At accidental disposition of grains and proper their concentration c_2 they can form clusters. The correlation radius is determined by relation (99), in which for 3D case $c_{2p} = 0.16$ and $v = 0.85$ [27, 28, 127]. For composite with superconductive inclusions, whose fraction aspires to percolation threshold from below ($\sigma_2 = \infty$, $c_2 \to c_{2p} - 0$) effective conductivity σ_0 in approximate form may again be calculated by using formula (101). At small concentration of grains ($c_2 \to 0$) the dimensionless effective conductivity λ_0 is also defined by using Garnet formula [36] as

$$\lambda_0 = \frac{\sigma_0}{\sigma_1} = 1 - 3c_2 \left[\frac{(2\sigma_1 + \sigma_2)}{(\sigma_1 - \sigma_2)} + c_2 \right]^{-1} \tag{115}$$

For superconductive inclusions ($\sigma_2 = \infty$) we obtain

$$\lambda_0 = \frac{\sigma_0}{\sigma_1} = \frac{1 + 2c_2}{1 - c_2}. \tag{116}$$

By joining relations (101) and (116), the approximate analytical solution coincides formally with relation (104), in which the factors are now found as $a_0 = 1$, $a_1 = 1 + 2c_{2p} + (1-c_{2p}^v)/(3c_{2p})$, $b_1 = 1 - c_{2p} + (1-c_{2p}^v)/(3c_{2p})$.

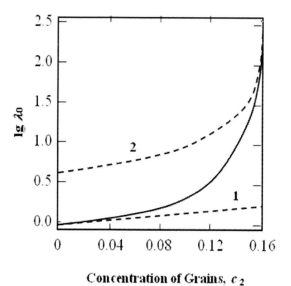

Figure 27. Effective conductivity of composite with superconductive grains.

The numerical results for composite with superconductive grains, whose concentration no greater than the percolation threshold, are shown in Figure 27. Here, the solid curve corresponds to formula (104), dashed curve 1 is stated by formula (116), and dashed curve 2 is defined by formula (101).

4.3. Effective Elastic Properties of Composites

During procedure defining longitudinal shear modulus G_L of filamentary composite, it is considered deformation of pure shear into planes x_1x_3 and x_2x_3 under action of tangential stresses σ_{13} and σ_{23} at zero other stresses. Then, the equilibrium equations into displacements u, v, w corresponding to coordinates x_1, x_2, x_3 for i-th component of the composite have the form:

$$G_i\left(\frac{\partial^2 w_i}{\partial x_1^2} + \frac{\partial^2 w_i}{\partial x_2^2}\right) = -f_i, i = 1, 2. \tag{117}$$

where f_i is the component (longitudinal) of bulk force along direction of x_3.

At the interface $\partial\Omega$ there are considered the conditions of ideal contact of the components of the composite consisting in equality their longitudinal displacements:

$$w_1 = w_2 \tag{118}$$

and tangential stresses along normal n to $\partial\Omega$:

$$\left\{G_1 \frac{\partial w_1}{\partial n} = G_2 \frac{\partial w_2}{\partial n}\right\}_{\partial\Omega}. \tag{119}$$

The boundary-value problem (117) − (119) is mathematically coincides with the conductivity problem (78), (79). Therefore, all results obtained in paragraph 4.1 for effective conductivity σ_0 directly carry over on effective shear modulus G_L at strain along longitudinal direction.

The effective bulk modulus K_T and effective shear modulus G_T at deformation in transverse direction may be obtained by using poly-dispersed model [47] and three-phase model [24], respectively. They do not take into account geometry of composite and lead to underestimated values for effective moduli in the case of high stiffness and high concentration of inclusions. Similar solutions for effective elastic moduli of granular composites could be obtained by using the same models [24, 44], no taking into account the type of spatial package of grains and assuming that the spherical inclusions are uniformly distributed into matrix and do not form clusters.

5. EFFECTS OF NEGATIVE STIFFNESS OF COMPONENTS

It should be assumed that materials near proper phase transitions will demonstrate negative stiffness in microscale due some causes. Superconductive ferroelastics [123] and ferroelectrics [22, 79] near structural phase transitions possess stiffness components which achieve minimal values or aspire to zero at critical temperature. Below critical temperature the domain or band structure forms. That phase transformations have been analyzed by using modified Landau theory [32, 33], in which free energy depends on strain, temperature and material constants. It has been shown that at low temperatures effective stiffness may be negative. At decreasing temperature initiates spontaneous deformation which corresponds to shear from unstable equilibrium state to one of stable points of minimal energy, and the material acquires domain or band structure.

Poly-domain material demonstrates positive stiffness. The formation of that band structure has been studied into framework of elastic model of the solid medium [68], in which dependence of stress on strain has unmonotonous section. Domain walls requires energy for their formation, therefore as single domains may be particles of sufficiently small sizes (from micrometer up to centimeter scale in dependence on material) [123]. It should be assumed that the single domains will demonstrate negative stiffness. Experimental results leading to conclusion about existence of negative stiffness have been published [74, 75]. Foam materials under strong compression demonstrate macroscopic positive stiffness. By this they create band structures under strong compression and that band structure corresponds to stability of individual cells [121]. Transformation to band structure is similar to bulging edges of the foam cell which also leads to non-linear dependence $\sigma - \varepsilon$. High-temperature superconductive (HTSC) foams[4] could find sufficiently wide application in modern electro-energetics and demand detailed studies their conductive and mechanical properties.

As it has been shown, for elastic two-phase composite consisting of isotropic phases the stiffness of Voigt-Reuss composite presents strong upper and lower bounds for Young modulus at given phase concentration. Hashin – Shtrikman equations [51] state strong bounds for elastic stiffness of macroisotropic composites. These bounds are narrower than Voigt-Reuss bounds.

The structure corresponding to Voigt composite (upper bound of Voigt-Reuss composite) has phases disposed along direction of applied loading, so each phase is subjected to the same strain. The Voigt bound (mixture rule) is defined in 1D model as

$$E_0 = E_1 V_1 + E_2 V_2, \tag{120}$$

where E_0, E_1, E_2 are Young moduli (stiffness) of composite and the both phases, V_1 and V_2 determine volume fractions of phases ($V_1 + V_2 = 1$). It is evident, that for Voigt structure

---

[4]The foam synthesized from HTSC-ceramic ($YBa_2Cu_3O_x$) in 2002 [119] may be ideal material for limiters of dangerous currents in electro-energetics. Their advantages compared with devices of tape and bulk forms from bismuth-containing high-temperature superconductors are caused by the next causes: (i) at T = 77 K the foam sustains critical currents which are significantly stronger than for superconductive phase Bi-2212 demonstrating big problems with pinning centers of magnetic flux at that 'high' temperatures, (ii) one has sufficiently high electric resistance at room temperature in order to scatter in heat an energy of superconducting current, (iii) one switches rapidly from superconductive state and into superconductive state. The last two properties in significant degree are forced due to the unique fact that the superconductive foam contains open pores filled by refrigerant. This circumstance provides continuously direct contact with liquid nitrogen compensating unsatisfactory heat capacity of the ceramic in bulk form.

defined by Equation (120), the phase with negative stiffness simply decreases in whole the composite stiffness.

The structure corresponding to Reuss composite (lower bound of Voigt-Reuss composite) has phases disposed in perpendicular direction to applied loading, so each phase is subjected to the same stress. The Reuss bound is defined in 1D model as

$$\frac{1}{E_0} = \frac{V_1}{E_1} + \frac{V_2}{E_2} \text{ or } S_0 = S_1 V_1 + S_2 V_2, \tag{121}$$

where S_0, S_1, S_2 are the compliance of the composite and the both phases. If stiffness one of the phase this composite is negative (that is the phase has also negative compliance), then at addition the negative compliance of the phase to positive compliance with similar value of other phase, it may be achieved near-to-zero compliance of the whole composite (or aspiring to infinity stiffness of the composite). Thus, the great values of stiffness should initiates in the composite structure in which the phases with positive and negative stiffness are almost balanced owing to that the interface between components moves far faster than applied loading providing large strain energy in each phase.

Reuss model including one element of each phase is stable only in conditions of restraint the phase deformations. Moreover, great negative stiffness is possible in the restrained system however a region for which stiffness of Reuss composite aspires to positive infinity to be unstable [74].

Investigate into framework of elasticity theory behavior of composite with required negative stiffness. With this aim, we consider spherically symmetric problem presenting a spherical matrix of isotropic material containing a spherical inclusion other phase at spherically symmetric boundary conditions. Into linear-elastic statement of the problem [55] effective volume modulus K_0 for the composite sphere is determined as

$$K_0 = \frac{a_1^3 K_1 \alpha_1 - 4 G_1 \beta_1 / 3}{a_1^3 \alpha_1 + \beta_1}, \tag{122}$$

where a_i are the radii, K_i and G_i are the volume and shear moduli, α_i and β_i are the unknown constants calculated from boundary conditions, $i = 1, 2$ (here indices 1 and 2 define matrix and inclusion, respectively).

Consider two types of the problems.

(1) *Homogeneous radial stress applied to the boundary of spherical matrix* $\sigma_{rr}(a_1) = \sigma$. Other conditions: (i) radial displacement in center of spherical inclusion $u_r(0) = 0$, and (ii) continuity of σ_{rr} and u_r on matrix-inclusion interface $r = a_2$. We define from boundary conditions the unknown constants (in particular, α_1 and β_1), and then taking into account dependence $2G = 3K(1 - 2v)/(1 + v)$, where v is Poisson's ratio, obtain K_0 as

$$\alpha_1 = a_1^3 \sigma (3K_2 + 4G_1)/\gamma ; \ \beta_1 = -3 a_1^3 a_2^3 \sigma (K_2 - K_1)/\gamma ; \tag{123}$$

$$\gamma = 12 a_2^3 G_1 (K_2 - K_1) + 3 a_1^3 K_1 (3K_2 + 4G_1);$$

$$K_0 = \frac{K_2(1+2V_2) + 2K_1(1-V_2)(1-2v_1)/(1+v_1)}{(1-V_2)K_2/K_1 + 2(1-2v_1)/(1+v_1) + V_2},$$
(124)

From here, the bulk modulus of the composite becomes infinite at negative bulk modulus of inclusion determined as

$$K_2 = -\frac{K_1}{(1-V_2)}\left[\frac{2(1-2v_1)}{(1+v_1)} + V_2\right].$$
(125)

When $\gamma \to 0$, the stress-strain state and displacement field become infinite. Due to Equation (123) it takes place at the next relation between moduli:

$$K_2 = -\frac{2(1-V_2)(1-2v_1)}{[2V_2(1-2v_1)+1+v_1]}K_1.$$
(126)

(2) *Homogeneous radial displacement applied to the boundary of spherical matrix $u_r(a_1)$* = u. Other conditions coincide with conditions (i) and (ii) of first problem. In this case the bulk modulus of the composite is also calculated from (124), becomes infinite at condition (125), and the unknown constants have the forms:

$$\alpha_1 = a_1^2 u(3K_2 + 4G_1)/\delta ; \ \beta_1 = -3a_1^2 a_2^3 u(K_2 - K_1)/\delta ;$$
(127)

$$\delta = -3a_2^3(K_2 - K_1) + a_1^3(3K_2 + 4G_1).$$

The relation for moduli is obtained from condition $\delta = 0$ as

$$K_2 = -\frac{K_1}{(1-V_2)}\left[\frac{2(1-2v_1)}{(1+v_1)} + V_2\right].$$
(128)

One coincides with relation (125), defining the condition of infinity for bulk modulus of the composite. So, in intrinsic more stable the boundary-value problem into displacements, the extremely great values of effective bulk modulus may be achieved without condition that obtained elastic fields become singular. Moreover, relation (124) shows that into conditions of given displacements, it may be obtained maximally great negative bulk modulus of the composite at corresponding selection of negative values of the inclusion bulk modulus. In order to achieve maximal positive values of the composite bulk modulus (this first of all is expedient from practical view), it may be shown [74], that inclusion bulk modulus K_2 should be some more negative compared with bulk modulus K_0 of the composite from Equation (124). However, the determined linear-elastic equation states that all elastic fields become infinite in the composite sphere already at the value of bulk modulus achieving the value of the composite bulk modulus K_0.

By considering the same problem into framework formulating finite strains with using constitutive equations presenting generalization to 3D statement of problem [109] of the constitutive equations for 'harmonic material' (firstly considered in [60]), it may be proved [74] that in difference from the linear-elastic case, the fields calculated into framework the finite strain conditions will always remain finite in the whole composite. This result exists even in the case when values of modulus K_2 achieve critical negative value which defines switch of effective volume modulus of the composite from very great negative values to very big positive ones.

This proposes that in actual elastic material in which could be expected existence some non-linearity in stress-strain state, the value of K_2 may be decreased below critical negative value predicted by linear-elastic analysis without forming singular behavior of the composite. In this case the composite will possess great positive bulk modulus at infinity-small deformations. Similar result occurs for the problem with given boundary stresses.

For 3D composites subjected to Hashin – Shtrikman formulation [51] and satisfactory for known microstructures for which the exact analytical solutions are fulfilled into framework of elastic theory, we obtain

$$G_0 = G_1 + \frac{V_2}{1/(G_2 - G_1) + 6(K_1 + 2G_1)V_1/[5(3K_1 + 4G_1)G_1]}; \tag{129}$$

$$K_0 = K_1 + \frac{V_2(K_2 - K_1)(3K_1 + 4G_1)}{(3K_1 + 4G_1) + 3V_1(K_2 - K_1)}. \tag{130}$$

Exact solution (129) for shear modulus is achieved into framework of hierarchical layer morphology [35, 98]. At the same time, formula (130) for bulk modulus is exact solution for hierarchical microstructure of covered spheres with various sizes for which ratio of radii of external layer and coverage has proper value for all inclusions [44]. The both exact solutions remain to be satisfactory also for negative stiffness of components. As it is known, in the case of rigid inclusions a small their concentration increases weakly the composite stiffness. However, if the inclusions have proper negative stiffness then the stiffness of that composite rises arbitrary aspiring to infinity.

Then, it may be shown that the variation principle providing estimations for all elastic moduli of arbitrary composites remains to be satisfactory in the case when one of phases of the composite has negative stiffness. The estimations obtained on the base of the variation principle also confirm that the composite stiffness may be made maximally great at corresponding selection of moduli of the phase with negative stiffness. In the variation approach developed in References [49, 50, 146], it has been introduced homogeneous 'comparison body' with moduli independing on coordinates. The stated variation principle satisfied even in the case when phases of the composite had negative values of stiffness until tensors of elastic moduli of actual composite and comparative one had usual symmetry of indices and in condition that the tensor of comparative moduli was selected so that for this solid existed some Green's function. In Reference [147] has been stated tensor of effective moduli for arbitrary composite with isotropic distribution of phases. The tensor presents variation estimation which is satisfactory for composites possessing phase with negative stiffness. In the case of two-phase composite consisting of isotropic matrix containing arbitrary distribution of isotropic inclusions of arbitrary forms, the variation estimations for bulk and shear moduli of the composite are determined in the forms [74]:

$$K_0 = \frac{4G(V_1 K_1 + V_2 K_2) + 3K_1 K_2}{4G + 3(V_1 K_2 + V_2 K_1)};$$ (131)

$$G_0 = \frac{G(9K + 8G)(V_1 G_1 + V_2 G_2) + 6(K + 2G)G_1 G_2}{G(9K + 8G) + 6(K + 2G)(V_1 G_2 + V_2 G_1)},$$ (132)

where the parameters without lower indices present corresponding comparative moduli. Note, that for composites with positive-definite phases the standard Hashin – Shtrikman lower (upper) bound is defined from relations (131) and (132) by selecting the comparative moduli to be equal minimal (maximal) values of the component moduli.

By considering a composite consisting of positive-definite matrix containing inclusions with negative stiffness the obvious selection for the comparative moduli is to select the matrix moduli. In this case Equations (131) and (132) show that the variation estimations of bulk and shear moduli can become arbitrary great at proper (negative value) selection of inclusion moduli achieving infinite values at

$$K_2 = -\frac{4G_1 + 3V_2 K_1}{3V_1}; \ G_2 = -G_1 \frac{3(3 + 2V_2)K_1 + 4(2 + 3V_2)G_1}{6(K_1 + 2G_1)V_1},$$ (133)

Note, that these results suggest that, if search for material with maximal bulk modulus then inclusions should be subjected to the condition of strong ellipticity. In this case the bulk modulus of inclusions should has value near to that which is defined by Equation (131), and shear modulus should be selected positive and sufficiently great in order to carry out the condition of strong ellipticity ($K_2 > -4G_2/3$, that will be demonstrated below).

As it easily saw the above-obtained formulae (124), (130), (131) are identical taking into account that the comparative modulus in the last formula is equal to corresponding matrix modulus. This conclusion fulfills also for formulae (129) and (132), again if the comparative moduli in the last formula are equal to corresponding matrix moduli. Thus, these formulae describe the same behavior of effective moduli in different situations. Therefore, the conclusion on design of the composites with high stiffness possessing phase with proper negative stiffness is applied to all considered cases with in fact quantitatively the same selection of negative moduli obtained in various formulae.

The negative stiffness leads to violation of usual dependence between force and displacement in deformed solid because in the last case the direction of force applied to the solid coincides with the strain direction by assuming that the restoration force will restore neutral state of the deformed solid. The negative stiffness defines unstable equilibrium and hence a positive supply of energy at equilibrium. The negative stiffness differs from negative

Poisson's ratio[5]. For isotropic materials Poisson's ratio changing into range from −1 up to 0.5 associates with stability of the material, at the same time, bulk materials with negative stiffness are unstable.

The examples of negative stiffness are characterized by proper structures [9]. In particular, the sequence of elements constrained by S-shaped configuration losing stability under loading relates to that structure (see Figure 28a) [74]. A proper longitudinal loading of that structure should lead to snapping this element. The condition of negative stiffness is unstable, but all system could be stabilized at a longitudinal constraint, for example due to its unification in rigid block. The negative stiffness forms also in single-cell model foam materials. For example, flexible tetrakaidecahedral models demonstrate non-monotonous dependence of strain on applied force under compression [121]. In this case, the cells bulge inwardly during high compressive deformation causing geometrical non-linearity. The negative stiffness forms also in cellular structure shown in Figure 28b [74], consisting of rigidly rotatable nodes connected by pre-deformed springs. At proper pre-deforming effective shear modulus of that cell becomes negative. That 2D lattice structure can demonstrate negative Poisson's ratio [71] and even negative shear modulus in condition of sufficient pre-straining.

Instability of materials including components with negative stiffness may be understood following way. If influence by using small force on block of materials possessing positive stiffness then the block will resist to strain due to initiation of opposite force. At the same

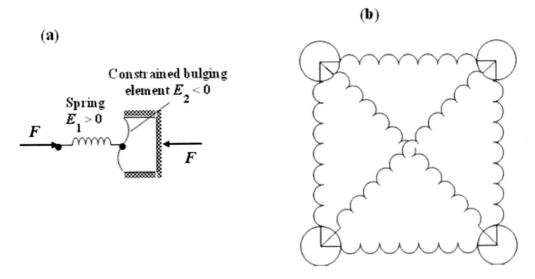

Figure 28. (a) Element with negative stiffness (effective modulus E_2) presented by constrained bulging element consecutively disposed with spring possessing positive stiffness (effective modulus E_1) in Reuss model; (b) cell consisting of rigid-rotatable nodes with pre-strained link springs. At sufficient pre-deformation shear modulus of the cell is negative.

---

[5]For example, the foams which expand in transverse cross-section when they are subjected to tension relates to the materials with negative Poisson's ratio. Negative Poisson's ratio is demonstrated by also several mono-crystals and layered structures [72]. Hierarchical layered structures [137] can demomstrate values of Poisson's ratio achieving −1 [100]. The foam materials with negative Poisson's ratio possessing non-convex cellular structure have been already designed and processed [70]. Moreover, there have been developed and prepared similar 2D structures [78, 148]. The value of Poisson's ratio which is equal to −1 could be achieved in chiral cellular structure [118].

time, in the case of negative stiffness the block carries the force in direction of the influence forming divergent unstable state.

Consider infinite-small homogeneous isotropic linear-elastic deformations from non-constrained state. The strain energy is positive-definite if and only if shear modulus G and Poisson's ratio ν satisfy to the conditions [130]:

$$G > 0; -1 < \nu < 0.5. \tag{134}$$

The materials subjected to conditions (134) lead to unique solutions for mixed boundary-value problems in which is found any proper combinations of surface stresses and displacements. It has been shown [54] that the uniqueness is the necessary condition for incremented stability, and that materials lead to stable solutions for broad class of mixed boundary-value problems. Based on this, it may be assumed a possibility to process materials with negative Poisson's ratio which should be stable in absence of external constraints.

For boundary-value problems under given displacements the uniqueness of solution [26] defines incremented stability [80] at strong ellipticity of elastic moduli:

$$G > 0; -\infty < \nu < 0.5 \text{ or } 1 < \nu < \infty. \tag{135}$$

These conditions are lesser constraining compared with conditions (134). By using Lame elastic constant λ conditions of strong ellipticity (135) are determined as

$$G > 0; \lambda + 2G > 0. \tag{136}$$

Then for bulk modulus $K = \dfrac{2G(1+\nu)}{3(1-2\nu)} = \lambda + 2G/3$ the condition of strong ellipticity takes the form: $-4G/3 < K < \infty$. These inequalities assume that the condition of strong ellipticity allows bulk modulus to be negative. In violation of strong ellipticity a material can demonstrate instability connected with formation of bands of the heterogeneous deformations [68].

Achievement of extremely large shear modulus of composite requires negative shear modulus of inclusions which does not violate the condition of strong ellipticity in the inclusions. However, violation of the strong ellipticity does not guarantee a stability loss of inclusions. The experiments show than in superconductive ferroelastics [123] and ferroelectrics [22, 79] expenditure of energy due to formation band (or domain) structure suppress the process forming that structure when the material particles to be sufficiently small. Thus, the instability criterion based only on theory of elasticity may predict instability in regimes its actual absence. Hence, negative stiffness is not excluded at all by physical laws. The solids with negative stiffness are unstable if they have free surfaces. They can be stabilized if they are constrained rigidly or by matrix of elastic composite. At the same time solid medium with negative volume modulus will stable under deformation constraints. Negative shear modulus of solid medium can form domain (band) instability associated with loss of strong ellipticity but that instability does not form always when violates strong ellipticity.

The composites with inclusions of negative stiffness may be used by investigating properties of single domains of the superconductive ferroelastics and ferroelectrics. In this case no need that that inclusions to be temperature-sensitive superconductive ferroeleastics because pre-stressed or pre-strained elements also may be used for this aim.

Thus, sharp (anomalous) changes of composite properties could be predicted in the case when inclusions have negative modulus close in value to the matrix modulus. Similar anomalies, in particular have place in hierarchical structures of Hashin – Shtrikman type and in various composites of 'matrix-inclusion' type [137]. The anomalies assume singular behavior of composites with linear-elastic components and also containing components in which a non-linearity is determined by geometric shape of the inclusions. If matrix demonstrates non-linearity in constitutive equations then anomaly of the composite modulus becomes finite. If components are linear-viscoelastic [73] then the anomaly in behavior of the composite modulus also to be finite and accompanied by high peak in viscoelastic damping $\tan\delta$.

The present analysis into framework theory of elasticity does not allow one to divide effects of various instability types in the composites possessing phase with negative stiffness. For example, in relation to the band instability it is known from theory of elasticity that in violation of strong ellipticity a formation of the band structure is predicted based on classic theory of elasticity. However, as it has been pointed above the tests show that this band instability suppresses in particles of sufficiently small sizes due to energetic conditions of domain formation. Thus, there is necessary additional physical basis outside frameworks theory of elasticity (for example, connected with experimental measurement of energy expenditure) to theoretically predict possible regimes of that instability.

Extreme elastic and viscoelstic behaviors of composites (elastic moduli and viscoelastic damping) possessing the component with negative stiffness have been demonstrated for composites with dispersed ferroelastic inclusions in quasi-static regime at frequencies in one order of magnitude lower than natural frequency of oscillations [75]. In the case of small concentration of inclusions there were observed extreme values of mechanical damping $\tan\delta$ and stiffness of composite (1% by volume of vanadium dioxide inclusions into tin matrix) in temperatures near of phase transition of the ferroelastic inclusions far higher than corresponding values for each component [75]. However, theory predicts in this case for particles with infinite stiffness, the composite would be only 2.1% stiffer than the matrix. If the particle stiffness aspires to zero then the composite would soften by 1.9% and have no change in $\tan\delta$.

ACKNOWLEDGMENTS

This research was supported in part by the Russian Foundation for Basic Research (Grants Nos. 10-08-00136, 10-08-13300) and National Science Council of Taiwan under contract NSC 99-2923-E-022-001-MY3.

REFERENCES

[1] Akopyan, V. A.; Soloviev, A. N.; Parinov, I. A.; Shevtsov S. N. *Definition of Constants for Piezoceramic Materials*; Nova Science: New York, 2010; pp 1 – 205.

[2] Andrianov, I. V.; Awrejcewicz, J.; Manevitch, L. I. *Asymptotical Mechanics of Thin-Walled Structures: a Handbook.*; Springer-Verlag: Berlin: Heidelberg, 2004; pp 1 – 535.

[3] Andrianov, I. V.; Danishevs'kyy, V. V.; Tokarzewski, S. *Int. J. Heat Mass Transfer*, 1996, *vol. 39*, 2349 – 2352.

[4] Andrianov, I. V.; Danishevs'kyy, V. V.; Tokarzewski, S. *Acta Appl. Math.* 2000, *vol. 61*, 29 – 35.

[5] Baker, G. A.; Graves-Morris, P. *Pade Approximants.* Two Volume Set; Addison-Wesley Publishing Company: New York, 1981; pp 1 – 572.

[6] Bakhvalov, N. S.; Panasenko, G. *Homogenisation: Averaging Processes in Periodic Media. Mathematical Problems in the Mechanics of Composite Materials*; Kluwer Academic Publishers: New York, 1989; pp. 1 – 408.

[7] Batchelor, G. K. *J. Fluid Mech.*, 1972, *vol. 52*, 245 – 268.

[8] Batchelor, G. K.; O'Brien, R. W. *Proc Roy. Soc. Lond. A*, 1977, *vol. 355*, 313 – 333.

[9] Bazant, Z.; Cedolin, L. *Stability of Structures: Elastic, Inelastic, Fracture and Damage Theories*; Oxford University Press: Oxford, 2003; pp. 1 – 1011.

[10] Berdichevski, V. L. *Variational Principles of Continuum Mechanics*; Nauka: Moscow; 1983; pp. 1 – 447.

[11] Bergman, D. J. *Phys. Rep. C.* 1978, *vol. 43*, 377 – 407.

[12] Berryman, J. G. *J. Acoust. Soc. Am.* 1980, *vol. 68*, 1809 – 1819.

[13] Berryman, J. G.; Berge, P. A. *Mech. Mater.*, 1996, *vol. 22*, 149 – 164.

[14] Berryman, J. G.; Milton, G. W. *J. Phys. D: Appl. Phys.* 1988, *vol. 21*, 87 – 94.

[15] *Bogoliubov, N. N.; Mitropolski, Yu. A. Asymptotic Methods in the Theory of Nonlinear Oscillations*; Gordon and Breach: New York, 1981; pp. 1 – 504.

[16] Bol'shakov, V. I.; Andrianov, I. V.; Danishev'skyy, V. V. *Asymptotical Methods for Calculation of Composite Materials Taking into Account Internal Structure*; Porogi: Dneropetrovsk; 2008; pp. 1 – 196.

[17] Boucher, S. *J. Composite Mater.* 1974, *vol. 8*, 82 – 89.

[18] Brown, W. F. *J. Chem. Phys.* 1955, *vol. 23*, 1514 – 1517.

[19] Bruggeman, D. *Ann. Physik (Liepzig).* 1935, *vol. 24*, 636 – 679.

[20] Bruno, O. P. *Proc R. Soc. Lond. A.* 1991, *vol. 433*, 353 – 381.

[21] Budiansky, B. *J. Mech. Phys. Solids.* 1965, *vol. 13*, 223 – 227.

[22] Cady, W. G. *Piezoelectricity*; Dover: New York, 1964; pp. 1 – 822.

[23] Cheng, H. W.; Greengard, L. *J. Comput. Phys.* 1997, *vol. 136*, 629 – 639.

[24] Christensen, R. M.; Lo K. H. *J. Mech. Phys. Solids.* 1979, *vol. 27*, 315 – 330.

[25] *Comprehensive Composite Materials.* Six Volume Set; A. Kelly, C. Zweben; Eds.; Pergamon Press: Oxford, 2000.

[26] Cosserat, E. F. *C. R. Acad. Sci. Paris.* 1898, *vol. 126*, 1089 – 1091.

[27] De Gennes, P. G. *J. Phys. Lett.* (Paris). 1976, *vol. 37L*, 1 – 14.

[28] Dykhne, A. M. *Soviet Phys. JETP*, 1971, *vol. 32*, 63 – 65.

[29] Efros, A. L. *Physics and Geometry of Disorder: Percolation Theory*; Mir: Moscow, 1987; pp. 1 – 259.

[30] Eischen, J. W.; Torquato S. *J. Appl. Phys.* 1993, *vol. 74*, 159 – 170.

[31] Espinosa, H. D.; Dwivedi, S. K.; Lu H.-C. *Comp. Meth. Appl. Mech. Eng.* 2000, *vol. 183*, 259 – 290.

[32] Falk, F. *Acta Metall.* 1980, *vol. 28*, 1773 – -1780.

[33] Falk, F. *Z. Phys. B.* 1983, *vol. 51*, 177 – 185.

[34] Feng, S.; Halperin, B. I.; Sen, P. N. *Phys. Rev. B.* 1987, *vol. 35*, 197 – 214.

[35] Francfort, G. A.; Murat, F. *Arch. Rational Mech. Analysis.* 1986, *vol. 94*, 307 – 334.

[36] Garnet, J. C. M. *Phil. Trans. Roy. Soc. Lond. A.* 1904, *vol. 203*, 385 – 420.

[37] Gibiansky, L. V.; Torquato, S. *Phys. Rev. Lett.* 1993, *vol. 71*, 2927 – 2930.

[38] Gibiansky, L. V.; Torquato, S. *J. Mech. Phys. Solids*, 1995, *vol. 43*, 1587 – 1613.

[39] Gibiansky, L. V.; Torquato, S. *Proc. Roy. Soc. Lond. A.* 1996, *vol. 452*, 253 – 283.

[40] Gibiansky, L. V.; Torquato, S. *Int. J. Eng. Sci.* 1996, *vol. 34*, 739 – 760.

[41] *Grigolyuk, E. I.; Fil'shtinski, L. A. Periodic Piecewise Homogeneous Elastic Structures*; Nauka: Moscow, 1992; pp. 1 – 288.

[42] Gubernatis, J. E.; Krumhansl, J. A. *J. Appl. Phys.* 1975, *vol. 46*, 1875 – 1883.

[43] *Guz', A. N.; Nemish, Yu. N. Perturbation Method of Boundary Shapes in Mechanics of Solid Media*; Vishcha Shkola: Kiev, 1989; pp. 1 – 352.

[44] Hashin, Z. *J. Appl. Mech.* 1962, *vol. 29*, 143 – 150.

[45] Hashin, Z. *J. Mech. Phys. Solids.* 1965, *vol. 13*, 119 – 134.

[46] Hashin, Z. *J. Appl. Phys.* 2001, *vol. 89*, 2261 – 2267.

[47] Hashin, Z.; Rosen, B. W. *J. Appl. Mech.*, 1964, *vol. 31*, 223 – 232.

[48] Hashin, Z.; Shtrikman, S. *J. Appl. Phys.* 1962, *vol. 33*, 1514 – 1517.

[49] Hashin, Z.; Shtrikman, S. *J. Mech. Phys. Solids.* 1962, *vol. 10*, 335 – 342.

[50] Hashin, Z.; Shtrikman, S. *J. Mech. Phys. Solids.* 1962, *vol. 10*, 343 – 352.

[51] Hashin, Z.; Shtrikman, S. *J. Mech. Phys. Solids.* 1963, *vol. 11*, 127 – 140.

[52] Helsing, J. *J. Stat. Phys.* 1998, *vol. 90*, 1461 – 1473.

[53] Hill, R. *Proc. Phys. Soc. Lond. A.* 1952, *vol. 65*, 349 – 355.

[54] Hill, R. *J. Mech. Phys. Solids.* 1957, *vol. 5*, 229 – 251.

[55] Hill, R. *J. Mech. Phys. Solids.* 1963, *vol. 11*, 357 – 372.

[56] Hill, R. *J. Mech. Phys. Solids.* 1965, *vol. 13*, 189 – 198.

[57] Hill, R. *J. Mech. Phys. Solids.* 1965, *vol. 13*, 213 – 222.

[58] Jeffrey, D. J. *Proc. Roy. Soc. Lond. A.* 1973, *vol. 335*, 355 – 367.

[59] Jeffrey, D. J. *Proc. Roy. Soc. Lond. A.* 1974, *vol. 338*, 503 – 516.

[60] John, F. *Commun. Pue Appl. Math.* 1960, *vol. 13*, 239 – 296.

[61] Kantor, Y.; Bergman D. J. *J. Mech. Phys. Solids.* 1984, *vol. 32*, 41 – 62.

[62] Keller, J. B. *J. Appl. Phys.* 1963, *vol. 34*, 991 – 993.

[63] Keller, J. B. *J. Math. Phys.* 1964, *vol. 5*, 548 – 549.

[64] Keller, J. B.; Rubenfeld, L.; Molyneux, J. *J. Fluid Mech.* 1967, *vol. 30*, 97 – 125.

[65] Kerner, E. H. *Proc. Phys. Soc. B.* 1956, *vol. 69*, 808 – 813.

[66] Kim, I. C.; Torquato, S. *J. Appl. Phys.* 1990, *vol. 68*, 3892 – 3903.

[67] Kim, I. C.; Torquato, S. *J. Appl. Phys.* 1991, *vol. 69*, 2280 – 2289.

[68] Knowles J. K., Sternberg E. *J. Elasticity*, 1978, *vol. 8*, 329 – 379.

[69] Kozlov, S. M. *Uspekhi Math. Nauk*, 1989, *vol. 44(2)*, 79 – 120.

[70] Lakes, R. S. *Science*, 1987, *vol. 235*, 1038 – 1040.

[71] Lakes, R. S. *J. Mater. Sci.* 1991, *vol. 26*, 2287 – -2292.

[72] Lakes, R. S. *Adv. Mater.* 1993, *vol. 5*, 293 – 296.

[73] Lakes, R. S. *Phys. Rev. Lett.* 2001, *vol. 86*, 2897 – 2900.

[74] Lakes, R. S.; Drugan, W. J. *J. Mech. Phys. Solids.* 2002, *vol. 50*, 979 – 1009.

[75] Lakes, R. S.; Lee, T.; Bersie, A.; Wang, Y. C. *Nature.* 2001, *vol. 410*, 565 – 567.

[76] Landauer, R. *J. Appl. Phys.*, 1952, *vol. 23*, 779 – 784.

[77] Landauer, R. In *Electrical, Transport and Optical Properties of Inhomogeneous Media*, J. C. Garland, D. B. Tanner; Eds.; AIP: New York, 1978, 2 – 43.

[78] Larsen, U. D.; Sigmund, O.; Bouwstra, S. *J. Vicroelectromech. Sys.* 1997, *vol. 6*, 99 – 106.

[79] Lines, M. E.; Glass, A. M. *Principles and Applications of Ferroelectrics and Related Materials.* Clarendon Press: Oxford, 1979; pp. 1 – 336.

[80] Lord Kelvin (Thomson, W.) *Philos. Mag.* 1888, *vol. 26*, 414 – 425.

[81] Lord Rayleigh. *Phil. Mag.* 1892, *vol. 34*, 481 – 502.

[82] Lurie, K. A.; Cherkaev, A. V. *Proc. R. Soc. Edinburgh.* 1984, *vol.* 99A, 71 – 87.

[83] Lurie, K. A.; Cherkaev, A. V. *Proc. R. Soc. Edinburgh.* 1986, *vol.* 104A, 21 – 38.

[84] *Marchenko, V. A.; Khruslov, E. Y.* Boundary-Value Problems in Regions with Fine-Grain Boundary; Naukova Dumka: Kiev; 1974; pp. 1 – 280.

[85] *Marchenko, V. A.; Khruslov, E. Y. Averaged Models of Microheterogeneous Media*; Naukova Dumka: Kiev, 2005; pp. 1 – 550.

[86] Maxwell, J. C. *Treatise on Electricity and Magnetism.* Clarendon Press: Oxford, 1873.

[87] Mendelson, K. S. *J. Appl. Phys.* 1975, *vol. 46*, 917 – 918.

[88] McHenry, K. D.; Koepke, B. G. In *Fracture Mechanics of Ceramics, vol. 5*; Bradt, R. C.; Evans, A. G.; Hasselman, D. P. H.; Lange, F. F.; Eds.; Plenum Press: New York, 1983; pp. 337 – 352.

[89] McKenzie, D. R.; McPhedran, R. C.; Derrick, G. H. *Proc. Roy. Soc. Lond. A.* 1978, *vol.* 362, 211 – 232.

[90] McLaughlin, R. *Int. J. Eng. Sci.* 1977, *vol.* 15, 237 – 244.

[91] McPhedran, R. C.; McKenzie, D. R. *Proc. Roy. Soc. Lond. A.* 1978, *vol.* 359, 45 – 63.

[92] McPhedran, R. C.; Milton, G. W. *Appl. Phys. A.* 1981, *vol.* 26, 207 – 220.

[93] *Mechanics of Composites.* Twelve Volume Set; Guz', A. N.; Ed.; Naukova Dumka: Kiev (*vols. 1-4*); ASK: Kiev (*vols. 5-12*), 1993 – 2003.

[94] Miller, C. A.; Torquato, S. *J. Appl. Phys.* 1990, *vol. 68*, 5486 – 5493.

[95] Milton, G. W. *Phys. Rev. Lett.* 1981, *vol. 46*, 542 – 545.

[96] Milton, G. W. *J. Mech. Phys. Solids.* 1982, *vol. 30*, 177 – 191.

[97] Milton, G. W. In *Physics and Chemistry of Porous Media*; Johnson, D. L.; Sen, P. N.; Eds.; AIP: New York, 1984; pp. 66 – 77.

[98] Milton, G. W. In *Homogenezation and Effective Moduli of Materials and Media*; Ericksen, J. L.; Kinderlehrer, D.; Kohn, R.; Lions, J. L.; Eds.; Springer-Verlag: New York, 1986; pp. 150 – 175.

[99] Milton, G. W. *Commun. Pure Appl. Math.*, 1990, *vol. 43*, 63 – 125.

[100] Milton, G. W. *J. Mech. Phys. Solids*, 1992, *vol. 40*, 1105 – 1137.

[101] Milton, G. W. *The Theory of Composites.* Cambridge University Press: Cambridge, 2002; pp. 1 – 719.

[102] Milton, G. W., Phan-Thien N. *Proc. Roy. Soc. Lond. A*, 1982, *vol. 380*, 305 – 331.

[103] *Mitropolski, Yu. A. Averaging Method in Nonlinear Mechanics*; Naukova Dumka: Kiev, 1971; pp. 1 – 440.

[104] Molyneux, J. E. *J. Math. Phys.* 1970, *vol. 11*, 1172 – 1184.

[105] Mori, T., Tanaka K. *Acta Metall.* 1973, *vol. 21*, 571 – 574.

[106] Needleman, A. *J. Mech. Phys. Solids.* 1990, *vol. 38*, 289 – 324.

[107] Norris, A. N. *Mech. Mat.* 1985, *vol. 4*, 1 – 16.

[108] Numan, K. C.; Keller, J. B. *J. Mech. Phys. Solids.* 1984, *vol. 32*, 259 – 280.

[109] Ogden, R. W. *Non-Linear Elastic Deformation.* 1997, Dover: New York; pp. 1 – 544.

[110] Parinov, I. A. *Microstructure and Properties of High-Temperature Superconductors, Two-Volume Set*; Rostov State University Press: Rostov-on-Don, 2004; pp 1 – 783.

[111] Parinov, I. A. *Microstructure and Properties of High-Temperature Superconductors*; Springer-Verlag: Berlin, Heidelberg, New York, 2007; pp 1 – 583.

[112] Parinov, I. A. *Superconductors and Superconductivity, Three-Volume Set: Vol. 2 Theory and Properties*; Southern Federal University Press: Rostov-on-Don, 2008; pp 1 – 924.

[113] Perrins, W. T.; McKenzie, D. R.; McPhedran, R. C. *Proc. Roy. Soc. Lond. A*, 1979, *vol. 369*, 207 – 225.

[114] Peterson, J. M., Hermans J. J. *J. Composite Mater.* 1969, *vol. 3*, 338 – 354.

[115] *Piezoelectric Materials and Devices*; Parinov, I. A.; Ed.; Nova Science Publishers: New York, 2011; pp 1 – 325.

[116] *Piezoceramic Materials and Devices*; I. A. Parinov; Ed.; Nova Science Publishers: New York, 2010; pp 1 – 335.

[117] *Pobedrya, B. E. Mechanics of Composite Materials;* Moscow State University Press: Moscow, 1984; pp. 1 – 336.

[118] Prall, D.; Lakes, R. S. *Int. J. Mech. Sci.* 1997, *vol. 39*, 305 – 314.

[119] Reddy, E. S.; Schmitz, G. J. *Supercond. Sci. Technol.* 2002, *vol. 15*, L21

[120] Reuss, A. *Z. Angew. Math. Mech.* 1929, *vol. 9*, 49 – 58.

[121] Rosakis, P.; Ruina, A.; Lakes, R. S. *J. Mater. Sci.* 1993, *vol. 28*, 4667 – 4672.

[122] Rubenfeld, L. A.; Keller, J. B. *SIAM J. Appl. Math.* 1969, *vol. 17*, 495 – 510.

[123] Salje, E. *Phase Transitions in Ferroelastic and Co-elastic Crystals*; Cambridge University Press: Cambridge, 1990; pp 1 –276.

[124] Sanchez-Palencia, E. *Non-homogenous Media and Vibration Theory*; *Springer*-Verlag: Berlin, New York, 1980; pp. 1 – 398.

[125] Sheng, P.; Kohn, R. V. *Phys. Rev. B.* 1982, *vol. 26*, 1331 – 1335.

[126] Shermergor, T. D. *Elasticity Theory of Macroheterogeneous Media*; Nauka: Moscow, 1977. pp. 1 – 400.

[127] Shklovskii, B. I.; Efros, A. L.; Luryi, S. *Electronic Properties of Doped Semiconductors*; Nauka: Moscow; 1979; pp. 1 – 406.

[128] Slepyan, L. I.; Yakovlev, Yu. S.; *Integral Transformations in Non-stationary Problems of Mechanics*; Sudostroenie: Leningrad; 1980; pp. 1 – 344.

[129] Telega, J. J.; Tokarzewski, S.; Galka, A. *Acta Appl. Math.* 2000, *vol. 61*, 295 – 315.

[130] Timoshenko, S. P.; Goodier, J. N. *Theory of Elasticity*; McGrow-Hill: New York, 1970; pp. 1 – 438.

[131] Tokarzewski, S.; Blawzdziewicz, J.; Andrianov, I. In *Advances in Structured and Heterogeneous Continua;* Alterton Press: New York, 1994; pp. 263 – 267.

[132] Torquato, S. *J. Appl. Phys.* 1985, *vol. 58*, 3790 – 3797.

[133] Torquato, S. *J. Chem. Phys.* 1985, *vol. 83*, 4776 – 4785.

[134] Torquato, S. *Appl. Mech. Rev.* 1991, *vol. 44*, 37 – 76.

[135] Torquato, S. In *Macroscopic Behavior of Heterogeneous Materials from the Microstructure*; Torquato, S.; Krajcinovic, D.; Eds.; *vol. 147*. Am. Soc. Mech. Eng. – AMD, 1992; pp. 53 – 65.

[136] Torquato, S. *J. Mech. Phys. Solids*, 1997, *vol. 45*, 1421 – 1448.

[137] Torquato, S. *Random Heterogeneous Materials. Microstructure and Macroscopic Properties*; Springer-Verlag: New York, 2002; pp. 1 – 701.

[138] Torquato, S.; Beasley J. D. *Int. J. Eng. Sci.* 1986, *vol. 24*, 415 – 433.

[139] Torquato, S.,; Rubinstein, J. *J. Appl. Phys.* 1991, *vol. 69*, 7118 – 7125.

[140] Tvergaard, V. *Mater. Sci. Eng. A*, 1995, *vol. 90*, 215 – 222.

[141] Van der Pol, C. *Rheol. Acta*, 1958, *vol. 1*, 198 – 205.

[142] Van Fo Fy, G. A. *Theory of Reinforced Materials with Coverage*; Naukova Dumka: Kiev, 1971; pp. 1 – 231.

[143] Vanin, G. A. *Micromechanics of Composite Materials*; Naukova Dumka: Kiev, 1985; pp. 1 – 304.

[144] Voigt, W. *Annallen der Physik und Chemie.* 1889, *vol. 38*, 573 – 587.

[145] Weng, G. J. *Int. J. Eng. Sci.* 1984, *vol. 22*, 845 – 856.

[146] Willis, J. R. *J. Mech. Phys. Solids.* 1977, *vol. 25*, 185 – 202.

[147] Willis, J. R. In *Mechanics of Solids: The R. Hill Anniversary Volume*; Hopkins, H. G.; Sewell, M. J.; Eds.; Pergamon Press: Oxford, 1982; pp. 653 – 686.

[148] Xu, B.; Arias, F.; Brittain, S. T.; et al. *Adv. Mater.* 1999, *vol. 11*, 1186 – 1189.

[149] Yonezawa, F.; Cohen, M. H. *J. Appl. Phys.* 1983, *vol. 54*, 2895 – 2899.

In: Ferroelectrics and Superconductors
Editor: Ivan A. Parinov

ISBN: 978-1-61324-518-7
©2012 Nova Science Publishers, Inc.

Chapter 6

INVESTIGATION OF THE STRUCTURAL FEATURES OF HIGH TEMPERATURE SUPERCONDUCTING CRYSTALS BY THE CHANNELING METHOD AND RESONANT NUCLEAR REACTIONS

V. S. Malyshevsky[*]
Department of Physics, Southern Federal University, Rostov-on-Don, Russia

ABSTRACT

The special features of Rutherford's method of ion backscattering in combination with channeling effect in multicomponent crystals allow ones to obtain information on the crystal structure which cannot be obtained by other methods. In the present work, the structural features of the high- temperature superconducting crystals are discussed based on an analysis of the experimental data on the orientational dependence of the yield of the resonant nuclear reactions. The features of the oxygen sublattice in a crystal $YBa_2Cu_3O_7$ are investigated by channeling technique. It has been measured the angular dependence of resonant reaction yield of 3.055 MeV He^+ ions channeled from oxygen nuclei along $\langle 001 \rangle$ direction. The best agreement of the calculated angular dependence of the reaction yield with experimental data is achieved with the assumption that the oxygen sublattice is partly disordered. About 20 % of oxygen atoms are situated in the plane (110) randomly; the oxygen atoms in oxygen chains are displaced from crystal sites at ~ 0.3 Å in the (110) plane. From a comparison of the experimental yields with the calculated ones is found that the oxygen sublattice of La_2CuO_4 is partly disordered (about 15 – 20% of the oxygen atoms displaced from the crystal sites in the La-Cu-O rows). The thermal vibrational amplitude of the oxygen atoms in the plane perpendicular to the O-O rows is apparently less (0.07 Å) than derived from the Debye model (0.15 Å). The orientation dependences of the $^7Li(p,\alpha)^4He$ nuclear reaction yield from a single crystal of Nd_2CuO_4 containing Li atoms as impurities have been used to study the lattice location of Li. The comparison revealed that neither one nor two different Li sites together can give a satisfactory agreement. The best agreement has been achieved for three different Li-sites: (i) 15% of Li atoms are located in the centers of the octants of the unit cell, (ii) 50% of Li atoms occupy

---

[*]E-mail: vsmalyshevsky@sfedu.ru, 5, Zorge Street, 344090, Rostov-on-Don, Russia.

the centers of parallelepipeds in which three neighbor corners are occupied by Nd, Cu and O atoms, and (iii) 35% of Li atoms are situated inside the (001) Nd-Cu rows at distances of 1.873 and 1.45 Å from Nd and Cu, respectively. Comparing the theoretical angular dependences of the resonance elastic scattering He^+ yield from oxygen atoms calculated by a computer simulation with the experimental ones for $La_{2-x}Sr_xCuO_{4-\delta}$ and $Nd_2CuO_{4-\delta}$ single crystals, the static displacements of O atoms from their lattice sites perpendicular to the $\langle 001 \rangle$ direction were determined. The directions of the displacements are rotated by 90° related to the displacements of the neighboring oxygen atoms which form the O-O row along $\langle 001 \rangle$. For $Nd_2CuO_{4-\delta}$ the anisotropic thermal vibrations of the oxygen atoms were determined.

1. INTRODUCTION

The effect of the charged particles channeling for multicomponent crystals' research allows receiving unique information on distribution of impurity, radiation defects and displacement of atoms from crystal sites in a crystal lattice [1]. Moreover, the features of the channeling technique for multicomponent crystals allow receiving the information on structure of crystals inaccessible by other methods. For example, the studying of the $YBa_2Cu_3O_7$ crystal by the channeling method in [2] has shown the abnormal change of static (and/or dynamic) displacements of atoms from crystal sites near the temperature of transition to the high-T_c state. The shoulder method[2] of the Rutherford backscattering angular dependences has enabled defining the model of $YBa_2Cu_3O_7$ structure, containing static displacement of oxygen atoms [3].

Developed earlier the approach to describe the channeling effect of ions in multicomponent crystals in [4] allows calculating the basic channeling parameters for the crystal structure under the given experimental conditions. It has allowed receiving the detailed information in [5-8] on structural features of a crystal lattice of La_2CuO_4 and Nd_2CuO_4 in a small depth (down to 0.1 μm) by comparison of the calculated and experimental results for various crystallographic directions. In the present work the research of features of He^+ ions channeling effect in a crystal film of $YBa_2Cu_3O_7$ is carried out. Theoretical analysis of experimental data on the angular dependence of resonant elastic scattering yield He^+ from oxygen nuclei gives new data on structure features of oxygen sublattice.

2. THEORETICAL TREATMENT

The ion distribution, flux-peaking and angular yield curves for backscattering and for nuclear reaction have been calculated using the channeling model developed specially for single crystal of compounds in which the channels are formed by several different atomic rows [8]. The model is based on the continuum theory of channeling. The distribution $g(\mathbf{p}_\perp, z)$ is obtained from the solution of the diffusion equation:

---

[2]The "shoulders" are the regions of the close-impact processes angular dependence where the yield is higher than the random yield.

$$\frac{\partial g(\mathbf{p}_\perp, z)}{\partial z} = div \left[D(\mathbf{p}_\perp) grad_{\mathbf{p}_\perp} g(\mathbf{p}_\perp, z) \right]^-_-, \tag{1}$$

where $\mathbf{p}_\perp, z$ are the transverse momentum of channeling ions and crystal depth, respectively. $D(\mathbf{p}_\perp)$ is the diffusion coefficient due to scattering by electrons and lattice vibrations. The flux distribution in the transverse plane (x, y) and crystal depth z is calculated from the equation:

$$F(x, y, z) = \int_{S_1} \frac{S_0}{S(\mathbf{p}_\perp)} g(\mathbf{p}_\perp, z) d\mathbf{p}_\perp + 1 - \int_{S_2} g(\mathbf{p}_\perp, z) d\mathbf{p}_\perp, \tag{2}$$

where S_1 and S_2 are the integration regions, which can be determined by

$$U(x, y) - U_{\min} \leq \mathbf{p}_\perp^2 / 2M_1 \leq \mathbf{p}_c^2 / 2M_1,$$

and

$$\mathbf{p}_\perp^2 / 2M_1 \leq \mathbf{p}_c^2 / 2M_1,$$

respectively. Here $U(x, y)$ is the continuum potential, U_{min} is the minimum value of the potential, M_1 is the mass of incident ions, $\mathbf{p_c}$ is the critical momentum, $S(\mathbf{p}_\perp)$ is the square of the accessible area in the transverse plane, S_0 is the square of the transverse elementary cell. For instance the resonant elastic scattering yield of ions with oxygen atoms at the incident angle ψ_{in} can be expressed as

$$\chi(z, \psi_{in}) = n_r + \sum_i n_i F(x_i, y_i, z), \tag{3}$$

where $F(x_i, y_i, z)$ is the flux of ions in the transverse plane points (x_i, y_i), n_i is the relative concentration of oxygen atoms located in the i-position, n_r is the relative concentration of the random fraction oxygen atoms.

3. AXIAL CHANNELING OF IONS IN YBA2CU3O7 CRYSTAL

The elementary cell of $YBa_2Cu_3O_7$ crystal in the orthorhombic phase is shown in Figure 1. Parameters of a lattice are $a = 3.82$ Å, $b = 3.86$ Å, $c = 11.67$ Å. The channeling of ions along a direction $\langle 001 \rangle$ is caused by interaction of ions with atom chains of four kinds: (i) by chains consisting of Y and Ba atoms, (ii) by chains consisting of Cu and O(1) atoms, (iii) by oxygen chains consisting of O(3) and O(4) atoms, (iv) by oxygen chains consisting of O(2) atoms.

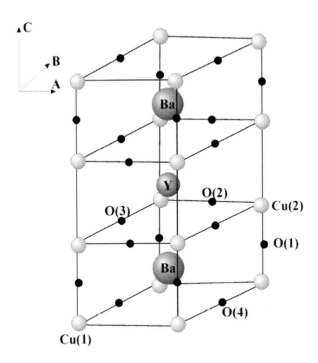

Figure 1. Structure of the YBa$_2$Cu$_3$O$_7$ crystal.

The calculated diffusion coefficient $D(\mathbf{p}_\perp)$ for ions He$^+$ with energy 3.055 MeV in channel $\langle 001 \rangle$ is shown in Figure 2. At small transverse momentum diffusion is defined by scattering with electrons and diffusion coefficient practically does not depend on a momentum. The increase of the diffusion coefficient at some value of a transverse momentum is corresponding to the contribution of scattering with atoms of different atomic chains. The first one is increased due to the scattering with atoms of oxygen in a "weak" oxygen chain O(2)-O(2). Ions with this momentum do not channeling along these oxygen chains and the accessible area covers the part of the channel, in which oxygen chains are located. So, the transverse momentum at the first step on the dependence $D(\mathbf{p}_\perp)$ is critical for a "weak" oxygen chain. The second step corresponds to a critical momentum, at which there is no channeling of ions along "strong" oxygen chains O(3)-O(4). The following features in behaviour of the diffusion coefficient are connected to scattering with mixed Cu-O(1) and Y-Ba atomic chains. At the transverse momentum larger than critical for Y-Ba atomic chains, all particles pass in a chaotic component of a beam. It is possible to say that this momentum is the critical for channeling along this crystallographic direction. The flux distribution (2) in the transverse plane for ions He$^+$ with energy 3.055 MeV for different incident angle in respect to the crystal axis $\langle 001 \rangle$ is shown in Figure 3. It should be noted that at the incident angle $\psi_{in} \neq 0$ and on the depths about (or more than) 0.1 μm the flux distribution is practically flat and is mainly formed by "heavy" atomic chains Y-Ba and Cu-O(1).

Experimental researches [3, 9] were carried out with the beam of ions He$^+$ with energy 3.055 MeV chosen so that the elastic resonant scattering of ions with oxygen atoms (^{16}O$(\alpha,\alpha)^{16}$O at $E = 3.045$ MeV) were occurred closely to the surface of a target (less than

0.05 μm). The crystal film of YBa$_2$Cu$_3$O$_7$ was put on a monocrystal substrate of Al$_2$O$_3$. The thickness of a film on the data of the Rutherford backscattering was approximately 0.4 μm. The value of the minimal yield of the Rutherford backscattering of ions by Ba atoms close to the surface did not exceed 0.06, that testified to sufficient perfection of crystal structure (Figure 4). The ions of energy spectra were measured at room temperature at various orientation of a beam in respect to the axis ⟨001⟩. In the spectra the energy range of elastic resonant scattering by oxygen atoms was selected.

The angular dependence of the elastic resonant scattering yield of He$^+$ ions from atoms of oxygen in respect to the axial direction ⟨001⟩ has "steps" at the incident angle in a crystal $\psi_{in} \approx \pm 0.4°$ (the smooth curves in Figure 4 were calculated according to the b-spline method). The absence of the similar step for barium dip confirms that this step is not

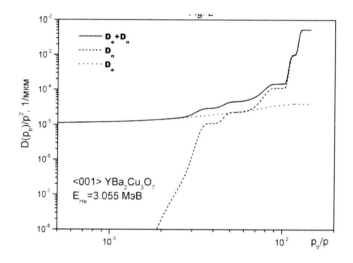

Figure 2. Diffusion coefficients of 3.055 MeV He$^+$ ions in the ⟨001⟩ channel of YBa$_2$Cu$_3$O$_7$ crystal

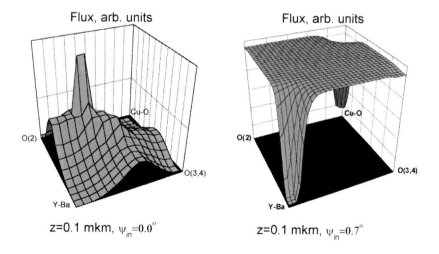

Figure 3. The flux distributions of ions He$^+$ with energy 3.055 MeV in the ⟨001⟩ channel of YBa$_2$Cu$_3$O$_7$ crystal for different incident angle in respect to the crystal axis ⟨001⟩.

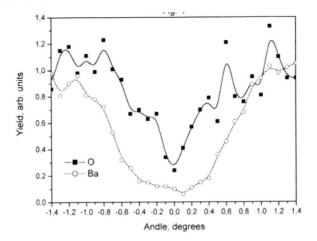

Figure 4. Orientation dependence of the relative yield of the Rutherford backscattering yield of He$^+$ ions by Ba atoms and elastic resonant scattering yield of ions with oxygen atoms in the ⟨001⟩ channel of YBa$_2$Cu$_3$O$_7$. The smooth curves were calculated according to the *b*-spline method.

connected with the influence of planes during angular scanning across ⟨001⟩ axes, and is connected with the oxygen sublattice structure. The similar form of angular dependence of the nuclear reaction yield of protons from atoms of oxygen in La$_2$CuO$_4$ crystal was discussed earlier in [6 – 8]. There was shown that position and width of the "step" are very sensitive to static or dynamic displacement of oxygen from crystal sites.

These "steps" in angular dependence of processes of close interactions can be explained as follows [4]. At small incident angle atomic chains of all four types form the flux distributions of channeling ions. At increase of the incident angle the transverse energy becomes more than potential barriers of "weak" O(3)-O(4) and O(2)-O(2) oxygen chains and the flux distribution are formed by more "strong" Y-Ba and Cu-O(1) atomic chains. The channeling ions in respect to the chains of Y-Ba and Cu-O(1) have the greater probability of scattering with atomic chains of O(3)-O(4) and O(2)-O(2).

Because the atoms of oxygen are located in three various chains, the angular dependence of processes of close interaction with atoms of oxygen is a superposition of dependences caused by interaction of ions with atoms of oxygen in chains of three types: Cu-O(1), O(3)-O(4) and O(2)-O(2). Such superposition with the weight factors of atoms of oxygen in each chain from its total one in a crystal defines the observable form of angular dependence of a yield of close interaction processes. Width of angular dependence and position of "steps" are defined by average potential of oxygen chains, position of oxygen atoms in a lattice and relative concentration of oxygen in various chains. As it was shown in [4] diffusion in transverse plane leads to the disappearance of "steps" on the depths 0.1 µm and more (see Figure 5 in Reference [4]). For this reason such "steps" were not observed in [10] because the resonant nuclear reaction of ions with energy 3.06 МэВ was observed from the greater depth than in [3, 9]. By comparing experimental angular dependence of the yield of close interaction process on atoms of oxygen and the calculated one for various amplitude thermal vibrations or static displacement from crystal sites it is possible to define their value.

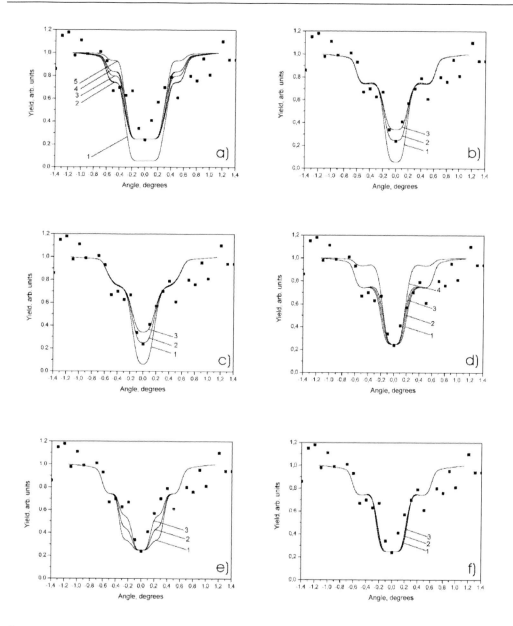

Figure 5. Calculated orientation dependence of the yield from oxygen atoms in the ⟨001⟩ channel of YBa$_2$Cu$_3$O$_7$: (a) Yield from the depth 0.01 μm; all atoms of oxygen are in a crystal sites; 1 – n_r = 0, 2 – n_r = 20 % in equal parts from chains O(3,4) and O(2), 3 – n_r = 20 % in equal parts from chains O(3,4), O(2) and Cu-O, 4 – n_r = 20 % in equal parts from chains O(3,4) and Cu-O or from chains O(2) and Cu-O, 5 – n_r = 20 % only from chains Cu-O. (b) Yield from the depth 0.01 μm; the atoms of oxygen from chains O(3,4) and O(2) are displaced from crystal sites in the (110) plane on distance 0.3 Å; 1 – n_r = 0, 2 – n_r = 20 % in equal parts from chains O(3,4) and O(2), 3 – n_r = 30 % in equal parts from chains O(3,4) and O(2). (c) Yield from the depth 0.05 μm; the atoms of oxygen from chains O(3,4) and O(2) are displaced from crystal sites in the (110) plane on distance 0.3 Å; 1 – n_r = 0, 2 –

$n_r = 20\%$ in equal parts from chains O(3,4) and O(2), $3 - n_r = 30\%$ in equal parts from chains O(3,4) and O(2). (d) Yield from the depth 0.01 μm, the atoms of oxygen from chains O(3,4) and O(2) are displaced from crystal sites in the (110) plane on distance 0.3 Å; $1 - n_r = 20\%$ only from chains O(3,4), $2 - n_r = 20\%$ in equal parts from chains O(3,4) and O(2), $3 - n_r = 20\%$ only from chains O(2), $4 - n_r = 20\%$ only from chains Cu-O. (e) Yield from the depth 0.01 μm, the atoms of oxygen only from chains O(2) are displaced from crystal sites in the (110) plane on distance 0.3 Å; $1 - n_r = 20\%$ only from chains O(2), $2 - n_r = 20\%$ in equal parts from chains O(3,4) and O(2), $3 - n_r = 20\%$ only from chains O(3,4). (f) Yield from the depth 0.01 μm; the atoms of oxygen only from chains O(3,4) are displaced from crystal sites in the (110) plane on distance 0.3 Å; $1 - n_r = 20\%$ only from chains O(2), $2 - n_r = 20\%$ in equal parts from chains O(3,4) and O(2), $3 - n_r = 20\%$ only from chains O(3,4).

The calculated angular dependences of the close-impact process yield along axial direction ⟨001⟩ from the depth 0.01 μm are shown in Figure 5, a with the assumption that all atoms of oxygen are situated in the lattice sites. The agreement of the calculated and experimental minimal yield is achieved if it is assumed that some part (no more than 20 %) of oxygen atoms is situated in random positions in plane (110). The calculated angular position and width of "steps" do not correspond to the experimental data. The increase of thermal vibration amplitude from 0.15 Å up to the abnormal large 0.40 Å reduces angular width of the dip, but it is not enough for the satisfactory agreement with experimental data. It is reasonable to assume the existence of static displacements of oxygen atoms. The value of displacements in the plane (110) (other orientations have not been tested in the present research) strongly influences on the angular width of close-impact processes, and in order to estimate it is necessary to achieve the best consent of the calculated and experimental data. In Figure 5, b the results of calculations with the assumption of such static displacements of oxygen atoms from crystal sites are shown. The calculations for the greater depth (Figure 5, c) show that diffusion processes lead to the easing of steps, and for depth more than 0.05 μm to their complete disappearance [4]. The satisfactory agreement with the experimental data is achieved with the assumption that the oxygen atoms from chains O(3)-O(4) and O(2)-O(2) are displaced from crystal sites at distance 0.3 Å in the (110) plane and the disordered part of oxygen is about 20 %. Thus the atoms of oxygen O(1) in chains Cu-O occupy equilibrium crystal sites.

The "steps" positions in the angular distribution correspond to experimental data (Figure 5, d) by assuming that the random part of oxygen in a crystal lattice is formed only by atoms from chains O(3)-O(4) and O(2)-O(2). Let's note, that the assumption that displaced atoms are only from chains O(3)-O(4) or only from chains O(2)-O(2) does not give such agreement with experimental results as the assumption of displacement of the atoms from both chains (see Figure 5, e, f). The reduction of the displacement value from 0.3 Å leads to increasing of the angular distribution width and the increasing leads to reduction one accordingly.

Comparison of the measured and calculated yield of elastic resonant ions of He^+ scattering with energy 3.055 MeV from oxygen atoms in $YBa_2Cu_3O_7$ crystal along the axial direction ⟨001⟩ leads to conclusion that oxygen atoms in chains O(3)-O(4) and O(2)-O(2) are displaced in the plane (110). The analysis of the angular distribution of the resonant elastic scattering yield features allows making a conclusion about presence of a random part of oxygen atoms. They occupy chaotic sites in the plane (110), and the random oxygen subsystem is formed by O(3)-O(4) and O(2)-O(2) chains. The value of the disordered oxygen is no more than 20 %. Let's notice, that the similar estimation for the disordered oxygen subsystem was discussed for the crystal La_2CuO_4, which also was investigated by a channeling technique [5].

Earlier in [3, 9] from the analysis of shoulder areas in angular dependence of the elastic resonant scattering yield of He^+ ions from atoms of oxygen the assumption on oxygen atoms displacements was made. Our results confirm and specify this conclusion. The maximal displacement is equal to 0.3 Å, and the vacancies arise in chains O(3)-O(4) and O(2)-O(2). The revealed properties of an oxygen subsystem of $YBa_2Cu_3O_7$ crystal are no consequence of the chosen channeling model and are connected with the specificity of crystal field of lattice.

Using for measurements single axial direction does not allow one to determine a direction of displacement of oxygen atoms in the projection to plane (110). It is possible to use the assumption about "two-by-two" displacements in one direction, and the directions of displacement change on opposite for the following pairs along chains $\langle 001 \rangle$. The similar model of displacement of the "goffered" type was discussed also in [3]. Such model of displacements of oxygen atoms and value of the displacements does not contradict to results of $YBa_2Cu_3O_7$ structure investigations with the X-ray methods and neutron diffraction. These methods are rather tolerant to local distortions of oxygen crystal sublattice structure and even the rather large static or dynamic displacements of oxygen atoms from crystal sites are hardly determined, if these displacements do not break of lattice symmetry.

4. PROTON-CHANNELING INVESTIGATION OF La_2CuO_4 CRYSTAL

The resonance nuclear reaction $O^{18}(p, \alpha)N^{15}$ yields (initial proton energy 0.65 MeV) in single crystals of $La_{2-x}Sr_xCuO_4$ have been obtained [11]. In this section we report the analysis of these experimental results. As can be seen (Figure 6), ions incident along $\langle 001 \rangle$ direction travel parallel to the four La-Cu-O rows and the one O-O row that construct the unit cell. The analysis of the channeling phenomenon in such a polyatomic crystal is complex, since each row contains different atoms and different interatomic spacing. Incident ions with transverse energy $\varepsilon_\perp > \varepsilon_\perp^{(1)}$ and $\varepsilon_\perp < \varepsilon_\perp^{(2)}$ can be expected to remain channeled with respect to the strong rows of La-Cu-O atoms, while being essentially not channeled by the weak rows of O-O atoms. The transverse energies $\varepsilon_\perp^{(1)}$ and $\varepsilon_\perp^{(2)}$ are the critical transverse energies with respect to the O-O rows and La-Cu-O rows respectively. Incident protons with transverse energy less than $\varepsilon_\perp^{(1)}$ can channeled with respect to the O-O rows and La-Cu-O rows. The value of $\varepsilon_\perp^{(1)}$ is very important and defines the peculiarity in the nuclear reaction dip at $0.4° - 0.8°$ (see Reference [11] and Figure 7).

The flux distribution (2) is calculated from the distribution in the transverse momentum which is derived by numerically solving the diffusion equation (1). The orientation dependences of the yield from oxygen atoms calculated according to the model developed are shown in Figure 7. In the calculations the contributions of protons

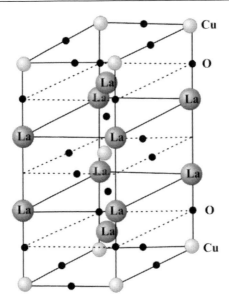

Figure 6. Structure of La_2CuO_4 crystal.

with energy less and more than the resonance one (0.63 MeV) are taken into account, so the contributions from crystal depths less and more than 0.15 – 0.20 µm give some increase of the calculated yield.

The calculated angular dependences show the characteristic features, i.e. a step at 0.4 - 0.8 degrees. These features are in best agreement with the experimental results when $\varepsilon_\perp^{(1)}$ is approximately 23 eV. This corresponds to the distance of closest approach to the O-O row $r_{min} = 0.24$ Å. By using the fundamental qualitative aspect of the Lindhard theoretical treatment [1] that $r_{min}^2 = a_{TF}^2 + u_\perp^2$ we have found the thermal vibration amplitude $u_\perp$ of oxygen atoms in O-O strings which is approximately equal to 0.07 Å. The similar estimation follows from the Barrett equation for the FWHM of the channeling dip based upon computer simulation studies of ion channeling [12].

Quite good agreement between the theoretical and experimental yields from oxygen atoms can be obtained by assuming that some part of the oxygen atoms was randomly distributed across the channel. As seen from Figure 7, d the best agreement was achieved when the random fraction of oxygen atoms displaced from the La-Cu-O strings was about 15 – 20%.

So, from a comparison of the experimental yields with the calculated ones is found that the oxygen sublattice is partly disordered (about 15 – 20% of the oxygen atoms are displaced from the crystal sites in the La-Cu-O rows). The thermal vibrational amplitude of the oxygen atoms in the plane perpendicular to the O-O rows is apparently less (0.07 Å) than derived from the Debye model (0.15 Å). We assume that this unusual behavior of the oxygen sublattice is not caused by the model assumptions, but is directly related to the specific crystal lattice fields within $La_{2-x}Sr_xCuO_4$, to which the strong electron-electron correlations within the Cu-O plane make the essential contribution. It is the electron-electron correlations that provide both the small thermal vibrational amplitude of oxygen atoms located in the O-O rows and the disorder of the oxygen atoms for the La-Cu-O rows.

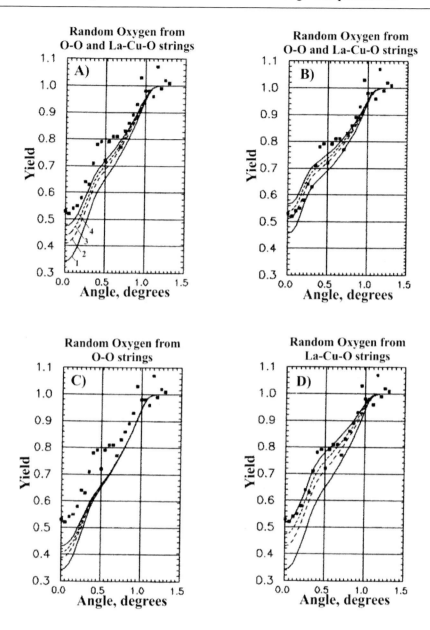

Figure 7. Calculated orientation dependences of the yield from oxygen atoms for different densities of the randomly distributed oxygen atoms: (1) 0%; (2) 10%; (3) 15%; (4) 20%; (A), (C), (D) $\varepsilon_\perp^{(1)} = 23$ eV, and (B) $\varepsilon_\perp^{(1)} = 15$ eV; (■) experiment [11].

5. LATTICE LOCATION OF LI IN ND$_2$CUO$_4$ CRYSTAL

It is well known that many physical properties of the high-T_c superconductors depend strongly on the atomic number and concentration of impurities [13]. For example, the temperature of the tetra-orto transition and low-temperature peculiarities change essentially

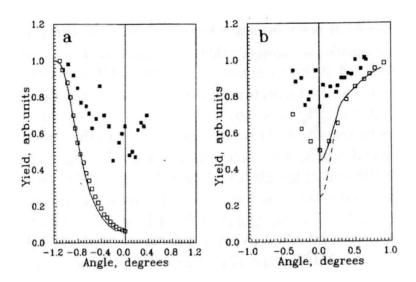

Figure 8. Angular scans across (a) (001) and (b) (111) axes in Nd$_2$CuO$_4$ single crystal containing Li [15]; E_P = 1.0 MeV. The open squares represent the backscattering yield from Nd atoms, the full squares the reaction yield from Li atoms. The solid lines represent calculations for Nd. The dashed line represents the calculations turning off the scattering by interstitial Li into the (111) channel.

with the concentration of Li and Cu [14]. So, it seems to be very interesting to determine the lattice positions of these impurities.

The orientation dependence of the yields for close encounter processes (backscattering, nuclear reaction, etc.) between light energetic ions and atoms of a single crystal has been widely used to obtain information about distribution of impurity atoms between substitutional and specific interstitial positions. Complete angular scans across the major crystal channeling directions and comparison between experiments and calculations offer the possibility to locate the lattice sites of impurity atoms more exactly. In the case of interstitial foreign atoms close to the center of a particular channel a peak or multiple peak is observed while for interstitial foreign atoms which lie within the forbidden "shadow region" of this channel the yield of close-impact processes is strongly reduced. A detailed knowledge about the ion distribution in the channel is essential for an interpretation of the results. For high-T_c superconductors the use of the channeling technique to locate impurities is complicated due to a number of experimental and theoretical problems (e.g. low concentrations and small atomic numbers of impurities, complex crystal structure, etc.).

The yield of the nuclear reaction ^{7}Li(p, α)^{4}He induced by protons with an initial energy E = 1 MeV has been measured at the Kharkov Institute of Physics and Technology [15] for high-quality single crystals of Nd$_2$CuO$_{4-\delta}$ containing Li (about 3%). Examples of the angular scans across the (001) and (111) axes are shown in Figure 8. When the ⟨001⟩ axis is aligned the proton backscattering yield is reduced to approximately 6% of the random yield, indicating a high-quality single crystal. The reaction yield is much higher and the angular dependence of the reaction yield reveals few small peaks superimposed on a dip. The similar angular-yield curves are measured for the (111) direction. A possible explanation of these results is the existence of several different positions of Li atoms. The angular dependencies indicate the some fraction of the Li atoms must occupy the interstitial sites close to the center of (001) channel, while the others locate within ⟨001⟩ and ⟨111⟩ atomic rows. In order to

Investigation of the Structural Features of High Temperature ... 219

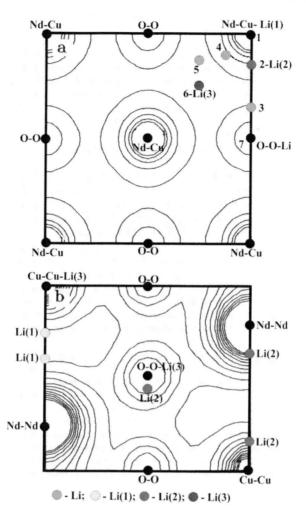

Figure 9. Equipotential contour plot for continuum Thomas-Fermi-Moliere potential in the (a) ⟨001⟩, and (b) ⟨111⟩ axial channels. The black circles represent the atomic rows; the color circles denoted by numbers 1 – 7 represent the various Li positions in the ⟨001⟩ channel.

determine the exact Li positions, it is necessary to know the ion flux distribution (2) as a function of the incidence angle and as a function of the crystal depth.

The axial channeling along ⟨111⟩ and ⟨001⟩ directions is formed by "mixed" and "pure" atomic rows. Along the ⟨001⟩ direction there are two different rows, namely: Nd-Cu atomic rows and O-O ones, while along ⟨111⟩ there are three different "pure" rows: Nd-Nd, Cu-Cu and O-O rows. The contours of the constant potential for the ⟨001⟩ and ⟨111⟩ channels are shown in Figure 9. Since the experimental widths of the nuclear reaction dips for ⟨001⟩ and ⟨111⟩ directions are approximately equal to those of the Nd dips (taking into account that the experimental Nd data were obtained from the surface region while the Li data were measured from a much deeper layer) we have to assume that some Li -atoms are located within the Nd-Cu rows. For the ⟨001⟩ channel there is a central maximum of the Li angle-yield curve, while for the ⟨111⟩ channel there is a double peak with a central minimum. Hence one can assume that some of the Li-atoms along ⟨001⟩ are located in the position where the flux distribution

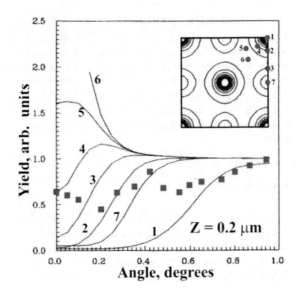

Figure.10. Variation of the relative reaction yield from depth of 0.2 μm below the surface of Nd_2CuO_4; numbers 1 – 7 have the same meaning as in Figure 9.

has a maximum (i.e. in the minima of the averaged potential), but along ⟨111⟩ no Li atoms are in the center of the channel.

Theoretical angle-yield curves for different locations of Li, indicated in Figure 9, a are shown in Figure 10. Calculations show that an agreement between the Li yield curves calculated for only two Li positions and experimental angular scans is impossible for all combinations of positions and concentrations. However, it is reasonable to assume that the agreement will be achieved if the Li atoms occupy three different positions and the theoretical Li curve is a sum of the yields from these Li positions with appropriated weighting coefficients. In this case the best fitting with the experimental results has been achieved when Li atoms are distributed in the plane orthogonal to the ⟨001⟩ direction in the following way: (i) 35% of Li atoms are situated inside Nd-Cu rows (Li(1)), (ii) 50% of Li atoms lie on the equipotential of 28 eV at a distance of about 0.73 Å from the Nd-Cu rows (Li(2)), and (iii) 15% of Li atoms are at the half of the distance between the neighboring Nd-Cu rows (Li(3)).

The results of calculations for Li(1), Li(2) and Li(3) in the ⟨001⟩ axial channel are shown in Figure 10 (curves 1, 2 and 6). The calculations for any four positions (or more) have not given good agreement with experimental results. The total sum of the theoretical angle-yield curves for the assumed Li positions is shown in Figure 11. The dotted line corresponds to the crystal surface (depth $z = 0$), the dashed line was calculated for $z = 0.4$ μm and the solid line was obtained as a sum of the yields from the all crystal (from $z = 0$ up to $z = 0.5$ μm). The summation was made with steps of 0.05 μm and the weighting coefficient chosen was proportional to the value of the $^7Li(p,\alpha)^4He$ cross-section for the proton energy at the given step. The yield from the crystal surface where the calculated peaks are stronger, gives maximum contribution to the angular-yield curve while contributions from the deeper layers (when the peaks are smoothed due to the change of the flux-peaking) are essentially less important.

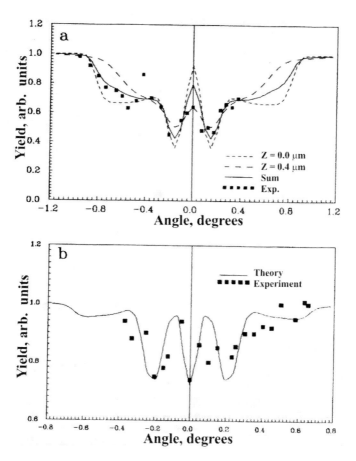

Figure.11. Calculation of summed angle-yield curves for Li atoms in the Li(1), Li(2) and Li(3) sites across (a) ⟨001⟩ and (b) ⟨111⟩ channel. Solid lines represent the total yield from 0 up to 0.5 μm, calculated taking into account the dependence of reaction cross-section on proton energy. Dashed lines refer to the surface (short dash) and the crystal depth of 0.4 μm (long dash).

Figure 9, b shows the possible Li positions on the plane normal to the ⟨111⟩ direction. From experimental data (Figure 8b) it is clear that some fraction of Li must be located inside ⟨111⟩ rows. The Li(1) atoms cannot be substitutional because in this case it is impossible to approach the agreement with the experimental data for the ⟨111⟩ direction (the fraction of the Li(1) obtained from the ⟨001⟩ data is too large for such agreement).

Therefore either Li(2) or Li(3) have to situate there. By considering all possible Li(1), Li(2) and Li(3) projections on the (111) plane and taking into account their fractions the best fitting has been obtained for the following case: (i) Li(1) atoms lie on the 42 eV equipotential at distances of 0.5 and 0.33 Å from the Nd-Nd and Cu-Cu rows, respectively; (ii) Li(2) atoms occupy the position near the center of the ⟨111⟩ channel at distances of 0.87 and 0.67Å from Nd-Nd and Cu-Cu rows, respectively; (iii) Li(3) atoms are situated inside the Cu-Cu or O-O rows. These positions are indicated in Figure 9b.

Now it is clear that the difference between the calculated and experimental Nd backscattering yield for the ⟨111⟩ direction is due to stronger scattering of the proton beam by impurities located near the center of the ⟨111⟩ channel (35% Li for the ⟨111⟩ while only 15% Li for the ⟨001⟩). This additional scattering leads to the higher value of the minimum Nd

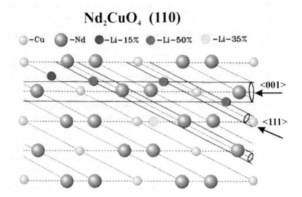

Figure.12. Cross section of the crystal structure along the Nd-Cu (110) plane. Cylinders around ⟨001⟩ and ⟨111⟩ axes are shown. The intersections of these cylinders determine the possible Li(2) positions; Li(1) and Li(3) positions are shown as well.

backscattering yield for the ⟨111⟩ axis. The angle-yields curve calculated with the modified diffusion coefficient [16] which takes into account the beam scattering by Li atoms is shown in Figure 8, b (solid line). It is obvious that in this case the agreement with experiments is much better. The theoretical angle-yield curve for the obtained Li positions in (111) direction is shown in Figure 11, b.

Thus Li(2) and Li(3) are located at the intersection lines of the cylinders around the ⟨001⟩ and the ⟨111⟩ rows (Figure 12). The radii of these cylinders are equal to the Li location distances from the appropriate rows. Hence, using some crystallographic considerations, one can conclude that the Li sites in the Nd_2CuO_4 structure are the following (Figure 13): (i) 15% of Li atoms are located in the centers of the octants of the unit cell; (ii) 50% of Li atoms occupy the centers of some parallelepipeds, one of which is shown in Figure 13; (iii) 35% of Li atoms can be located in two different positions (Li(1') and Li(1")) inside the ⟨001⟩ rows. Li(1') and Li(1") are situated at distances 1.873 and 1.45 Å from Nd and Cu, respectively.

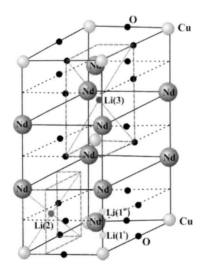

Figure.13. The unit cell of Nd_2CuO_4 single crystal. The Nd-Cu and O-O rows of atoms along the ⟨001⟩ direction are shown separately. For clarity only one of the physically equivalent Li sites is shown in each case.

6. THE OXYGEN DISPLACEMENTS IN $LA_{2-x}SR_xCUO_{4-\Delta}$ AND $ND_2CUO_{4-\Delta}$ CRYSTALS

In the previous paragraphs the common channeling approximations (averaged potential, critical energy etc.) have been used which allowed one to obtain information only about "averaged" distortions of the atomic rows or planes and gave no possibility to distinguish the kind of distortions (e.g. static displacements or uncorrelated thermal vibrations or correlated thermal vibrations of the atoms into row etc.). In References [17, 18] it has been demonstrated that the dip shoulders (i.e. the regions of the angular dependences of the small impact parameter events (SPE) where the yield is higher than normal) are very sensitive to the composition and the structure of atomic rows and planes. Furthermore, the shoulder angles ψ_S depend strongly on the kind of row distortions.

As the shoulders are formed by the ions which can move "inside" the rows or planes where the SPE probabilities are very high, it is not correct to use the continuous potential approximations. To calculate the shoulder parameters the most reasonable technique seems to be the computer simulation of the ion trajectories. Comparing the experimental angular dependences with the simulated ones it is possible to extract some quantitative information about the crystalline structure. Due to the rigorous consideration of the ion-atom interaction the computer simulation permits one to examine compound crystals with complex structure taking into account any kind of static or dynamic displacement of the atoms from the lattice sites.

The $\langle 001 \rangle$ angular dependences of the resonance elastic scattering yield (E_{He} = 3.045 MeV) from oxygen atoms were measured in [19] for small depths (< 50 nm) in $La_{2-x}Sr_xCuO_4$ and Nd_2CuO_4 monocrystals. For the first crystal two strong shoulders were obtained at ψ_S = 0.3° and ψ_S = 1.3° while for the second only one shoulder was measured as usual. Along $\langle 001 \rangle$ in La_2CuO_4 the oxygen atoms are located in "mixed" La-Cu-O rows and in "pure" O-O rows while in Nd_2CuO_4 oxygen atoms are located in O-O rows, only. The presence of two kinds of oxygen rows explains [5] the peculiarities of the measured dip for $La_{2-x}Sr_xCuO_4$. The maximum measured at the smaller ψ_S was due to the "pure" O-O rows.

The experimental results [19] are compared with data calculated according to the computer simulation model which was described in detail elsewhere [17]. The main features of this model are the following: (i) the elastic interactions of the incident ion with the lattice atoms are described as ordered sequences of binary collisions, (ii) the angular divergence of the incident beam and the ion energy losses are neglected, (iii) it is supposed that SPE occurs when the angle between the ion trajectory and the channeling direction is greater than $10\psi_L$ where ψ_L is the Lindhard angle, (iv) the thermal vibrations are simulated either as independent isotropic displacements or as independent anisotropic ones. For the last case it is assumed that the directions of the thermal displacements are uniformly distributed and the thermal amplitudes are direction dependent.

6.1. The Oxygen Displacements in La_2CuO_4

In the tetragonal phase of La_2CuO_4 single crystal the $\langle 001 \rangle$ channels are formed by "pure" O_I-O_{II} rows with atomic spacing d_0 = 6.58 Å and "mixed" La-Cu-O_{III} rows (Figure 14, a). The

SPE yields from the "pure" oxygen rows calculated with the various amplitudes of thermal vibrations u_{rms} are shown in Figure 15. The calculated maximum yield was obtained at $\psi_S = 0.44°$. The calculated value of ψ_S was 30% greater than the experimental one. When the isotropic thermal vibrations were taken into account in the calculations, the agreement with the experimental data did not improve. By calculating deeper ion penetrations into the crystal (> 50 nm) the discrepancy between calculation and experiment increased. The angular divergence of the incident ion beam smoothed the calculated angular dependences but the values of ψ_S did not decrease.

In order to provide an attractive fit to the experimental data one can suppose that the O_I and O_{II} atoms are displaced from their usual lattice sites to the neighboring Cu atoms (see Figure 14, a) along normal to $\langle 001 \rangle$ so that the angles between the directions of the O_I and O_{II} displacements are equal to 90°. In this case if the oxygen displacements $|\mathbf{b}|$ are large enough (for 3.06 MeV He ions and crystal depth < 50 nm, $|\mathbf{b}|$ must be more than 0.07 Å) the incident ions interact with the O_I-O_{II} row as with two separate rows which have interatomic spacing $d_1 = 2d_0$ or $d_2 = 4d_0$ in dependence on the sequence of displacement directions $\mathbf{b}$ of the O_I (or O_{II}) atoms.

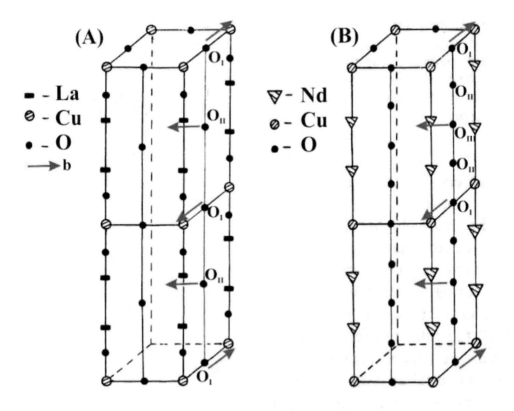

Figure.14. Crystal structures of La_2CuO_4 (A) and Nd_2CuO_4 (B) and the oxygen rows along the $\langle 001 \rangle$ axis. Only the visible atoms of the crystal structures are shown. The displacement directions $\mathbf{b}$ of the O atoms are shown only for one row.

Investigation of the Structural Features of High Temperature ... 225

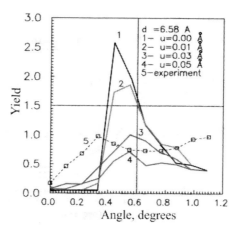

Figure.15. Angular dependences of the SPE yield from oxygen atoms which form O_I-O_{II} rows along the ⟨001⟩ direction in La_2CuO_4 (E_{He} = 3.045 MeV) . The curves are calculated with various *rms* amplitudes u; □ show experimental data [19].

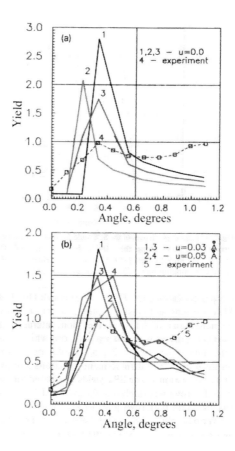

Figure 16. The ⟨001⟩ angular dependences of the SPE yield from displaced O_I and O_{II} atoms in La_2CuO_4 which form rows with different interatomic spacing: (a) d_1 row (curve 1), d_2 row (curve 2), d_1 and d_2 rows (curve 3); (b) d_1 row (curves 1 and 2), d_1 and d_2 rows (curves 3 and 4); □ show experimental data [19].

The SPE yields calculated for the d_1 and d_2 rows separately and the averaged SPE yield calculated for both rows together are shown in Figure 16 for various values of the amplitudes of thermal vibrations. One can see that very good agreement with the experimental results was achieved when the SPE yield was calculated either for rows with interatomic space d_1 or for two different rows with d_1 and d_2. In order to provide the best fit it is necessary to assume that the oxygen atoms located at their new equilibrium positions vibrate with $u_{rms} <$ 0.03 Å.

6.2. The Oxygen Displacements in Nd$_2$CuO$_4$

The $\langle 001 \rangle$ channels of the Nd$_2$CuO$_4$ single crystals are formed by "pure" O$_I$-O$_{II}$-O$_{III}$ rows with d_0 = 3.03 Å and "mixed" Nd-Cu rows (see Figure 14, b). The calculated yr_s for the "frozen" oxygen rows (u_{rms} = 0.0) was ~ 20% greater than the experimental one (see Figure 17). The thermal vibrations of the oxygen atoms have been simulated as follows: (i) as isotropic vibrations with $u_1 = u_2 = u_3$, where u_1, u_2 and u_3 are the rms-amplitudes of the O$_I$, O$_{II}$ and O$_{III}$ atoms, respectively; (ii) as isotropic vibrations with $u_1 = u_2$ and $u_2 > u_1$; (iii) as isotropic vibrations of O$_I$ and O$_{III}$ atoms with $u_1 = u_3$ and anisotropic vibrations of O$_{II}$ atoms. The results of set of calculations carried out in these three ways with varied values of u_1, u_2 and u_3 are shown in Figure 17. The experimental data disagree with these results in any case. Like for La$_{2-x}$Sr$_x$CuO$_4$, the corrections of the calculations which take into account the beam angular divergence and the deeper ion penetration into the crystal have not eliminated this discrepancy.

Therefore, similarly as it has been done for La$_{2-x}$Sr$_x$CuO$_4$ we must involve the static displacements of the oxygen atoms along normal to the $\langle 001 \rangle$ axis (see Figure 14, b) which result in the following: (A) The O$_{II}$, atoms are displaced. Then along $\langle 001 \rangle$ the oxygen

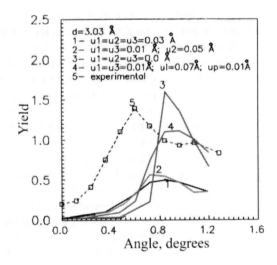

Figure 17. The $\langle 001 \rangle$ angular dependences of the SPE yield from O$_I$, O$_{II}$ and O$_{III}$ atoms in Nd$_2$CuO$_4$. The curves are calculated with thermal vibrations determined in three ways as described in the text; □ show experimental data [19].

sublattice forms O_I-O_{II} rows with non-equidistant atoms and O_{III}-O_{III} rows. For the O_{III}-O_{III} row the interatomic spacing is $d_1 = 4d_0$ or $d_2 = 8d_0$ in dependence on the sequence of displacement directions **b** of O_{III} atoms. For the O_I-O_{II} row there are two interatomic spacing, $d_3 = d_0$ and $d_4 = 2d_0$. (B) The O_I and O_{III} atoms are displaced. The oxygen sublattice forms O_I-O_{II}, O_{II}-O_{II} and O_{III}-O_{III} rows along $\langle 001 \rangle$ (see Figure 14, b). Then, if the displacements $|\mathbf{b}| > 0.1\text{Å}$ (under given experimental conditions) the incident ions interact with every row separately.

The calculations have been carried out for both situations (A and B). The three following cases have been considered: (i) the displaced atoms form rows with $d_1 = 4d_0$; (ii) they form rows with $d_2 = 8d_0$; (iii) there are two rows with $d_1 = 4d_0$ and $d_2 = 8d_0$. For every situation (A or B) the thermal vibrations have been simulated in the three ways described above with varied values of the *rms*-amplitudes.

The results of such a series of calculations have demonstrated that agreement with the experiment would be achieved in two cases: (i) The static displacements normal to the $\langle 001 \rangle$ direction ($|\mathbf{b}| > 0.1\text{Å}$) take place only for half of the oxygen atoms belonging to the Cu-O planes (i.e. only O_{III} atoms are displaced). Both the displaced O_{III} atoms and the undisplaced O_I atoms vibrate isotropically with $u_1 = u_3$. The undisplaced O_{II} atoms belonging to "pure" oxygen planes vibrate anisotropically with the "longitudinal" $u_L \gg u_T$ (see Figure 18, a). (ii) All oxygen atoms in the Cu-O planes are displaced normal to $\langle 001 \rangle$. The thermal vibrations of the displaced O_I and O_{III} atoms are isotropic with $u_1 = u_3$. The O_{II} atoms be longing to the "pure" oxygen (001) planes are not displaced but they vibrate anisotropically with the "longitudinal" $u_L \gg u_T$ (see Figure 18, b). In every case the displaced oxygen atoms form either two rows with interatomic spacing $d_1 = 4d_0$ or two different rows with $d_1 = 4d_0$ and $d_2 = 8d_0$.

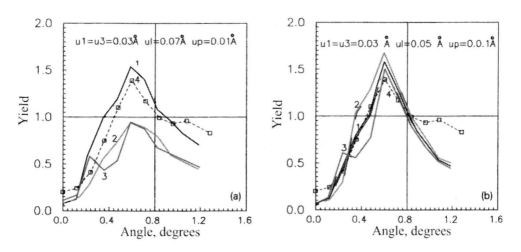

Figure 18. The $\langle 001 \rangle$ angular dependences of the SPE yield from displaced oxygen atoms in Nd_2CuO_4 which form rows with different interatomic spacings: (a) the non-equidistant row together with the d_1 and d_2 rows (curve 1), with the d_1 row (curve 2) and with the d_2 row (curve 3); (b) d_3 row together with the d_1 and d_2 rows (curve 1), with the d_1 row (curve 2) and with the d_2 row (curve 3); □ show experimental data [19].

Thus by comparing the experimental angular dependences of the resonance elastic scattering yield from oxygen atoms with the calculated ones obtained by a computer simulation it has been found that in $La_{2-x}Sr_xCuO_{4-\delta}$ and $Nd_2CuO_{4-\delta}$ single crystals the oxygen atoms are displaced from their usual lattice sites along normal to the $\langle 001 \rangle$ direction (> 0.07Å for $La_{2-x}Sr_xCuO_{4-\delta}$ and > 0.1 Å for $Nd_2CuO_{4-\delta}$). The angles between the directions of the displacements of the neighboring oxygen atoms which form $\langle 001 \rangle$ rows are equal to $90°$. For $La_{2-x}Sr_xCuO_{4-\delta}$ the *rms*-amplitudes of the oxygen atom thermal vibrations have been obtained to be < 0.03 Å. For $Nd_2CuO_{4-\delta}$ crystal the oxygen atoms located in the Cu-O planes vibrate isotropically with $u_{rms} < 0.05$ Å while the oxygen atoms which form the oxygen (00 1) planes vibrate anisotropically: the u_{rms} along the $\langle 001 \rangle$ direction are much larger than the u_{rms} along normal to this direction. The results obtained here and also previously reported in paragraph 4 related to randomly distributed oxygen atoms (15% – 20%) in $La_{2-x}Sr_xCuO_{4-\delta}$ allow us to suggest that the unit cells of the studied crystals have a much larger lattice unit cell volume than has been usually considered.

REFERENCES

[1] Lindhard, J. Kgl. Danske Videuskab. *Selskab. Mat.-Fis. Medel.* 1965, *vol. 34, No 14.*

[2] Sharma, R.P.; Rehn, L.E.; Baldo, P.M. *Phys. Rev. B.* 1991, *vol. 43,* 13711.

[3] Borovik, A.S.; Kobzev, A.P.; Kovaleva, E.A. *Fizika i Himiya Obrabotki Materialov,* 1998, *No 5,* 69 (in Russian).

[4] Borovik, A.S.; Epifanov, A.A.; Korneev, D.A.; Malyshevsky, V.S. *Preprint JINR.* 1992, P14-92-396.

[5] Makarov, V.I.; Slabospitskii, R.P.; Skakun, N.A.; Borovik, A.S.; Voronov, A.P.; Grinchenko, A.Y.; Malyshevsky, V.S.; Oleink, V.A. Sov. J. *Low Temp. Phys.* 1991, *vol. 17(4),* 251.

[6] Borovik, A.S.; Epifanov, A.A.; Malyshevsky, V.S.; Makarov, V.I. *Phys. Lett.* 1992, *vol. 161A,* 523.

[7] Borovik, A.S.; Kovaleva, E.A.; Malyshevsky, V.S.; Makarov, V. I. *Phys. Lett.* 1992, *vol. 171A,* 397.

[8] Borovik, A.S.; Epifanov, A.A.; Malyshevsky, V.S.; Makarov, V.I. *Nucl. Instr. And Meth. In Phys. Res.* 1993, *vol. B73,* 512.

[9] Borovik, A.S.; Kobzev, A.P.; Kovaleva, E.A.; Potapov, S.N. *Poverhnost.* 1997, *No 2,* 116 (in Russian).

[10] Remmel, J.; Meyer, O.; Geerk, J.; Reiner, J.; Linker, G.; Erb, A.; Muller-Vogt, G., *Phys. Rev. B.* 1993, *vol. 48,* 16168.

[11] Skakun, N.A.; et al. In*: Proc. 20th Natl. Conf. on Atomic Collisions in Solids,* May 1990, Moscow, 1990, Moscow Univ. Publ. House: Moscow, p. 135 (in Russian).

[12] Barrett, J.H. *Phys. Rev.* 1971, *vol. B3,* 1527.

[13] *Physical Properties of High Temperature Superconductors*; D. M. Ginsberg; Ed.; World Scientific: London; 1989; pp. 1 – 550.

[14] Zavaritski, N.V.; Makarov, V.I.; Klochko, V.S.; Molchanov, V.N.; Tamazjan, R.A.; Jurgens, A.A. *Zh. Eksp. Teor. Fiz.* 1991, *vol. 100,* 1987 (in Russian).

[15] Skakun N.A.; et al., *Private communication.*

[16] Borovik, A.S.; Shipatov, E.T. *Radiat. Eff.* 1984, *vol. 81*.

[17] Kovaleva, E.A.; Shipatov, E.T. *Radiat. Eff.* 1985, *vol. 84*, 301.

[18] Kovaleva, E.A.; Shipatov, E.T. *Radiat. Eff.* 1986, *vol. 86*, 141.

[19] Grinchenko, A. Yu.; et al. In: *Proc. 21st Natl. Conf. on Atomic Collisions in Solids*, May 1991, Moscow, 1991, Moscow Univ. Publ. House: Moscow, p. 115 (in Russian).

In: Ferroelectrics and Superconductors
Editor: Ivan A. Parinov

ISBN: 978-1-61324-518-7
©2012 Nova Science Publishers, Inc.

Chapter 7

IMPROVED FINITE ELEMENT APPROACHES FOR MODELING OF POROUS PIEZOCOMPOSITE MATERIALS WITH DIFFERENT CONNECTIVITY

A. V. Nasedkin[*] and M. S. Shevtsova[‡]

Department of Mathematic Modelling, Southern Federal University,
Rostov-on-Don, Russia

ABSTRACT

Porous piezoceramic materials have received considerable attention due to their application in ultrasonic transducers, hydrophones, pressure sensors and other piezoelectric devices. The classification of piezoelectric composites was initiated by Newnham's connectivity theory. In accordance with this theory porous piezoceramic can be classified as two-phase composite. As is well known, the 3-0 and 3-3 piezocomposites in the form of porous PZT materials show considerably improved transducer characteristics. Porous piezocomposites have great potential for low acoustic impedance and higher efficiency compared to conventional dense PZT piezoceramic materials. However, their depth-handling capability and the stability to hydrostatic pressure have not yet to be proved, particularly for 3-3 porous structures.

Porous composite piezoceramic may be accepted as a quasi-homogeneous medium with some effective moduli for most applications. At present, there are many publications in which the effective properties of 3-0, 3-3 piezocomposite media have been analyzed. The approximate formulae of engineering character for piezocomposites with various type of connectivity have been received. However, the strict mathematical approaches of mechanics of composites were used only in a small number of papers.

In the present work we have developed the effective moduli method and finite-element technique. Theoretical aspects of the effective moduli method for inhomogeneous piezoelectric media were examined. Four static piezoelectric problems for a representative volume that allow finding the effective moduli of an inhomogeneous body were specified. These problems differ by the boundary conditions which were set on a representative volume surfaces, namely: (i) mechanical displacements and electric

---

[*]E-mail: nasedkin@math.rsu.ru., 8a, Milchakova Street, 344090 Rostov-on-Don, Russia.
[‡]E-mail: mariamarcs@bk.ru., 8a, Milchakova Street, 344090 Rostov-on-Don, Russia.

potential, (ii) mechanical displacements and normal component of electric displacement vector, (iii) mechanical stress vector and electric potential, and (iv) mechanical stress vector and normal component of electric displacement vector. Respective equations for calculation of effective moduli of piezoelectric media with arbitrary anisotropy were derived.

Based on these equations the full set of effective moduli for porous composite piezoceramics having wide porosity range was calculated with help of finite element method (FEM). Different models of representative volume were considered: piezoelectric cubes with one cubic and one spherical pore inside, cubic volume evenly divided on partial cubic volumes a part of which randomly declared as pores for 3-0 and 3-0 − 3-3 connectivity. For the modeling of porous piezoceramics with 3-3 connectivity the representative volume having skeleton structure was considered. For taking into consideration inhomogeneous or incomplete ceramics polarization the preliminary modeling of polarization process was performed. In order to determine the zones that have different polarization the FEM calculations of the electrostatic problem were executed.

The results of the FEM modeling were compared with the experimental date for different porous ceramics in the relative porosity range of 0 − 70 %.

1. INTRODUCTION

Recently porous piezoceramic materials have received considerable attention due to their application in ultrasonic transducers, hydrophones, pressure sensors and other piezoelectric devices. The microstructural characterization, analysis and classification of piezoelectric composites were initiated by Newnham's connectivity theory [1]. In compliance with this theory porous piezoceramic can be classified as two phase composite. Depending on whether their pores are unconnected or connected, such piezoceramics are said to have 3-0 or 3-3 connectivity. Here, the first number denotes three-dimensional interconnection of the ceramic material and the second number marks the dimensions in which the pore volume is interconnected.

As is well known, the 3-0 and 3-3 piezocomposites in the form of porous Lead Zirconate Titanate (PZT) materials show considerably improved transducer characteristics. Porous piezocomposites have great potential for low acoustic impedance and higher efficiency compared to conventional dense PZT piezoceramic materials. However, their depth-handling capability and the stability to hydrostatic pressure have not yet to be proved, particularly for 3-3 porous structures. The mechanical strength of the devices can be improved by filling with polymer material as the second phase. Note, that the original concept of microstructural designing of polymer-free polycrystalline composite materials is suggested in [2]. This concept is based on controllable substitution during composite formation of separate crystallites by pores (making a porous polycrystal), crystallites with other compositions and/or structure or amorphous substances, and on preliminary FEM modeling of polycrystalline composite properties.

Porous or polycrystalline composite piezoceramics having pore or inclusion sizes lesser that 100 μm may be accepted as a quasi-homogeneous medium with some effective moduli for most applications. At present, there are many publications in which the effective properties of 3-0, 3-3 piezocomposite media have been analyzed. The approximate formulae of engineering character for piezocomposites with various type of connectivity have been received. However, the strict mathematical approaches of mechanics of composites were used only in a small number of papers.

The material properties of porous or polymer–crystalline piezocomposites with mixed connectivities (3-0, 3-3 and 0-3) have been evaluated using different theoretical models [3 – 20]. The use of Marutake's and Bruggeman's approximations for calculation of effective moduli of piezocomposites was offered in [3]. The approximate equations for elastic, dielectric and piezoelectric constants of diphasic piezocomposite from 3-0 to 3-3 connectivity were obtained in [4] based on simplified model combing cubes and 3-3 models. Some cubic and other models also were used in [5 – 8]. The modified cubes matrix method was proposed for analysis of the piezocomposite with different connectivity in [9] and other papers of the same authors. The dilute, self-consistent, Mori-Tanaka and differential micromechanics theories were extended in [10] to consider the effective characteristic of piezocomposite materials. The application of each theory was based on three-dimensional static solution of an ellipsoidal inclusion in an infinite piezoelectric media. Theoretical models including optimization techniques and homogenization methods have also been proposed for piezocomposite in [11] and methods of statistical mechanics have been applied for piezocomposite in [12, 13]. Direct finite element modeling for porous piezocomopite was carried out in [14 – 16].

In the present work we developed a metodology to predict the effective moduli using the finite element technique in accordance with [17 – 20].

2. MATHEMATICAL ASPECTS OF EFFECTIVE MODULI METHOD

Let Ω be a region occupied by a heterogeneous body with piezoelectric properties, $\Gamma = \partial\Omega$ is the boundary of the region, $\mathbf{n}(\mathbf{x})$ is the vector of the external unit normal to Γ ($\mathbf{x} \in \Gamma$), $\mathbf{u}(\mathbf{x})$ is the displacement vector-function, $\varphi(\mathbf{x})$ is the electric potential function. We denote by ε the deformation tensor, σ the stress tensor, $\mathbf{E}$ the electric field vector, $\mathbf{D}$ the electric displacement vector. The fields ε and $\mathbf{E}$ are determined from the functions $\mathbf{u}$ and φ by the following form:

$$\varepsilon = \frac{1}{2}(\nabla\mathbf{u} + \nabla\mathbf{u}^*); \; \mathbf{E} = -\nabla\varphi, \tag{1}$$

where superscript "*" denotes transposition operation.

On the boundary of region Γ we will consider the mechanical stress vector $\mathbf{p}$ and surface density of electric charges q :

$$\mathbf{p} = \mathbf{n} \cdot \boldsymbol{\sigma}; \tag{2}$$

$$q = -\mathbf{n} \cdot \mathbf{D}. \tag{3}$$

As usually, we will denote the volume-averaged quantities in the broken brackets as

$$\langle...\rangle = \frac{1}{\Omega} \int_{\Omega} (...)d\Omega. \tag{4}$$

Following a substantiation of a method of effective moduli for elastic media, for piezoelectric body we will formulate the auxiliary statements. These statements are proved under similar techniques, as for an elastic body [21]. The following states are found on the basis of effective moduli theory.

Lemma 1. These representations take place for volume averaging field characteristic by means of appropriate values on the boundary Γ:

$$(i) \ \varepsilon = \frac{1}{2}(\nabla \mathbf{u} + \nabla \mathbf{u}*) \Rightarrow \langle \varepsilon \rangle = \frac{1}{2\Omega} \int (\mathbf{un}* + \mathbf{nu}*)\,d\Gamma, \tag{5}$$

$$(ii) \ \mathbf{E} = -\nabla \varphi \Rightarrow \langle \mathbf{E} \rangle = -\frac{1}{\Omega} \int \mathbf{n}\varphi\,d\Gamma, \tag{6}$$

$$(iii) \ \mathbf{p} = \mathbf{n} \cdot \boldsymbol{\sigma}; \ \nabla \cdot \boldsymbol{\sigma} = 0 \Rightarrow \langle \boldsymbol{\sigma} \rangle = \frac{1}{2\Omega} \int (\mathbf{px}* + \mathbf{xp}*)\,d\Gamma, \tag{7}$$

$$(iv) \ q = -\mathbf{n} \cdot \mathbf{D}; \ \nabla \cdot \mathbf{D} = 0 \Rightarrow \langle \mathbf{D} \rangle = -\frac{1}{\Omega} \int \mathbf{x}q\,d\Gamma. \tag{8}$$

◀ In order to prove statements (5) – (8) we can use component-wise notations, formulae (1) and Gauss divergence theorem:

$$\langle \varepsilon_{ij} \rangle = \frac{1}{2\Omega} \int_{\Omega} (u_{i,j} + u_{j,i})\,d\Omega = \frac{1}{2\Omega} \int (n_j u_i + n_i u_j)\,d\Gamma;$$

$$\langle E_i \rangle = -\frac{1}{\Omega} \int_{\Omega} \varphi_{,i}\,d\Omega = -\frac{1}{2\Omega} \int n_i \varphi\,d\Gamma.$$

In addition to prove (7), (8) we use the formulae:

$$(\sigma_{ik} x_j)_{,k} = \sigma_{ik,k} x_j + \sigma_{ij}; \ (\sigma_{jk} x_i)_{,k} = \sigma_{jk,k} x_i + \sigma_{ji}; \ \sigma_{ik,k} = 0;$$

$$\sigma_{ij} = \sigma_{ji} \Rightarrow \sigma_{ij} = [(\sigma_{ik} x_j)_{,k} + (\sigma_{jk} x_i)_{,k}]/2;$$

$$(D_i x_j)_{,i} = D_{i,i} x_j + D_i; \ D_{i,i} = 0.$$

Therefore,

$$\langle \sigma_{ij} \rangle = \frac{1}{2\Omega} \int_\Omega (\sigma_{ik} x_j + \sigma_{jk} x_i)_{,k} \, d\Omega = \frac{1}{2\Omega} \int (n_k \sigma_{ik} x_j + n_k \sigma_{jk} x_i) \, d\Gamma =$$

$$= \frac{1}{2\Omega} \int (p_i x_j + p_j x_i) \, d\Gamma \, ;$$

$$\langle D_i \rangle = \frac{1}{\Omega} \int_\Omega (D_i x_j)_{,i} \, d\Omega = \frac{1}{\Omega} \int n_i D_i x_j \, d\Gamma = -\frac{1}{\Omega} \int q x_j \, d\Gamma \, . \blacktriangleright$$

Lemma 2.

(i) Let $\mathbf{u} = \mathbf{x} \cdot \boldsymbol{\varepsilon}_0 \big|_\Gamma$ for $\mathbf{x} \in \Gamma$, where $\boldsymbol{\varepsilon}_0 = \boldsymbol{\varepsilon}_0{}^* = \text{const}$. Then we have $\langle \boldsymbol{\varepsilon} \rangle = \boldsymbol{\varepsilon}_0$.

(ii) Let $\varphi = -\mathbf{x} \cdot \mathbf{E}_0 \big|_\Gamma$ for $\mathbf{x} \in \Gamma$, where $\mathbf{E}_0 = \text{const}$. Then we have $\langle \mathbf{E} \rangle = \mathbf{E}_0$.

(iii) Let $\mathbf{p} = \mathbf{n} \cdot \boldsymbol{\sigma}_0 \big|_\Gamma$ for $\mathbf{x} \in \Gamma$, where $\mathbf{p}$ is the stress vector from (2) and $\boldsymbol{\sigma}_0 = \text{const}$. Then we have $\langle \boldsymbol{\sigma} \rangle = \boldsymbol{\sigma}_0$.

(iv) Let $q = -\mathbf{n} \cdot \mathbf{D}_0 \big|_\Gamma$ for $\mathbf{x} \in \Gamma$, where q is the surface density of electric charges from (3) and $\mathbf{D}_0 = \text{const}$. Then we have $\langle \mathbf{D} \rangle = \mathbf{D}_0$.

◀ In order to prove statements (i) – (iv) we apply the corresponding statements (i) – (iv) from Lemma 1 and transfer from integration on Γ to integration on Ω by Gauss divergence theorem.▶

Lemma 3.

(i) Let $\mathbf{u} = \mathbf{x} \cdot \boldsymbol{\varepsilon}_0 \big|_\Gamma$ for $\mathbf{x} \in \Gamma$, where $\boldsymbol{\varepsilon}_0 = \boldsymbol{\varepsilon}_0{}^* = \text{const}$, and the equilibrium equation $\nabla \cdot \boldsymbol{\sigma} = 0$ takes place. Then we have $\langle \boldsymbol{\varepsilon} \cdot\cdot \boldsymbol{\sigma} \rangle / 2 = \langle \boldsymbol{\varepsilon} \rangle \cdot\cdot \langle \boldsymbol{\sigma} \rangle / 2$.

(ii) Let $\varphi = -\mathbf{x} \cdot \mathbf{E}_0 \big|_\Gamma$ for $\mathbf{x} \in \Gamma$, where $\mathbf{E}_0 = \text{const}$, and the equation of quasi-electrostatic $\nabla \cdot \mathbf{D} = 0$ takes place. Then we have $\langle \mathbf{D} \cdot \mathbf{E} \rangle / 2 = \langle \mathbf{D} \rangle \cdot \langle \mathbf{E} \rangle / 2$.

(iii) Let $\mathbf{p} = \mathbf{n} \cdot \boldsymbol{\sigma}_0 \big|_\Gamma$ for $\mathbf{x} \in \Gamma$, where $\boldsymbol{\sigma}_0 = \text{const}$, $\mathbf{p}$ is the stress vector from (2), and the equilibrium equation $\nabla \cdot \boldsymbol{\sigma} = 0$ takes place. Then we have $\langle \boldsymbol{\varepsilon} \cdot\cdot \boldsymbol{\sigma} \rangle / 2 = \langle \boldsymbol{\varepsilon} \rangle \cdot\cdot \langle \boldsymbol{\sigma} \rangle / 2$.

(iv) Let $q = -\mathbf{n} \cdot \mathbf{D}_0\big|_\Gamma$ for $\mathbf{x} \in \Gamma$, where $\mathbf{D}_0 = \mathrm{const}$, q is the surface density from (3), and the equation of $\nabla \cdot \mathbf{D} = 0$ takes place. Then we have $\langle \mathbf{D} \cdot \mathbf{E} \rangle / 2 = \langle \mathbf{D} \rangle \cdot \langle \mathbf{E} \rangle / 2$.

◄ (i) If $\mathbf{u} = \mathbf{x} \cdot \boldsymbol{\varepsilon}_0\big|_\Gamma$ for $\mathbf{x} \in \Gamma$, $\nabla \cdot \boldsymbol{\sigma} = 0$, then it is correct the following sequence of identities:

$$\frac{1}{2}\langle \boldsymbol{\varepsilon} \cdot\cdot \boldsymbol{\sigma} \rangle = \frac{1}{2\Omega} \int_\Omega \nabla \mathbf{u} \cdot\cdot \boldsymbol{\sigma} \, d\Omega = \frac{1}{2\Omega} \int_\Gamma (\mathbf{n} \cdot \boldsymbol{\sigma}) \cdot \mathbf{u} \, d\Gamma = \frac{1}{2\Omega} \int_\Gamma (\mathbf{n} \cdot \boldsymbol{\sigma}) \cdot \boldsymbol{\varepsilon}_0 \cdot \mathbf{x} \, d\Gamma =$$

$$= \frac{1}{2\Omega} \int_\Omega \nabla \cdot (\boldsymbol{\sigma} \cdot \boldsymbol{\varepsilon}_0 \cdot \mathbf{x}) \, d\Omega = \frac{1}{2\Omega} \int_\Omega \boldsymbol{\sigma} \cdot\cdot \boldsymbol{\varepsilon}_0 \, d\Omega = \frac{1}{2\Omega} \int_\Omega \boldsymbol{\sigma} \, d\Omega \cdot\cdot \boldsymbol{\varepsilon}_0 = \frac{1}{2}\langle \boldsymbol{\sigma} \rangle \cdot\cdot \langle \boldsymbol{\varepsilon} \rangle,$$

since $\boldsymbol{\varepsilon}_0 = \langle \boldsymbol{\varepsilon} \rangle$ from Lemma 2, (i).

(ii) If $\varphi = -\mathbf{x} \cdot \mathbf{E}_0\big|_\Gamma$ for $\mathbf{x} \in \Gamma$, $\nabla \cdot \mathbf{D} = 0$, then

$$\frac{1}{2}\langle \mathbf{D} \cdot \mathbf{E} \rangle = -\frac{1}{2\Omega} \int_\Omega \mathbf{D} \cdot \nabla \varphi \, d\Omega =$$

$$= -\frac{1}{2\Omega} \int_\Omega \nabla \cdot (\mathbf{D}\varphi) \, d\Omega = -\frac{1}{2\Omega} \int_\Gamma \mathbf{n} \cdot \mathbf{D}\varphi \, d\Gamma = \frac{1}{2\Omega} \int_\Gamma \mathbf{n} \cdot \mathbf{D}(\mathbf{x} \cdot \mathbf{E}_0) \, d\Gamma =$$

$$= \frac{1}{2\Omega} \int_\Omega \nabla \cdot (\mathbf{D}(\mathbf{x} \cdot \mathbf{E}_0)) \, d\Omega = \frac{1}{2\Omega} \int_\Omega \mathbf{D} \cdot \mathbf{E}_0 \, d\Omega = \frac{1}{2}\langle \mathbf{D} \rangle \cdot \mathbf{E}_0 = \frac{1}{2}\langle \mathbf{D} \rangle \cdot \langle \mathbf{E} \rangle,$$

since $\mathbf{E}_0 = \langle \mathbf{E} \rangle$ from Lemma 2, (ii).

(iii) If $\mathbf{p} = \mathbf{n} \cdot \boldsymbol{\sigma}_0\big|_\Gamma$, $\nabla \cdot \boldsymbol{\sigma} = 0$, then

$$\frac{1}{2}\langle \boldsymbol{\varepsilon} \cdot\cdot \boldsymbol{\sigma} \rangle = \frac{1}{2\Omega} \int_\Omega \nabla \mathbf{u} \cdot\cdot \boldsymbol{\sigma} \, d\Omega = \frac{1}{2\Omega} \int_\Gamma (\mathbf{n} \cdot \boldsymbol{\sigma}) \cdot \mathbf{u} \, d\Gamma =$$

$$= \frac{1}{2\Omega} \int_\Gamma (\mathbf{n} \cdot \boldsymbol{\sigma}_0) \cdot \mathbf{u} \, d\Gamma = \frac{1}{2\Omega} \int_\Omega \nabla \mathbf{u} \cdot\cdot \boldsymbol{\sigma}_0 \, d\Omega = \frac{1}{2\Omega} \int_\Omega \boldsymbol{\varepsilon} \cdot\cdot \boldsymbol{\sigma}_0 \, d\Omega = \frac{1}{2}\langle \boldsymbol{\varepsilon} \rangle \cdot\cdot \boldsymbol{\sigma}_0 = \frac{1}{2}\langle \boldsymbol{\varepsilon} \rangle \cdot\cdot \langle \boldsymbol{\sigma} \rangle,$$

since $\boldsymbol{\sigma}_0 = \langle \boldsymbol{\sigma} \rangle$ from Lemma 2, (iii).

(iv) If $q = -\mathbf{n} \cdot \mathbf{D}_0\big|_\Gamma$, $\nabla \cdot \mathbf{D} = 0$, then

$$\frac{1}{2}\langle \mathbf{D} \cdot \mathbf{E} \rangle = -\frac{1}{2\Omega}\int_{\Omega}\mathbf{D} \cdot \nabla \varphi \, d\Omega = -\frac{1}{2\Omega}\int_{\Omega}\nabla \cdot (\mathbf{D}\varphi) \, d\Omega =$$

$$= -\frac{1}{2\Omega}\int \mathbf{n} \cdot \mathbf{D}\varphi \, d\Gamma = -\frac{1}{2\Omega}\int \mathbf{n} \cdot \mathbf{D}_0\varphi \, d\Gamma = -\frac{1}{2\Omega}\int_{\Omega}\nabla \cdot (\mathbf{D}_0\varphi) \, d\Gamma =$$

$$= -\frac{1}{2\Omega}\int_{\Omega}\mathbf{D}_0 \cdot \nabla \varphi \, d\Omega = \frac{1}{2\Omega}\int_{\Omega}\mathbf{D}_0 \cdot \mathbf{E} \, d\Omega = \frac{1}{2}\langle \mathbf{D} \rangle \cdot \langle \mathbf{E} \rangle,$$

since $\mathbf{D}_0 = \langle \mathbf{D} \rangle$ from Lemma 2, (iv). ▶

In accordance with four equivalent fundamental forms of constitutive equations [22] we will introduce the moduli of piezoelectric medium $\mathbf{c}^E$, $\mathbf{e}$, $\boldsymbol{\varepsilon}^S$, and so on:

– in independent variables $\boldsymbol{\varepsilon}$, $\mathbf{E}$ as

$$\boldsymbol{\sigma} = \mathbf{c}^E \cdot\cdot \boldsymbol{\varepsilon} - \mathbf{e}^* \cdot \mathbf{E}, \ \mathbf{D} = \mathbf{e} \cdot\cdot \boldsymbol{\varepsilon} + \boldsymbol{\varepsilon}^S \cdot \mathbf{E}; \tag{9}$$

– in independent variables $\boldsymbol{\sigma}$, $\mathbf{E}$ as

$$\boldsymbol{\varepsilon} = \mathbf{s}^E \cdot\cdot \boldsymbol{\sigma} + \mathbf{d}^* \cdot \mathbf{E}, \ \mathbf{D} = \mathbf{d} \cdot\cdot \boldsymbol{\sigma} + \boldsymbol{\varepsilon}^T \cdot \mathbf{E}; \tag{10}$$

– in independent variables $\boldsymbol{\varepsilon}$, $\mathbf{D}$ as

$$\boldsymbol{\sigma} = \mathbf{c}^D \cdot\cdot \boldsymbol{\varepsilon} - \mathbf{h}^* \cdot \mathbf{D}, \ \mathbf{E} = -\mathbf{h} \cdot\cdot \boldsymbol{\varepsilon} + \boldsymbol{\beta}^S \cdot \mathbf{D}; \tag{11}$$

– in independent variables $\boldsymbol{\sigma}$, $\mathbf{D}$ as

$$\boldsymbol{\varepsilon} = \mathbf{s}^D \cdot\cdot \boldsymbol{\sigma} + \mathbf{g}^* \cdot \mathbf{D}, \ \mathbf{E} = -\mathbf{g} \cdot\cdot \boldsymbol{\sigma} + \boldsymbol{\beta}^T \cdot \mathbf{D}. \tag{12}$$

Here $\mathbf{c}^E$, $\mathbf{c}^D$ are the fourth rank tensors of elastic stiffness moduli at constant electric field and at constant electric displacement, respectively; $\mathbf{e}$, $\mathbf{d}$, $\mathbf{h}$, $\mathbf{g}$ are the third rank tensors of piezoelectric moduli (stress coefficients, charge coefficients, strain coefficients, voltage coefficient, respectively); $\boldsymbol{\varepsilon}^S$, $\boldsymbol{\varepsilon}^T$ are the second rank tensors of dielectric permittivity moduli at constant mechanical strain and at constant mechanical strain, respectively; $\mathbf{s}^E$, $\mathbf{s}^D$ are the fourth rank tensors of elastic compliance moduli at constant electric field and at constant electric displacement, respectively; $\boldsymbol{\beta}^S$, $\boldsymbol{\beta}^T$ are the second rank tensors of dielectric impermittivity moduli at constant mechanical strain and at constant mechanical strain, respectively. These different material constants are connected by the relations:

$$\mathbf{s}^E = (\mathbf{c}^E)^{-1}, \qquad \mathbf{s}^D = (\mathbf{c}^D)^{-1}, \qquad \boldsymbol{\beta}^S = (\boldsymbol{\epsilon}^S)^{-1}, \qquad \boldsymbol{\beta}^T = (\boldsymbol{\epsilon}^T)^{-1}, \qquad \mathbf{c}^D = \mathbf{c}^E + \mathbf{e} * \cdot \mathbf{h},$$

$$\mathbf{s}^D = \mathbf{s}^E - \mathbf{d} * \cdot \mathbf{g}, \qquad \boldsymbol{\epsilon}^T = \boldsymbol{\epsilon}^S + \mathbf{d} \cdot \cdot \mathbf{e} *, \qquad \boldsymbol{\beta}^T = \boldsymbol{\beta}^S - \mathbf{g} \cdot \cdot \mathbf{h} *, \qquad \mathbf{e} = \mathbf{d} \cdot \cdot \mathbf{c}^E, \qquad \mathbf{d} = \boldsymbol{\epsilon}^T \cdot \mathbf{g},$$

$$\mathbf{g} = \boldsymbol{\beta}^T \cdot \mathbf{d}, \mathbf{h} = \mathbf{g} \cdot \cdot \mathbf{c}^D.$$

For inhomogeneous body these moduli will be functions of coordinates $\mathbf{x}$: $\mathbf{c}^E = \mathbf{c}^E(\mathbf{x})$; $\mathbf{e} = \mathbf{e}(\mathbf{x})$; $\boldsymbol{\epsilon}^S = \boldsymbol{\epsilon}^S(\mathbf{x})$, etc.

Let Ω is the representative volume of heterogeneous piezoelectric materials. We will determine the effective moduli $\mathbf{c}^{Eeff}$, $\mathbf{e}^{eff}$, $\boldsymbol{\epsilon}^{Seff}$ by the following technique [17 – 20].

We consider the static piezoelectric problem for representative volume Ω:

$$\nabla \cdot \boldsymbol{\sigma} = 0, \ \nabla \cdot \mathbf{D} = 0, \ \mathbf{x} \in \Omega; \tag{13}$$

$$\mathbf{u} = \mathbf{x} \cdot \boldsymbol{\epsilon}_0, \ \varphi = -\mathbf{x} \cdot \mathbf{E}_0, \ \mathbf{x} \in \Gamma = \partial\Omega. \tag{14}$$

We call the problem (1), (9), (13), (14) the $u\varphi$-problem and denote the solution of this problem by $\mathbf{u}^{u\varphi}$, $\varphi^{u\varphi}$. From the obtained solution and (1), (9) we find $\boldsymbol{\epsilon}^{u\varphi}$, $\mathbf{E}^{u\varphi}$, $\boldsymbol{\sigma}^{u\varphi}$ and $\mathbf{D}^{u\varphi}$, where $\boldsymbol{\epsilon}^{u\varphi} = \boldsymbol{\epsilon}(\mathbf{u}^{u\varphi})$, $\mathbf{E}^{u\varphi} = \mathbf{E}(\varphi^{u\varphi})$, $\boldsymbol{\sigma}^{u\varphi} = \boldsymbol{\sigma}(\mathbf{u}^{u\varphi}, \varphi^{u\varphi})$, $\mathbf{D}^{u\varphi} = \mathbf{D}(\mathbf{u}^{u\varphi}, \varphi^{u\varphi})$.

By virtue of linearity of $u\varphi$-problem (13), (14) the tensor functions $\mathbf{A}^{u\varphi}$, $\mathbf{B}^{u\varphi}$, $\mathbf{F}^{u\varphi}$ and $\mathbf{G}^{u\varphi}$ depending on coordinates exist. By using these functions we join $\boldsymbol{\epsilon}^{u\varphi}$ and $\mathbf{E}^{u\varphi}$ with $\boldsymbol{\epsilon}_0$ and $\mathbf{E}_0$ from (14) as

$$\boldsymbol{\epsilon}^{u\varphi} = \mathbf{A}^{u\varphi} \cdot \cdot \boldsymbol{\epsilon}_0 + \mathbf{B}^{u\varphi} \cdot \mathbf{E}_0; \ \mathbf{E}^{u\varphi} = \mathbf{F}^{u\varphi} \cdot \cdot \boldsymbol{\epsilon}_0 + \mathbf{G}^{u\varphi} \cdot \mathbf{E}_0. \tag{15}$$

By substituting (15) into constitutive equations (9), we obtain the next relations $\boldsymbol{\sigma}^{u\varphi}$, $\mathbf{D}^{u\varphi}$ with $\boldsymbol{\epsilon}_0$ and $\mathbf{E}_0$:

$$\boldsymbol{\sigma}^{u\varphi} = (\mathbf{c}^E \cdot \cdot \mathbf{A}^{u\varphi} - \mathbf{e} * \cdot \mathbf{F}^{u\varphi}) \cdot \cdot \boldsymbol{\epsilon}_0 + (\mathbf{c}^E \cdot \cdot \mathbf{B}^{u\varphi} - \mathbf{e} * \cdot \mathbf{G}^{u\varphi}) \cdot \mathbf{E}_0; \tag{16}$$

$$\mathbf{D}^{u\varphi} = (\mathbf{e} \cdot \cdot \mathbf{A}^{u\varphi} + \boldsymbol{\epsilon}^S \cdot \mathbf{F}^{u\varphi}) \cdot \cdot \boldsymbol{\epsilon}_0 + (\mathbf{e} \cdot \cdot \mathbf{B}^{u\varphi} + \boldsymbol{\epsilon}^S \cdot \mathbf{G}^{u\varphi}) \cdot \mathbf{E}_0.$$

We note from Lemma 2, that for $u\varphi$-problem $\langle \boldsymbol{\epsilon}^{u\varphi} \rangle = \boldsymbol{\epsilon}_0$ and $\langle \mathbf{E}^{u\varphi} \rangle = \mathbf{E}_0$.

We supply in conformity to the initial heterogeneous medium some "equivalent" homogeneous medium with effective moduli $\mathbf{c}^{Eeff}$, $\mathbf{e}^{eff}$ and $\boldsymbol{\epsilon}^{Seff}$. The constitutive equations for "equivalent" medium, similar to (9), are in the forms:

$$\boldsymbol{\sigma}_0 = \mathbf{c}^{Eeff} \cdot \cdot \boldsymbol{\varepsilon}_0 - \mathbf{e}^{eff} * \mathbf{E}_0, \ \mathbf{D}_0 = \mathbf{e}^{eff} \cdot \cdot \boldsymbol{\varepsilon}_0 + \boldsymbol{\varepsilon}^{Seff} \cdot \mathbf{E}_0. \tag{17}$$

For $u\varphi$-problem we accept the following equations such as relations for definition of effective moduli from (17)

$$\left\langle \boldsymbol{\sigma}^{u\varphi} \right\rangle = \boldsymbol{\sigma}_0; \ \left\langle \mathbf{D}^{u\varphi} \right\rangle = \mathbf{D}_0. \tag{18}$$

The moduli, found from these conditions, are marked with superscripts "$u\varphi$". As a result from (16) – (18) we obtain the formulae for $u\varphi$-moduli:

$$\mathbf{c}^{Eeff,u\varphi} = \left\langle \mathbf{c}^E \cdot \cdot \mathbf{A}^{u\varphi} - \mathbf{e} * \cdot \mathbf{F}^{u\varphi} \right\rangle; \ \boldsymbol{\varepsilon}^{Seff,u\varphi} = \left\langle \mathbf{e} \cdot \cdot \mathbf{B}^{u\varphi} + \boldsymbol{\varepsilon}^S \cdot \mathbf{G}^{u\varphi} \right\rangle,$$

$$\mathbf{e}^{Eeff,u\varphi} = \left\langle \mathbf{e} \cdot \cdot \mathbf{A}^{u\varphi} + \boldsymbol{\varepsilon}^S \cdot \mathbf{F}^{u\varphi} \right\rangle = -\left\langle (\mathbf{c}^E \cdot \cdot \mathbf{B}^{u\varphi} - \mathbf{e} * \cdot \mathbf{G}^{u\varphi}) * \right\rangle. \tag{19}$$

Note, that due to Lemma 3 the average energies to be equal for heterogeneous and for "equivalent" homogeneous piezoelectric media:

$$\left\langle \boldsymbol{\sigma}^{u\varphi} \cdot \cdot \boldsymbol{\varepsilon}^{u\varphi} + \mathbf{D}^{u\varphi} \cdot \mathbf{E}^{u\varphi} \right\rangle / 2 = (\boldsymbol{\sigma}_0 \cdot \cdot \boldsymbol{\varepsilon}_0 + \mathbf{D}_0 \cdot \mathbf{E}_0)/2. \tag{20}$$

This variant of equivalent moduli method for piezoelectric media follows from [17, 18]. However, for piezoelectric media it is possible to suggest the other ways of introducing effective moduli, by considering the problems with other mechanical and electric boundary conditions from Lemma 2. Just, we can consider the following problems [19]:

(i) $p\varphi$-problem with boundary condition for stress vector $\mathbf{p}$ and electric potential φ:

$$\mathbf{p} = \mathbf{n} \cdot \boldsymbol{\sigma}_0; \ \varphi = -\mathbf{x} \cdot \mathbf{E}_0; \ \mathbf{x} \in \Gamma; \tag{21}$$

(ii) uq-problem with boundary condition for displacement $\mathbf{u}$ and surface density of electric charges q:

$$\mathbf{u} = \mathbf{x} \cdot \boldsymbol{\varepsilon}_0; \ q = -\mathbf{n} \cdot \mathbf{D}_0; \ \mathbf{x} \in \Gamma; \tag{22}$$

(iii) pq-problem with boundary condition for stress vector $\mathbf{p}$ and surface density of electric charges q:

$$\mathbf{p} = \mathbf{n} \cdot \boldsymbol{\sigma}_0; \ q = -\mathbf{n} \cdot \mathbf{D}_0; \ \mathbf{x} \in \Gamma. \tag{23}$$

In the all these problems the field equations (13) of equilibrium and electrostatic are considered. For the $p\varphi$-problem constitutive equations (10) are used and originally the effective moduli $\mathbf{s}^{Eeff,p\varphi}$, $\mathbf{d}^{eff,p\varphi}$ and $\boldsymbol{\varepsilon}^{Teff,u\varphi}$ are defined; respectively, for the uq-problem constitutive equations (11) are used and moduli $\mathbf{c}^{Deff,uq}$, $\mathbf{h}^{eff,uq}$, $\boldsymbol{\beta}^{Seff,uq}$ are obtained; and for the pq-problem constitutive equations (12) are used and moduli $\mathbf{s}^{Deff,pq}$, $\mathbf{g}^{eff,pq}$, $\boldsymbol{\beta}^{Teff,pq}$ are defined.

Note, that from Lemma 3 for the all of these problems, similar for the $u\varphi$-problem, the average energy is conserved, i.e. the relation (20) satisfied with replace of the superscript "$u\varphi$" by "$p\varphi$", "uq" or "pq".

In any of these problems from obtained effective moduli we can find also other moduli from constitutive equations (9) – (12) for "equivalent" homogeneous medium. We note, that effective moduli, found from different problems, will differ, i.e. $\mathbf{c}^{Eeff,u\varphi} \neq \mathbf{c}^{Eeff,p\varphi} \neq \mathbf{c}^{Eeff,pq} \neq \mathbf{c}^{Eeff,uq}$. At that, in the problems "$p\varphi$" and "uq" constitutive relations (10) and (11) are used with the moduli forming positively defined matrixes in the corresponding two-index notations:

$$\begin{pmatrix} \mathbf{s}^E & \mathbf{d}* \\ \mathbf{d} & \boldsymbol{\varepsilon}^T \end{pmatrix}, \quad \begin{pmatrix} \mathbf{c}^D & -\mathbf{h}* \\ \mathbf{h} & \boldsymbol{\beta}^S \end{pmatrix}.$$

Important task for practical implementation of the effective moduli method consist in the solution of boundary electroelastic problems (13), (14); (13), (21); (13), (22) or (13), (23) and determination of such tensor functions as $\mathbf{A}^{u\varphi}$, $\mathbf{B}^{u\varphi}$, $\mathbf{F}^{u\varphi}$, $\mathbf{G}^{u\varphi}$ for $u\varphi$-problem. However, in practice it is obvious that defining these tensor functions is not required. Indeed, from (17), (18) it follows, that for $u\varphi$-problem the equations take place:

$$\mathbf{c}^{Eeff,u\varphi} \cdot\cdot\boldsymbol{\varepsilon}_0 - \mathbf{e}^{eff,u\varphi} * \cdot\mathbf{E}_0 = \langle \boldsymbol{\sigma}^{u\varphi} \rangle; \quad \mathbf{e}^{eff,u\varphi} \cdot\cdot\boldsymbol{\varepsilon}_0 + \boldsymbol{\varepsilon}^{Seff,u\varphi} \cdot \mathbf{E}_0 = \langle \mathbf{D}^{u\varphi} \rangle. \tag{24}$$

Now, by using Equations (24), we can select such boundary conditions, at which the obvious expressions for effective moduli are obtained. For example, we consider $u\varphi$-problem (13), (14) with

$$\boldsymbol{\varepsilon}_0 = \varepsilon_0 (\mathbf{e}_k \mathbf{e}_m + \mathbf{e}_m \mathbf{e}_k); \quad \mathbf{E}_0 = 0, \tag{25}$$

where k, m are some fixed numbers ($k,m = 1,2,3$); $\mathbf{e}_k$ are the unit vectors of Cartesian basis. Then, from (24), (25) we obtain

$$c_{ijkm}^{Eeff,u\varphi} = \langle \sigma_{ij}^{u\varphi} \rangle /(2\varepsilon_0); \quad e_{jkm}^{eff,u\varphi} = \langle D_j^{u\varphi} \rangle /(2\varepsilon_0). \tag{26}$$

If in $u\varphi$-problem (13), (14) we accept

$$\boldsymbol{\varepsilon}_0 = 0; \quad \mathbf{E}_0 = E_0 \mathbf{e}_k, \tag{27}$$

then from (24), (27) we find

$$e_{kij}^{eff,u\varphi} = -\left\langle \sigma_{ij}^{u\varphi} \right\rangle / E_0 \,; \; \varepsilon_{jk}^{Seff,u\varphi} = \left\langle D_j^{u\varphi} \right\rangle / E_0 \,. \tag{28}$$

Note, that the quantities $\left\langle \sigma_{ij}^{u\varphi} \right\rangle$ and $\left\langle D_j^{u\varphi} \right\rangle$ in (26) and (28) are different, since they are calculated from the solutions of the $u\varphi$-problems with different boundary conditions (14): (25) and (27).

Analogously, for $p\varphi$-problem (13), (21) we have

$$\mathbf{s}^{Eeff,p\varphi} \cdot\cdot\boldsymbol{\sigma}_0 + \mathbf{d}^{eff,p\varphi} * \cdot \mathbf{E}_0 = \left\langle \boldsymbol{\varepsilon}^{p\varphi} \right\rangle \,; \; \mathbf{d}^{eff,p\varphi} \cdot\cdot\boldsymbol{\sigma}_0 + \boldsymbol{\varepsilon}^{Teff,p\varphi} \cdot \mathbf{E}_0 = \left\langle \mathbf{D}^{p\varphi} \right\rangle. \tag{29}$$

Then for the boundary conditions with

$$\boldsymbol{\sigma}_0 = \sigma_0 (\mathbf{e}_k \mathbf{e}_m + \mathbf{e}_m \mathbf{e}_k) \,; \; \mathbf{E}_0 = 0 \tag{30}$$

from (29), (30) we have

$$s_{ijkm}^{Eeff,p\varphi} = \left\langle \varepsilon_{ij}^{p\varphi} \right\rangle / (2\sigma_0) \,; \; d_{jkm}^{eff,p\varphi} = \left\langle D_j^{p\varphi} \right\rangle / (2\sigma_0). \tag{31}$$

At the same time, for the boundary conditions with

$$\boldsymbol{\sigma}_0 = 0 \,; \; \mathbf{E}_0 = E_0 \mathbf{e}_k \tag{32}$$

we obtain

$$d_{kij}^{eff,p\varphi} = \left\langle \varepsilon_{ij}^{p\varphi} \right\rangle / E_0 \,; \; \varepsilon_{jk}^{Teff,p\varphi} = \left\langle D_j^{p\varphi} \right\rangle / E_0 \,. \tag{33}$$

For uq-problem (13), (22) the basic relations for effective moduli are given as

$$\mathbf{c}^{Deff,uq} \cdot\cdot\boldsymbol{\varepsilon}_0 - \mathbf{h}^{eff,uq} * \cdot \mathbf{D}_0 = \left\langle \boldsymbol{\sigma}^{uq} \right\rangle \,; \; -\mathbf{h}^{eff,uq} \cdot\cdot\boldsymbol{\varepsilon}_0 + \boldsymbol{\beta}^{Seff,uq} \cdot \mathbf{D}_0 = \left\langle \mathbf{E}^{uq} \right\rangle. \tag{34}$$

Therefore, for boundary conditions (22) with

$$\boldsymbol{\varepsilon}_0 = \varepsilon_0 (\mathbf{e}_k \mathbf{e}_m + \mathbf{e}_m \mathbf{e}_k) \,; \; \mathbf{D}_0 = 0 \tag{35}$$

from (34), (35) we obtain

$$c_{ijkm}^{Deff,uq} = \left\langle \sigma_{ij}^{uq} \right\rangle / (2\varepsilon_0) \,; \; h_{jkm}^{eff,uq} = -\left\langle E_j^{uq} \right\rangle / (2\varepsilon_0). \tag{36}$$

While for boundary conditions (22) with

$$\varepsilon_0 = 0; \; \mathbf{D}_0 = D_0 \mathbf{e}_k \tag{37}$$

we have

$$h_{kij}^{eff,uq} = -\left\langle \sigma_{ij}^{uq} \right\rangle / D_0; \; \beta_{jk}^{Seff,uq} = \left\langle E_j^{uq} \right\rangle / D_0. \tag{38}$$

Finally, for the effective moduli of the pq-problem (13), (23) we have the formulae:

$$\mathbf{s}^{Deff,pq} \cdot\cdot \boldsymbol{\sigma}_0 + \mathbf{g}^{eff,pq} * \mathbf{D}_0 = \left\langle \boldsymbol{\varepsilon}^{pq} \right\rangle; \; -\mathbf{g}^{eff,pq} \cdot\cdot \boldsymbol{\sigma}_0 + \boldsymbol{\beta}^{Teff,pq} \cdot \mathbf{D}_0 = \left\langle \mathbf{E}^{pq} \right\rangle. \tag{39}$$

Here for boundary conditions (23) with

$$\boldsymbol{\sigma}_0 = \sigma_0 (\mathbf{e}_k \mathbf{e}_m + \mathbf{e}_m \mathbf{e}_k); \; \mathbf{D}_0 = 0 \tag{40}$$

from (39), (40) we obtain

$$s_{ijkm}^{Deff,pq} = \left\langle \varepsilon_{ij}^{pq} \right\rangle / (2\sigma_0); \; g_{jkm}^{eff,pq} = -\left\langle E_j^{pq} \right\rangle / (2\sigma_0). \tag{41}$$

While for boundary conditions (23) with

$$\boldsymbol{\sigma}_0 = 0; \; \mathbf{D}_0 = D_0 \mathbf{e}_k \tag{42}$$

we obtain

$$g_{kij}^{eff,pq} = \left\langle \varepsilon_{ij}^{pq} \right\rangle / D_0; \; \beta_{jk}^{Teff,pq} = \left\langle E_j^{pq} \right\rangle / D_0. \tag{43}$$

Equations (25) – (43) allow us to obtain a full set of effective moduli for piezoelectric media with arbitrary anisotropy class. Use of different constitutive equations above four types of problems can be useful to determine effective moduli of the inhomogeneous structures by dealing with mainly one-dimensional movements with piezoinflexible and piezocompliant modes, for example, for piezoceramic rods, plates and disks, etc.

From the set of solutions of these problems it is possible to define all effective moduli of porous piezocomposites. Piezoceramic, homogeneously polarized along the axis x_3, is anisotropic material of crystallographic class $6mm$ and have 10 independent moduli (5 elastic, 3 piezoelectric and 2 dielectric moduli). Thus, for all piezoceramic moduli determination it is enough to consider five various problems for each type of problems ($u\varphi$, $p\varphi$, uq, pq). So for the $u\varphi$-problem we have the following problems and the formulae for calculation of the moduli in two-index tables of symbols ($i, j = 1,2,3$; $\alpha = 1,2,...6$):

I-$u\varphi$. $\boldsymbol{\varepsilon}_0 = \varepsilon_0 \mathbf{e}_1 \mathbf{e}_1$, ($\varepsilon_{011} = \varepsilon_0$), $\mathbf{E}_0 = 0 \Rightarrow$

$$c_{11}^{E\,\text{eff},u\varphi} = \langle \sigma_{11}^{u\varphi} \rangle / \varepsilon_0 ; \quad c_{12}^{E\,\text{eff},u\varphi} = \langle \sigma_{22}^{u\varphi} \rangle / \varepsilon_0 ;$$
$$c_{13}^{E\,\text{eff},u\varphi} = \langle \sigma_{33}^{u\varphi} \rangle / \varepsilon_0 ; \quad e_{31}^{\text{eff},u\varphi} = \langle D_3^{u\varphi} \rangle / \varepsilon_0 . \tag{44}$$

II-$u\varphi$. $\boldsymbol{\varepsilon}_0 = \varepsilon_0 \mathbf{e}_3 \mathbf{e}_3$, ($\varepsilon_{033} = \varepsilon_0$), $\mathbf{E}_0 = 0 \Rightarrow$

$$c_{33}^{E\,\text{eff},u\varphi} = \langle \sigma_{33}^{u\varphi} \rangle / \varepsilon_0 ; \quad e_{33}^{\text{eff},u\varphi} = \langle D_3^{u\varphi} \rangle / \varepsilon_0 ;$$

$$c_{13}^{E\,\text{eff},u\varphi} = \langle \sigma_{11}^{u\varphi} \rangle / \varepsilon_0 = \langle \sigma_{22}^{u\varphi} \rangle / \varepsilon_0 . \tag{45}$$

III-$u\varphi$. $\boldsymbol{\varepsilon}_0 = \varepsilon_0 (\mathbf{e}_2 \mathbf{e}_3 + \mathbf{e}_3 \mathbf{e}_2)$, ($\varepsilon_{023} = \varepsilon_{032} = \varepsilon_0$), $\mathbf{E}_0 = 0 \Rightarrow$

$$c_{44}^{E\,\text{eff},u\varphi} = \langle \sigma_{23}^{u\varphi} \rangle / (2\varepsilon_0) ; \quad e_{15}^{\text{eff},u\varphi} = \langle D_2^{u\varphi} \rangle / (2\varepsilon_0) . \tag{46}$$

IV-$u\varphi$. $\boldsymbol{\varepsilon}_0 = 0$, $\mathbf{E}_0 = E_0 \mathbf{e}_1$ ($E_{01} = E_0$) $\Rightarrow$

$$e_{15}^{\text{eff},u\varphi} = -\langle \sigma_{13}^{u\varphi} \rangle / E_0 ; \quad \varepsilon_{11}^{S\,\text{eff},u\varphi} = \langle D_1^{u\varphi} \rangle / E_0 . \tag{47}$$

V-$u\varphi$. $\boldsymbol{\varepsilon}_0 = 0$, $\mathbf{E}_0 = E_0 \mathbf{e}_3$ ($E_{03} = E_0$) $\Rightarrow$

$$e_{31}^{\text{eff},u\varphi} = -\langle \sigma_{11}^{u\varphi} \rangle / E_0 = -\langle \sigma_{22}^{u\varphi} \rangle / E_0 ; \tag{48}$$

$$e_{33}^{\text{eff},u\varphi} = -\langle \sigma_{33}^{u\varphi} \rangle / E_0 ; \quad \varepsilon_{33}^{S\,\text{eff},u\varphi} = \langle D_3^{u\varphi} \rangle / E_0 .$$

In (44) – (48), we use the vector-matrix forms for moduli [22]: $\mathbf{c}^E$ is the 6×6 matrix (pseudomatrix) of elastic moduli, $c_{\alpha\beta}^E = c_{ijkl}^E$; $\alpha, \beta = 1,2,\ldots 6$; $i,j,k,l = 1,2,3$ with correspondence law $\alpha \leftrightarrow (ij)$, $\beta \leftrightarrow (kl)$, $1 \leftrightarrow (11)$, $2 \leftrightarrow (22)$, $3 \leftrightarrow (33)$, $4 \leftrightarrow (23) = (32)$, $5 \leftrightarrow (13) = (31)$, $6 \leftrightarrow (12) = (21)$; $\mathbf{e}$ is the 3×6 matrix (pseudomatrix) of piezoelectric moduli ($e_{i\beta} = e_{ikl}$).

Note that from the first $u\varphi$-problem by using (44) it is possible to find the following essentially non-zero effective moduli $c_{11}^{Eeff,u\varphi}$, $c_{12}^{Eeff,u\varphi}$, $c_{13}^{Eeff,u\varphi}$ and $e_{31}^{eff,u\varphi}$. Analogously, from the second $u\varphi$-problem by using (45) we can determine the effective moduli $c_{33}^{Eeff,u\varphi}$, $c_{13}^{Eeff,u\varphi} = c_{23}^{Eeff,u\varphi}$, $e_{33}^{eff,u\varphi}$. From the third $u\varphi$-problem by using (46) we can find $c_{44}^{Eeff,u\varphi}$,

$e_{15}^{eff,u\varphi} = e_{24}^{eff,u\varphi}$. From the fourth $u\varphi$-problem by using (47) we can find $e_{15}^{eff,u\varphi}$, $\varepsilon_{11}^{Seff,u\varphi}$. Finally, from the fifth $u\varphi$-problem by using (48) we can find $e_{31}^{eff,u\varphi} = e_{32}^{eff,u\varphi}$, $e_{33}^{eff,u\varphi}$ and $\varepsilon_{33}^{Seff,u\varphi}$. As a result we shall have a full set of effective moduli of porous piezoceramic.

It is simple to obtain similar formulae for the $p\varphi$-, uq- and pq-problems (21), (22), (23), respectively. Thus, it is important to take into account, that at transfer to two-index notations the following formulae take place: $s_{\alpha\beta}^{E,D} = 2^p s_{ijkl}^{E,D}$; $d_{j\alpha} = 2^l d_{jkl}$; $g_{j\alpha} = 2^l g_{jkl}$; where $p = 0,1,2$; $l = 0,1$, depending on number of the values of indexes α, β, greater than 3.

Thus for the $p\varphi$-problem we have the following problems and the formulae for calculation of the moduli in two-index tables of symbols:

I-$p\varphi$. $\boldsymbol{\sigma}_0 = \sigma_0 \mathbf{e}_1 \mathbf{e}_1$, $(\sigma_{011} = \sigma_0)$, $\mathbf{E}_0 = 0 \Rightarrow$

$$s_{11}^{E\,\text{eff},p\varphi} = \langle \varepsilon_{11}^{p\varphi} \rangle / \sigma_0; \quad s_{12}^{E\,\text{eff},p\varphi} = \langle \varepsilon_{22}^{p\varphi} \rangle / \sigma_0; \tag{49}$$

$$s_{13}^{E\,\text{eff},p\varphi} = \langle \varepsilon_{33}^{p\varphi} \rangle / \sigma_0; \quad d_{31}^{\text{eff},p\varphi} = \langle D_3^{p\varphi} \rangle / \sigma_0$$

II-$p\varphi$. $\boldsymbol{\sigma}_0 = \sigma_0 \mathbf{e}_3 \mathbf{e}_3$, $(\sigma_{033} = \sigma_0)$, $\mathbf{E}_0 = 0 \Rightarrow$

$$s_{33}^{E\,\text{eff},p\varphi} = \langle \varepsilon_{33}^{p\varphi} \rangle / \sigma_0; \quad d_{33}^{\text{eff},p\varphi} = \langle D_3^{p\varphi} \rangle / \sigma_0; \tag{50}$$

$$s_{13}^{E\,\text{eff},p\varphi} = \langle \varepsilon_{11}^{p\varphi} \rangle / \sigma_0 = \langle \varepsilon_{22}^{p\varphi} \rangle / \sigma_0.$$

III-$p\varphi$. $\boldsymbol{\sigma}_0 = \sigma_0 (\mathbf{e}_2 \mathbf{e}_3 + \mathbf{e}_3 \mathbf{e}_2)$, $(\sigma_{023} = \sigma_{032} = \sigma_0)$, $\mathbf{E}_0 = 0 \Rightarrow$

$$s_{44}^{E\,\text{eff},p\varphi} = \langle 2\varepsilon_{23}^{p\varphi} \rangle / (\sigma_0); \quad d_{15}^{\text{eff},p\varphi} = \langle D_2^{p\varphi} \rangle / \sigma_0. \tag{51}$$

IV-$p\varphi$. $\boldsymbol{\sigma}_0 = 0$, $\mathbf{E}_0 = E_0 \mathbf{e}_1$ $(E_{01} = E_0) \Rightarrow$

$$d_{15}^{\text{eff},p\varphi} = \langle 2\varepsilon_{13}^{p\varphi} \rangle / E_0; \quad \varepsilon_{11}^{T\,\text{eff},p\varphi} = \langle D_1^{p\varphi} \rangle / E_0. \tag{52}$$

V-$p\varphi$. $\boldsymbol{\sigma}_0 = 0$, $\mathbf{E}_0 = E_0 \mathbf{e}_3$ $(E_{03} = E_0) \Rightarrow$

$$d_{31}^{\text{eff},p\varphi} = \langle \varepsilon_{11}^{p\varphi} \rangle / E_0 = \langle \varepsilon_{22}^{p\varphi} \rangle / E_0; \tag{53}$$

$$d_{33}^{\text{eff},u\varphi} = \langle \varepsilon_{33}^{p\varphi} \rangle / E_0; \quad \varepsilon_{33}^{T\,\text{eff},p\varphi} = \langle D_3^{p\varphi} \rangle / E_0.$$

Here from the first $p\varphi$-problem by using (49) we can find the effective compliances $s_{11}^{Eeff,p\varphi}$, $s_{12}^{Eeff,p\varphi}$, $s_{13}^{Eeff,p\varphi}$ and the effective piezoelectric module $d_{31}^{eff,p\varphi}$. Further, from the second $p\varphi$-problem by using (50) we can determine the effective compliances $s_{33}^{Eeff,p\varphi}$, $s_{13}^{Eeff,p\varphi} = s_{23}^{Eeff,p\varphi}$ and the effective piezoelectric module $d_{33}^{eff,p\varphi}$. From the third $p\varphi$-problem by using (51) we can find $s_{44}^{Eeff,p\varphi}$, $d_{15}^{eff,p\varphi} = d_{24}^{eff,p\varphi}$. From the fourth $p\varphi$-problem by using (52) we can find $d_{15}^{eff,p\varphi}$, $\varepsilon_{11}^{Teff,p\varphi}$. Finally, from the fifth $p\varphi$-problem by using (53) we can find $d_{31}^{eff,p\varphi} = d_{32}^{eff,p\varphi}$, $d_{33}^{eff,p\varphi}$ and $\varepsilon_{33}^{Teff,p\varphi}$. As a result, the solution of five $p\varphi$-problems allows us to obtain full set of effective moduli $s_{\alpha\beta}^{Eeff,p\varphi}$, $d_{i\alpha}^{eff,p\varphi}$, $\varepsilon_{ii}^{Teff,p\varphi}$ of porous piezoceramic.

Analogously for the uq-problem we have the following problems and the formulae to calculate a full set of the moduli in two-index notation:

I-uq. $\boldsymbol{\varepsilon}_0 = \varepsilon_0 \mathbf{e}_1 \mathbf{e}_1$, ($\varepsilon_{011} = \varepsilon_0$), $\mathbf{D}_0 = 0 \Rightarrow$

$$c_{11}^{D\,\mathrm{eff},uq} = \left\langle \sigma_{11}^{uq} \right\rangle / \varepsilon_0 ; \quad c_{12}^{D\,\mathrm{eff},uq} = \left\langle \sigma_{22}^{uq} \right\rangle / \varepsilon_0 ; \tag{54}$$

$$c_{13}^{D\,\mathrm{eff},uq} = \left\langle \sigma_{33}^{uq} \right\rangle / \varepsilon_0 ; \quad h_{31}^{\mathrm{eff},uq} = -\left\langle E_3^{uq} \right\rangle / \varepsilon_0 .$$

II-uq. $\boldsymbol{\varepsilon}_0 = \varepsilon_0 \mathbf{e}_3 \mathbf{e}_3$, ($\varepsilon_{033} = \varepsilon_0$), $\mathbf{D}_0 = 0 \Rightarrow$

$$c_{33}^{D\,\mathrm{eff},uq} = \left\langle \sigma_{33}^{uq} \right\rangle / \varepsilon_0 ; \quad h_{33}^{\mathrm{eff},uq} = -\left\langle E_3^{uq} \right\rangle / \varepsilon_0 ; \tag{55}$$

$$c_{13}^{D\,\mathrm{eff},uq} = \left\langle \sigma_{11}^{uq} \right\rangle / \varepsilon_0 = \left\langle \sigma_{22}^{uq} \right\rangle / \varepsilon_0 .$$

III-uq. $\boldsymbol{\varepsilon}_0 = \varepsilon_0 (\mathbf{e}_2 \mathbf{e}_3 + \mathbf{e}_3 \mathbf{e}_2)$, ($\varepsilon_{023} = \varepsilon_{032} = \varepsilon_0$), $\mathbf{D}_0 = 0 \Rightarrow$

$$c_{44}^{D\,\mathrm{eff},uq} = \left\langle \sigma_{23}^{uq} \right\rangle / (2\varepsilon_0) ; \quad h_{15}^{\mathrm{eff},uq} = -\left\langle E_2^{uq} \right\rangle / (2\varepsilon_0) . \tag{56}$$

IV-uq. $\boldsymbol{\varepsilon}_0 = 0$, $\mathbf{D}_0 = D_0 \mathbf{e}_1$ ($D_{01} = D_0$) $\Rightarrow$

$$h_{15}^{\mathrm{eff},uq} = -\left\langle \sigma_{13}^{uq} \right\rangle / D_0 ; \quad \beta_{11}^{S\,\mathrm{eff},uq} = \left\langle E_1^{uq} \right\rangle / D_0 . \tag{57}$$

V-uq. $\boldsymbol{\varepsilon}_0 = 0$, $\mathbf{D}_0 = D_0 \mathbf{e}_3$ ($D_{03} = D_0$) $\Rightarrow$

$$h_{31}^{\text{eff},uq} = -\left\langle \sigma_{11}^{uq} \right\rangle / D_0 = -\left\langle \sigma_{22}^{uq} \right\rangle / D_0 \; ; \tag{58}$$

$$h_{33}^{\text{eff},uq} = -\left\langle \sigma_{33}^{uq} \right\rangle / D_0 \; ; \quad \beta_{33}^{S\,\text{eff},uq} = \left\langle E_3^{uq} \right\rangle / D_0 \; .$$

At last for pq-problem we have the following problems and the formulae to calculate a full set of the moduli in two-index notation:

I-pq. $\boldsymbol{\sigma}_0 = \sigma_0 \mathbf{e}_1 \mathbf{e}_1$, $(\sigma_{011} = \sigma_0)$, $\mathbf{D}_0 = 0 \Rightarrow$

$$s_{11}^{D\,\text{eff},pq} = \left\langle \varepsilon_{11}^{pq} \right\rangle / \sigma_0 \; ; \quad s_{12}^{D\,\text{eff},pq} = \left\langle \varepsilon_{22}^{pq} \right\rangle / \sigma_0 \; ; \tag{59}$$

$$s_{13}^{D\,\text{eff},pq} = \left\langle \varepsilon_{33}^{pq} \right\rangle / \sigma_0 \; ; \quad g_{31}^{\text{eff},pq} = -\left\langle E_3^{p\varphi} \right\rangle / \sigma_0 \; .$$

II-pq. $\boldsymbol{\sigma}_0 = \sigma_0 \mathbf{e}_3 \mathbf{e}_3$, $(\sigma_{033} = \sigma_0)$, $\mathbf{D}_0 = 0 \Rightarrow$

$$s_{33}^{D\,\text{eff},pq} = \left\langle \varepsilon_{33}^{pq} \right\rangle / \sigma_0 \; ; \quad g_{33}^{\text{eff},pq} = -\left\langle E_3^{pq} \right\rangle / \sigma_0 \; ; \tag{60}$$

$$s_{13}^{D\,\text{eff},pq} = \left\langle \varepsilon_{11}^{pq} \right\rangle / \sigma_0 = \left\langle \varepsilon_{22}^{pq} \right\rangle / \sigma_0 \; .$$

III-pq. $\boldsymbol{\sigma}_0 = \sigma_0 (\mathbf{e}_2 \mathbf{e}_3 + \mathbf{e}_3 \mathbf{e}_2)$, $(\sigma_{023} = \sigma_{032} = \sigma_0)$, $\mathbf{D}_0 = 0 \Rightarrow$

$$s_{44}^{D\,\text{eff},pq} = \left\langle 2\varepsilon_{23}^{pq} \right\rangle / (\sigma_0) \; ; \quad g_{15}^{\text{eff},pq} = -\left\langle E_2^{pq} \right\rangle / \sigma_0 \; . \tag{61}$$

IV-pq. $\boldsymbol{\sigma}_0 = 0$, $\mathbf{D}_0 = D_0 \mathbf{e}_1$ $(D_{01} = D_0) \Rightarrow$

$$g_{15}^{\text{eff},pq} = \left\langle 2\varepsilon_{13}^{pq} \right\rangle / D_0 \; ; \quad \beta_{11}^{T\,\text{eff},pq} = \left\langle E_1^{pq} \right\rangle / D_0 \; . \tag{62}$$

V-pq. $\boldsymbol{\sigma}_0 = 0$, $\mathbf{D}_0 = D_0 \mathbf{e}_3$ $(D_{03} = D_0) \Rightarrow$

$$g_{31}^{\text{eff},pq} = \left\langle \varepsilon_{11}^{pq} \right\rangle / D_0 = \left\langle \varepsilon_{22}^{pq} \right\rangle / D_0 \; ; \tag{63}$$

$$g_{33}^{\text{eff},uq} = \left\langle \varepsilon_{33}^{pq} \right\rangle / D_0 \; ; \quad \beta_{33}^{T\,\text{eff},pq} = \left\langle E_3^{pq} \right\rangle / D_0 \; .$$

3. MODELS OF REPRESENTATIVE VOLUMES, FEM AND NON-UNIFORM POLARIZATION

Above-mentioned formulae for calculation of effective moduli assume the solution of corresponding boundary problems of electroelasticity in region Ω, which should be representative volume of porous piezocomposites. As representative volumes in an ideal it is necessary to take areas, big enough compared with the sizes of heterogeneity (i.e. pore), but small in comparison with distances on which averaged variables essentially vary. Nevertheless, as the first approximation it is possible to take volume Ω in the form of a cube with one cubic pore. The given model we shall call the model 1. For the analysis of influence of corners on the results it is possible to consider model without angular points, for example, spherical volume with a spheroidal cavity, or, more generalized model of ellipsoid with ellipsoidal cavity. The given model we shall name the model 2.

More adequate model for the real porous piezoceramic structure at low relative porosity is a model of piezoceramic cube evenly divided into lesser cubes, part of which randomly declared as pores (model 3). Note that the similar model was considered in [18].

Obviously that even in simplest cases (models 1 and 2) solving the boundary problems of electroelasticity (13), (14); (13), (21); (13), (22) or (13), (23) by using analytical methods is exceedingly difficult and for the model 3 these boundary problems can be solved only numerically. In accordance with modern standards the most effective method for numerical solution of the electroelasticity problems for representative volumes of heterogeneous media is the finite element method (FEM) [23, 24]. For solving these problems it is convenient to use finite element package ANSYS [24] with macro-language APDL.

Some finite element models for cube with cubic pore and for ellipsoid with ellipsoidal pore are shown on Figure 1 and Figure 2, respectively.

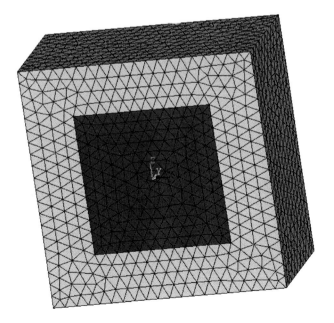

Figure 1. Finite element model for cube with cubic pore (20 % porosity).

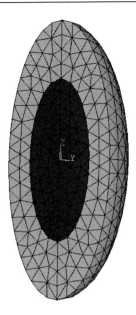

Figure 2. Finite element models for ellipsoid with ellipsoidal pore (20 % porosity).

Figure 3. Finite-element model 3 (20 % porosity).

In finite-element model 3 each small cube is accepted as a single finite element with material properties of piezoelectric body or pore. Pore is modelled as a body with negligible small elastic moduli and dielectric permittivities, equal to permittivity of vacuum. One variant of finite element model 3 is shown in Figure 3.

It is possible to examine more complex model 3 taking into account partial polarization of the ceramics in pores vicinity [25]. In this case, in first stage the process of partial polarization of porous ceramics along z-axis direction is modelled. With this aim, at first a quasi-electrostatic problem for inhomogeneous dielectric is solved in region Ω:

$$\nabla \cdot \mathbf{D} = 0, \; \mathbf{D} = \varepsilon_* \mathbf{E}, \; \mathbf{E} = -\nabla \varphi, \; \mathbf{x} \in \Omega; \tag{64}$$

$$\varphi = V_j, \; \mathbf{x} \in \Gamma_{\varphi j}; \tag{65}$$

$$\mathbf{n} \cdot \mathbf{D} = 0, \; \mathbf{x} \in \Gamma_q, \tag{66}$$

where $\Gamma = \Gamma_{\varphi 1} \cup \Gamma_{\varphi 2} \cup \Gamma_q$; $\Gamma_{\varphi 1}$, $\Gamma_{\varphi 2}$ are the electrodes $z = L$ and $z = 0$, respectively; $V_1 = -E_p L$, $V_2 = 0$, E_p is a preliminary value for polarization field.

Inhomogeneity of the dielectric can be defined by the following way. At fixed relative porosity p number of cubic finite elements n_p representing pores is calculated for the model 3. Than material properties with dielectric constant of vacuum are randomly assigned to n_p elements from macrocube ε_*. By using electrostatic finite elements with different material properties problem (64) – (66) is solved. Then the calculated average values of the field E_{zm} for each finite element that is not a pore are analyzed. Marks of polarization degree are assigned to the elements if $E_{zm} > E_c$, where $E_c = k_c E_p$, $k_c \leq 1$ is a some fixed value for polarization field E_c.

It is possible also propose a "hyperpolarization" hypothesis for individual elements for which polarization field E_{zm} exceeds some fixed value $E_{zm} > E_s$, where $E_c = k_c E_p$, $k_s \geq 1$ is the fixed value of "hyperpolarization" field. Additional factor $m_s \geq 1$ is assigned to "hyperpolarized" elements and used for piezoceramics moduli calculation by formulae: $c_{\alpha\beta}^{E,s} = (c_{\alpha\beta}^E - c_{\alpha\beta}^I)m_s + c_{\alpha\beta}^I$, $e_{i\alpha}^s = e_{i\alpha}m_s$, $\varepsilon_{ij}^{S,s} = (\varepsilon_{ij}^E - \varepsilon_{ij}^I)m_s + \varepsilon_{ij}^I$, where $c_{\alpha\beta}^I$, ε_{ij}^I are the moduli of non-polarized ceramics.

In the second modeling stage the electrostatic finite elements are modified to the elements with possibility of the piezoelectric analysis. Four types of material properties are assigned to the new elements: (i) properties of polarized piezoceramics for polarized elements, (ii) negligible elastic moduli and omissible piezoelectric moduli for pores, (iii) isotropic, elastic, and dielectric properties for non-polarized ferroelectric ceramics, (iv) properties of "hyperpolarized" piezoceramics for "hyperpolarized" elements. Then for effective moduli the stated electroelasticity problem for representative volume is solved by using methods described in previous section.

In order to describe fully closed 3-0 type porosity it is possible to complicate the model 3 by using partially controllable porosity structure. For example, it is possible to subdivide elementary compound cube into piezoelectric skeleton and lesser cube, as shown in Figure 4a, b. Thus, in Figure 4a the cube, which can be a pore, is shown by taupe color. This cubic cell can be pore or medium with the same piezoelectric properties as the skeleton ones. To obtain given porosity part of lesser cubes randomly declared as pores. As a result, we obtain new

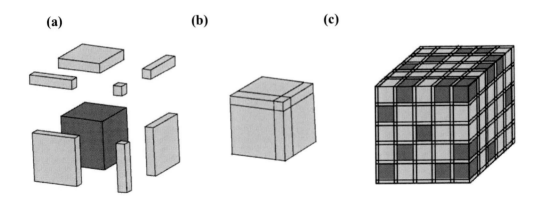

Figure 4. Model 4 is the model of 3–0 connectivity type: (a) configuration of a cell, (b) aggregated cell, and (c) representative volume.

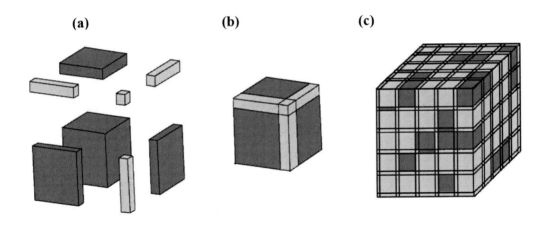

Figure 5. Model 5 is the model of 3–0 – 3–3 connectivity type: (a) configuration of a cell, (b) aggregated cell, and (c) representative volume.

model 4 with fully closed porosity of 3–0 connectivity type. One variant of the representative volume assembled from such cells is shown in Figure 4c. Note that in the model 4 besides percent of porosity there is one more important geometrical parameter, namely the relation of the skeleton thickness to the size of an elementary compound cube.

For 3–3 connectivity type modeling two additional models were developed. The model 5 named 3–0 – 3–3 connectivity model is a future development of the model 4. In this model additionally to a greater cube in elementary cell thin parallelepipeds also can be pores, as is shown in Figure 5a, b, where separate intrinsic cube and thin parallelepipeds, which can be a pore, are shown by taupe color. In this model exception of a situation when the volume does not represent the connected structure of solid-state components is algorithmically stipulated.

The model 6 (3–3 connectivity type) is constructed by the following manner. The cube constructed by translation of equal structured cells along three directions is considered as a representative volume. Each cell in its turn also represents cube consisting of 10×10×10 cubic

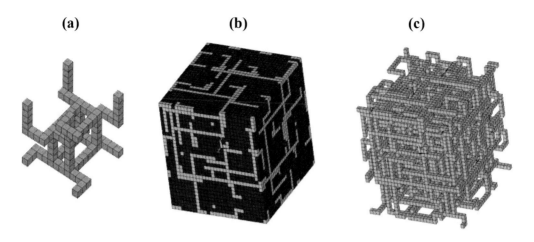

Figure 6. Model 6 is the model of 3–3 connectivity type: (a) elementary skeleton, (b) volume with voids, and (c) volume without voids.

elements. The connected skeleton with the elements in the cube corners always exist in the cell. The skeleton consist of parallelepiped represented by own edges (linear dimensions are pointed out by the randomizer) as well as the chains of elements connecting its corners with the corners of the main cell. The connecting elements of the chains are generated by the randomizer also. The skeleton occupied 10 % of cell volume (see Figure 6a). So, maximal possible porosity that can be achieved in this model attained up to 90 %. One variant of the representative volume of 3–3 connectivity is shown in Figure 6b (where black color is used for voids) and Figure 6c (where void is not presented).

4. COMPUTATIONAL AND EXPERIMENTAL RESULTS

Experimental samples of porous PZT ceramics in relative porosity range 0 – 70 % have been fabricated using specially developed technology [2]. Measurements of resonance impedance spectra as a function of frequency were made using Solartron 1260 frequency response analyzer. Piezoelectric Resonance Analysis Program (PRAP) has been used for analysis of impedance spectra to determine dielectric, piezoelectric and electromechanical properties of porous ceramics.

For the models 1 and 2 finite-element calculations of effective moduli of porous ceramics PCR-1 [26] were conducted for all four types of problems: $u\varphi$, $p\varphi$, uq and pq. It was found that the influence of mechanical boundary conditions on the effective moduli is stronger than the electrical ones. The $u\varphi$- and uq-models have the pointed out fixation conditions being more rigid than for the $p\varphi$- and pq-models in which mechanical stresses are fixed. The dielectric constants obtained on the base of the $u\varphi$- and uq-models are less than corresponding values found from the $p\varphi$- and pq-models. The $u\varphi$- and uq-models gives overvalued elastic stiffness moduli, ultrasonic velocities and electromechanical coupling factors. On the

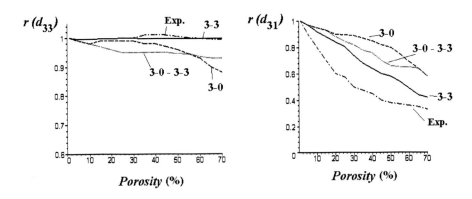

Figure 7. Dependencies of effective moduli on porosity.

contrary, piezoelectric moduli d_{33} and d_{31} are overvalued for the $p\varphi$- and pq-models. Dielectric constant ε_{33}^T is less dependent from the problem type.

Comparison of the calculating and measuring results [26, 27] for different porosities have shown that $p\varphi$-problem gives more reasonable values of ultrasonic velocities, stiffness moduli and piezoelectric moduli $e_{i\alpha}$ than $u\varphi$-problem. Note that in all cases piezoelectric modulus d_{31} decreases with the porosity no so fastly as in the experiment. By the way, $u\varphi$-problem is most convenient for FEM calculations, particularly for models with complex boundaries.

From the results of the models 1 and 2 follows, that corner points enters minor distortions to integral characteristics of porous ceramics. Piezoelectric modulus d_{31} in the model 3 at homogeneous polarization practically does not decrease with the porosity. Decrease of the piezoelectric modulus d_{31} can be reached by using partial polarization models however in this case piezoelectric modulus d_{33} also decreases. Serious disadvantage of the model 3 compared to more complex models 4 – 6 is a dislocation of ceramic skeleton connectivity. The results of these models are much better than for the model 3. At the same time, 3–3 connectivity type model gives calculated values of d_{33} that are very close to experimental ones for broad range of porosity. The calculated values of d_{31} for porous ceramics are in less conformity with the experiments. However, the model 6 for 3–3 connectivity type also gives more reasonable results.

Thus, in Figure 7 we can see the dependencies of piezomoduli relations for porous and dense ceramics $r(d_{3j}) = d_{3j}^{e\!f\!f}/d_{3j}$ ($j=1,3$) on porosity for all three models of 3–0; 3–0 – 3–3 and 3–3 connectivities and the evaluation of experimental data.

CONCLUSION

In this chapter different FEM models of porous piezoceramics based on the effective moduli method were developed and considered. For the models the properties of porous piezoceramics with well controllable porosity (0 – 70 %) were measured. The results of FEM modeling show that the representative volume model for 3–3 connectivity type with rigid skeleton gives better results compared to other examined models and describes very well experimental results for porous ceramics in the relative porosity range of 0 – 70 %.

ACKNOWLEDGMENTS

This work is supported by State contract P401 of the Federal Target Program "Scientific and Science-Teacher Personnel of Innovative Russia" (2009 – 2013 years) and in part by the Russian Foundation for Basic Research (Grant No. 10-08-13300).

REFERENCES

[1] Newnham, R. E.; Skinner, D. P.; Cross, L. E. *Vater. Res. Bull.* 1978, *vol. 13*, 525-536.
[2] Rybianets, A. N.; Nasedkin, A.V.; Turik, A. V. *Integrated Ferroelectrics.* 2004, *vol. 63*, 179-182.
[3] Wersing, W.; Lubitz, K.; Moliaupt, J. *Ferroelectrics.* 1986, *vol. 68*, 77-97.
[4] Banno, H. *Jpn. J. Appl. Phys.* 1993, *vol. 32*, 4214-4217.
[5] Rittenmyer, K.; Shrout, T.; Schulze, W.A.; Newnham, R.E. *Ferroelectrics.* 1982, *vol. 4*, 189-195.
[6] Perry, A.; Bowen, C. R.; Mahon, S. W. *Scripta Materialia.* 1999, *vol. 41*, 1001–1007.
[7] Bowen, C. R.; Perry, A.; Kara, H.; Mahon, S. W. *J. Eur. Ceram. Soc.* 2001, *vol. 21*, 1463-1467.
[8] Topolov, V. Yu.; Bowen, C. R. *Electromechanical Properties in Composites Based on Ferroelectrics.* Springer: London, 2009; pp. 1-202.
[9] Levassort, F.; Lethiecq, M.; Sesmare, R.; Tran-IIuu-IIuc, L.P. *IEEE Trans. Ultras. Ferr. Freq. Control.* 1999, *vol. 46*, 1028-1033.
[10] Dunn, H.; Taya, M. *J. Am. Ceram. Soc.* 1993, *vol. 76*, 1697-1706.
[11] Silva, E.C.N.; Fonseca, J.S.O.; Kikuchi, N. *Comput. Meth. Appl. Mech. Eng.* 1998, *vol. 159*, 49-77.
[12] Sokolkin, Yu. V.; Pan'kov, A. A. *Electroelasticity of Piezocomposites with Irregular Sstructure.* Fizmatlit: Moscow, 2003; pp. 1-176 (in Russian)
[13] Pan'kov, A. A. *Statistical Mechanics of Piezocomposites.* Perm State University Press: Perm, 2009; pp. 1-480. (in Russian)
[14] Ramesh, R.; Kara, H.; Bowen, C. R. *Comput. Materials Sci.* 2004, *vol. 30*, 397–403.
[15] Ramesh, R.; Kara, H.; Bowen, C. R. *Ultrasonics.* 2005, *vol. 43*, 173–181.
[16] Soloviev, A. N.; Vernigora, G.D. In: *Piezoceramic Materials and Devices.* I. A. Parinov; Ed.; Nova Publishers: N.-Y., 2010, pp 217-240.
[17] Khoroshun, L.P.; Maslov, B.P.; Leshchenko, P.V. *Prediction of Effective Pproperties of Piezoactive Composite Materials.* Naukova Dumka: Kiev, 1989. (in Russian)
[18] Getman, I., Lopatin, S. *Ferroelectrics.* 1996, *vol. 186*, 301-304.

[19] Nasedkin, A. V. *Modern Problems of Continuum Mechanics, Proc. VII Int. Conf., Rostov on Don, 22-24 Oct. 2001*. Novaja Kniga: Rostov on Don, 2002, vol. 1, pp 182-188. (in Russian)

[20] Nasedkin, A.; Rybjanets, A.; Kushkuley, L.; Eshel, Y.; Tasker, R. *Proc. 2005 IEEE Ultrason. Symp., Rotterdam, Sept. 18 -21, 2005*, pp 1648-1651.

[21] Pobedria, B.E. *Mechanics of Composite Materials*. Moscow State University Press: Moscow, 1984; pp. 1-336. (in Russian)

[22] Berlincourt, D.A.; Curran, D.R.; Jaffe, H. *Piezoelectric and Piezomagnetic Materials and Their Application in Transducers*. Part A, In: W.P. Mason (Ed.), *Physical Acoustics*, vol. 1, Academic Press: New York, 1964; pp. 1-592.

[23] Kaltenbacher M. *Numerical Simulation of Mechatronic Sensors and Actuators*. Springer: Berlin, Heidelberg, New York, 2010; pp. 1-446.

[24] *ANSYS. Theory Ref.* Rel.11.0. Ed. P. Kothnke. ANSYS Inc. Houston, 2007.

[25] Nasedkin, A. V.; Rybjanets, A. N. *Russian Izvestiya, North-Caucas. Region. Ser. Techn Sciences, Spec. Issue*. 2004, 91-95. (in Russian)

[26] Dantziger, A. J.; Razumovskaya, O. N.; Reznichenko, L. A.; et al. *Multicomponent Systems of Ferroelectric Solid Solutions: Physics, Crystallochemistry, Technology. Design Aspects of Piezoelectric Materials*. Rostov on Don: Rostov State University Press, *vol. 1-2*, 2001; pp. 1-808. (in Russian)

[27] Rybyanets, A.N. In: *Piezoceramic Materials and Devices*. I. A. Parinov; Ed.; Nova Publishers: N.-Y., 2010, pp 113-176.

In: Ferroelectrics and Superconductors
Editor: Ivan A. Parinov

ISBN: 978-1-61324-518-7
©2012 Nova Science Publishers, Inc.

Chapter 8

COMPUTER MODELING OF RESONANCE CHARACTERISTICS OF THE PIEZOELECTRIC CYLINDRICAL AND BIMORPF TRANSDUCERS

T. G. Lupeiko[*,1], *A. N. Soloviev*[2,3], *B. S. Medvedev*[1], *A. S. Pahomov*[1], *M. P. Petrov*[1], *V. S. Chernov*[1] *and D. V. Nasonova*[1]

[1]Department of Chemistry, Southern Federal University, Rostov-on-Don, Russia,
[2]Southern Scientific Center of Russian Academy of Sciences, Rostov-on-Don, Russia
[3]Department of Material Strength, Don State Technical University,
Rostov-on-Don, Russia

ABSTRACT

Computer modeling and experimental tests were used to study the influence of piezoelectric material nature and geometry of cylindrical hydro-acoustic transducers and piezobimorphs on their amplitude-frequency behavior. The outer diameter of the cylinders was changed within the range of 50 – 100 mm whereas their wall thickness was within the range of 5 – 15 mm. The diameter and the thickness of bimorph membranes were 32 mm and 100 μm, respectively. The diameter and the thickness of their piezoelements were 20 – 30 mm and 100 – 200 μm, respectively.

1. INTRODUCTION

Computer modeling of the transducers became possible quite recently judging by the publications [1 – 32]. The research in this sphere is mainly connected with the development and optimization of varies piezoelectric materials and transducer models.

A theory of a piezoelectric axisymmetric bimorph is presented in [1]. This theory is based on the shell model. The finite-element method (FEM) is applied to solve this boundary-value problem. An efficient and accurate approach to piezoelectric bimorph based on refined

---

*E-mail: lupeiko@rsu.ru, 7, Zorge Street, 344090 Rostov-on-Don, Russia.

expansion of the elastic displacement and electric potential is proposed in [2]. An axisymmetric electroelastic problem of hollow radially polarized piezoceramic cylinders made of functionally-graded materials (FGM) is analyzed in [3].

A high-order theoretical model based on the governing equations and the corresponding natural boundary conditions for functionally-graded piezoelectric generic shells has been derived in [4]. In [5] the authors examine the wave propagation in a piezoelectric coupled cylindrical shell affected by the shear effect and rotary inertia. A complete mathematical analysis of wave propagation solution in this piezoelectric coupled cylindrical shell is provided. Piezoelectric porous lead zirconate titanate (PZT) ceramics prepared by different methods has been examined in terms of microstructure in [6].

A micro-ultrasonic motor using a micro-machined bulk piezoelectric transducer is introduced in [7, 8]. A cylindrical micro-ultrasonic motor utilizing PZT thin film was fabricated and successfully operated. Theoretical and numerical results were investigated by using an equivalent circuit in terms of scale law.

The exact solutions of a simply supported functionally-graded piezoelectric plate/laminate under cylindrical bending are derived in [9]. An alternative method for the derivation of exact solutions of a simply supported rectangular functionally-graded piezoelectric plate or laminate is present in [10].

The design and modeling of FGM piezoelectric transducers in comparison with non-FGM ones are proposed in [11]. Analytical and finite element (FE) modeling of FGM piezoelectric transducers radiating a plane pressure wave in fluid medium are developed and their results are compared.

The governing equations of hollow cylindrical piezo-ceramics with axial polarization by using the force–velocity boundary conditions and equalizing the mechanical elements to the electrical ones are solved in [12]. Then the equivalent electro-mechanical impedance is represented via matrix notation.

An analytical solution in series form for the problem of a circularly cylindrical layered piezoelectric composite consisting of **n** dissimilar layers is given in [13] within the framework of linear piezoelectricity. The alternating technique in conjunction with the method of analytical continuation is applied to derive the general multilayered media solution in an explicit series form, whose convergence is guaranteed numerically.

The vibration of a cylindrical piezoelectric transducer induced by applied voltage, which can be used as the stator transducer of a cylindrical micro-motor, is studied based on shell theory in [14]. This work provides a general and precise theoretical modeling the dynamical movement of the transducer.

The coupled extensional and flexural vibrations of an annular corrugated shell piezoelectric transducer consisting of multiple circularly-annular surfaces smoothly connected along the interfaces were investigated in [15]. A theoretical solution was obtained by using the classical shell theory. The radial vibration characteristics of piezoelectric cylindrical transducers are present in [16].

The dynamic electromechanical behavior of a triple-layer piezoelectric composite cylinder with imperfect interfaces is investigated in [17]. The obtained solution is valid for analyzing the dynamic electromechanical behavior of composite cylinder with arbitrary thickness for both elastic and piezoelectric layers.

The radial vibration of a piezoelectric ceramic thin circular ring polarized in the thickness direction is studied in [18]. Its electro-mechanical equivalent circuit is derived, the resonance and anti-resonance frequency equations are obtained, and the relationship between the resonance frequencies, the material parameters and the geometrical dimensions is analyzed.

The transmission of electric energy through a circular cylindrical elastic shell by acoustic wave propagation and piezoelectric transducers is studied in [19]. Mechanical model consists of a circular cylindrical elastic shell with finite piezoelectric patches on both sides of the shell. A trigonometric series solution is obtained.

The radial vibration of piezoelectric cylindrical transducers polarized in the radial direction is studied in [20]. The purpose of this paper is to establish a formula for calculating the piezoelectric natural frequency of these transducers. The dependence of the piezoelectric natural frequency on the radius and thickness of the piezoelectric cylinder is discussed.

A new type of radial composite piezoelectric transducers with radial vibrations is developed and analyzed in [21, 22]. Their radial electro-mechanical equivalent circuits are obtained. The resonance and the anti-resonance frequencies, the electro-mechanical equivalent circuit parameters are measured. Nonlinear elastic vibrations of cylindrical piezoelectric transducers are investigated both experimentally and theoretically in [23].

Recently due to the new mathematic package based on the finite-element method [24 – 31], the computer modeling has become an effective instrument for transducers performance research. This gave the opportunity to carry out multifactor chemical-technological and dimension design of amplitude-frequency characteristics (AFC) for transducers of various types which presents both theoretical and practical interest.

In the present chapter the systematic calculation results using the software ACELAN and ANSYS to define transducer resonant characteristics of cylindrical and bimorph types depending on their geometry and piezomaterial nature are given.

2. CYLINDRICAL TRANSDUCERS

2.1. Introduction

The cylindrical transducers with electrodes on the generating surfaces and polarization along the wall thickness (Figure 1) occupy a special place among transducers. They can serve as a base for generating highly effective cylinder wall vibrations (Figure 2).

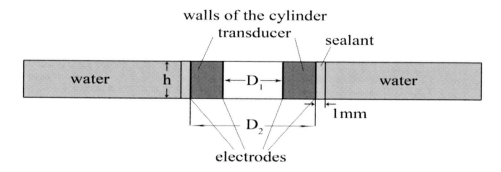

Figure 1. The section of a cylindrical transducer loaded against water with electrodes on the external surfaces.

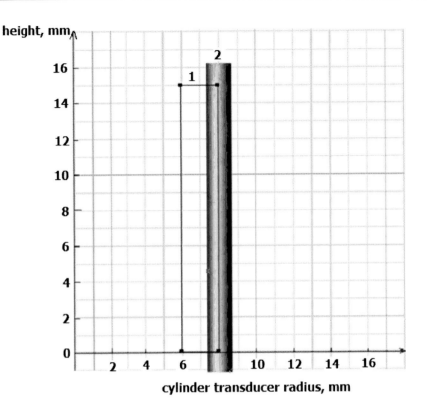

Figure 2. Schematic pattern of a cylindrical transducer wall section in initial (1) and displaced (2) position at radial vibration.

The above-mentioned emitters are of great interest for hydro-acoustics. They emit into the water and their outer surface is protected with a sealant (Figure 1).

2.2. Experimental Procedure

In the present work the results of the resonance characteristic research of the hydro-acoustic cylindrical transducers by using computer modeling are given. The emitters studied were made of two piezoelectric ceramics: PZT-19 and PZTB-3. Their principal electro-physical characteristics are presented in Table 1.

Table 1. Some characteristics of PZT-19 and PZTB-3 materials

Material parameters	PZT-19	PZTB-3
Permeability, ε_r (at 1 kHz)	1650	2325
Coupling coefficient, K_p	0.56	0.52
Piezoelectric constant, d_{31} (pC/N)	155	158
Velocity of sound, $V_E \cdot 10^{-3}$ m/s	2.95	3.47

The titan and zirconium ratio as well as the solidus of the system $PbZrO_3 - PbTiO_3$ [32] show that at the temperature lower than the Curie point the mentioned material structure is tetragonal. The material PZTB-3 is also widely used and also has the tetragonal structure at the temperature lower than the Curie point.

Cylindrical transducers with 10 mm height and external surfaces coated with 1 mm ebonite sealant were used as the model objects. During the procedure the applied alternating voltage was 100 Volts and the samples were acoustically loaded against water (Figure 1).

2.3. Results and Discussion

In this paper we give the results of computer modeling. The variable dimensions of characteristics were the next:

- outer diameter (D_1) changed within the range 50 – 150 mm with 10 mm increment
- inner diameter (D_2) also changed with 10 mm increment.

The inner diameter of transducers was taken from conditions that the wall thickness was no less than 5 and no more than 15 mm.

The typical pattern of AFC emitters fragment within the frequency band up to 200 kHz is shown in Figure 3.

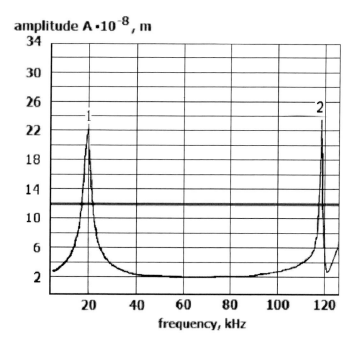

Figure 3. The example of estimated amplitude dependence for the cylindrical transducer made of PZTB-3 (the outer diameter is 120 mm, the height and thickness of the walls are 10 mm, the alternating field voltage is 100 V); 1 - radial vibration resonance, 2 – flexural vibration resonance.

The x-axis corresponds to the frequency and the y-axis defines the resonance vibration amplitude. In this partial case low-frequency vibration is radial whereas high-frequency vibration is flexural. The distance between the points at the amplitude plot correlates with the transducer's frequency working band width for radial vibration mode.

The results of the AFC research were supplemented with vibration analysis in animated run (Figure 2). This allowed one to find out the nature of the vibration resonances, identify the radial vibrations and estimate the degree of their "accuracy". These data allowed one to obtain the frequency values (f, kHz) and amplitude values (A, m) of hydro-acoustic transducer radial vibrations as well as to measure the working band width (Δf_{wb}, kHz) and vibration amplitude values corresponding to the working frequency band level (A_{wb}, m).

Table 2. Vibration resonance amplitude-frequency characteristics of cylindrical transducers made of PZT-19

D_1, mm	D_2, mm	f, kHz	$A \cdot 10^{-8}$, m	$A_{wb} \cdot 10^{-8}$, m	Δf_{wb}, kHz	f_f, kHz
50	20	28.7	8.4	5.9	2.6	101
50	30	24.3	13.3	9.3	3.3	128
50	40	20.3	17.4	12.2	5.3	109
60	30	21.9	11.0	7.7	2.5	99.3
60	40	19.0	13.4	9.4	3.8	125
60	50	16.4	17.0	11.9	6.5	108
70	40	18.0	12.0	8.4	2.5	97.3
70	50	15.4	14.0	9.8	3.8	127
70	60	12.8	16.0	11.7	8.2	143
80	50	14.4	11.6	8.2	3.1	97.3
80	60	12.8	15.1	10.6	3.9	118
80	70	10.2	17.1	11.9	7.3	143
90	60	12.8	13.0	9.1	2.7	96
90	70	11.5	15.3	10.7	3.7	127
90	80	8.9	19.6	13.7	5.6	109
100	70	11.5	13.2	9.3	3.0	94.7
100	80	10.2	15.8	11.8	3.2	127
100	90	7.6	21.1	14.8	5.6	109
110	80	1.,2	12.6	9.9	3.2	97.3
110	90	8.9	15.9	11.7	3.6	127
110	100	6.3	21.6	15.1	4.1	113
120	90	9.0	14.6	10.3	2.6	96.5
120	100	7.6	16.9	11.8	4.4	127
120	110	6.3	23.0	16.1	4.3	108
130	100	7.6	14.9	10.5	2.5	96.5
130	110	6.3	16.0	11.2	3.7	127
130	120	5.0	23.4	16.1	4.4	110
140	110	7.6	14.0	9.8	3.0	96.5
140	120	6.3	18.0	12.6	3.1	127
140	130	5.0	25.2	17.6	3.9	108
150	120	6.3	16.7	11.7	2.1	96.5
150	130	6.3	18.4	12.9	3.1	127
150	140	4.9	25.6	17.9	3.9	108

The radial vibration resonances of the cylindrical transducers in question are revealed in the frequency band limited from above by vibration resonances of flexural type. Moreover, the closer the value of this vibration frequency is to the radial resonance frequency, the stronger is the distortion of the latter. Consequently, by modeling we also determined the frequency and the amplitude of the cylindrical wall flexural vibrations, when their frequency being closest to its radial resonance. The calculated data set is given in Tables 2 and 3.

In order to control the computer estimations of transducer AFCs they were compared with the experimental data. With this purpose two cylindrical emitters were made of PZT-19 by the conventional solid-state sintering method. Their resonance characteristics are shown in

Table 3. Vibration resonance amplitude-frequency characteristics of cylindrical transducers made of PZTB-3

D_1, mm	D_2, mm	f, kHz	$A \cdot 10^{-8}$, m	$A_{wb} \cdot 10^{-8}$, m	Δf_{wb}, kHz	f_f, kHz
50	20	34.7	12.3	8.6	2.9	121
50	30	29.6	19.5	13.7	3.3	140
50	40	240	23.7	16.6	5.7	128
60	30	26.8	15.3	10.7	2.7	117
60	40	23.0	20.7	13.7	3.7	140
60	50	20.0	25.0	16.2	5.5	130
70	40	21.9	17.4	12.7	2.8	116
70	50	19.3	21.1	14.7	3.4	140
70	60	16.7	22.6	15.8	7.5	113
80	50	18.0	19.6	13.7	2.2	116
80	60	16.7	20.6	14.7	3.7	140
80	70	14.1	23.2	16.2	7.7	127
90	60	15.4	19.1	13.4	2.4	114
90	70	14.1	20.8	14.5	3.9	140
90	80	11.5	23.9	16.7	7.4	127
100	70	12.8	17.9	12.5	3.1	114
100	80	11.8	21.3	14.9	3.8	139
100	90	10.2	26.9	18.8	5.9	127
110	80	11.5	17.5	12.3	3.3	116
110	90	10.5	19.7	14.8	4.7	138
110	100	8.9	25.4	17.8	7.1	127
120	90	10.3	19.0	13.3	3.0	114
120	100	9.2	20.7	14.5	4.4	139
120	110	7.6	27.0	18.9	6.4	127

Table 4. The dimension characteristics and frequencies of cylindrical transducer radial vibration resonances

Diameter, mm		Height, mm	Resonance frequency, kHz (model)	Resonance frequency, kHz (experiment)
outer	inner			
51.5	46.5	10	18.3	18.6
73	64.5	10	11.7	11.9

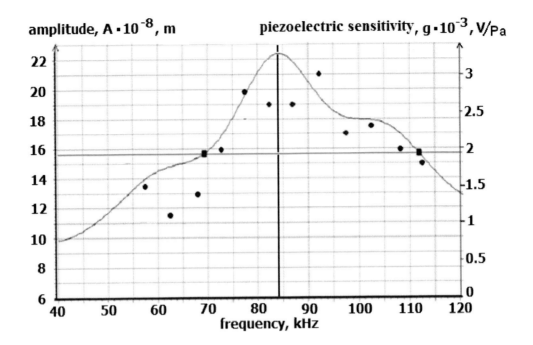

Figure 4. The comparison of AFC dependence of cylindrical transducer loaded against the water (the curve) and hydrophone sensitivity (points); Δf is the working band.

Table 4. As it can be seen the resonance frequencies which are calculated by modeling and obtained from experimental samples have similar values.

In addition, the process of creating broadband emitters resulted in modeling of a cylindrical hydro-acoustic emitter has great practical interest. Figure 4 shows the correlation between its AFC and a hydrophone created on its basis.

As it can be seen from the Figure 4 the estimated and the experimental data agree. It is also noteworthy that the above-mentioned hydro-acoustic transducer has the wide working band of frequencies from 70 up to 113 kHz.

By analyzing the type of vibration in animated mode (see also Tables 2 and 3) we can conclude that cylindrical transducers in the studied dimension range have radial vibration resonances within the following frequency ranges: (i) from 4.9 up to 28.7 kHz for PZT-19, and (ii) from 5.0 up to 34.7 kHz for PZTB-3.

The amplitudes with the following range of values (10^{-8}, m) correspond to the vibration resonances of this interval: (i) from 8.4 up to 25.6 for PZT-19, and (ii) from 12.3 up to 31.0 for PZTB-3

It is also clear from the tables that at the same thickness of cylindrical transducer walls when their diameters diminish then their radial vibration frequency of resonance increases, which well agrees with the known experimental data and long-term experience of cylindrical transducer application. In addition to these known experimental data for the first time there were obtained the general quantitative characteristics that changed. In particular, it shows that a rate change of the resonance frequency increases when the transducer diameter diminishes.

In this case the above-mentioned dependencies are the same for transducers made of PZT-19 (Figure 5) and of PZTB-3 (Figure 6).

At the same time, the vibration amplitude changes depending on samples' diameter at the same wall thickness are not so definite and demonstrate extremal values (see Table 2). Most probably these values appear due to the computer modeling inaccuracy by estimating such an extremal characteristic as vibration resonance amplitude. In fact, even at minimal frequency scan spacing within the process of modeling the experimental estimations there is no guarantee that the resonance frequency will be within the estimated range of frequencies.

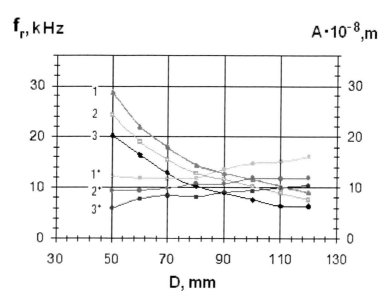

Figure 5. The dependence of resonance frequency (f_r) and vibration amplitude in the working band (A_{wb}) on the outer diameter of PZT-19 cylindrical transducers loaded against the water when the wall thickness is 5.0 (1), 10.0 (2) and 15.0 (3) mm.

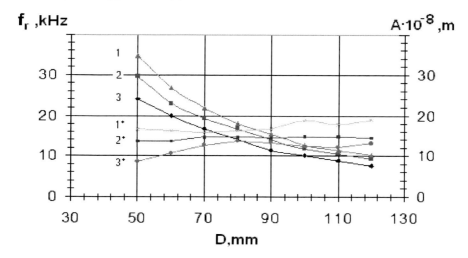

Figure 6. The dependence of resonance frequency (f_r) and vibration amplitude in the working band (A_{wb}) on the outer diameter of PZTB-3 cylindrical transducers loaded against the water when the wall thickness is 5.0 (1), 10.0 (2) and 15.0 (3) mm.

In connection with this, in order to estimate the amplitude values of vibration resonances their amplitude in the working band (A_{wb}) (see Table 2, 3 and Figures 4, 5) was taken. However the character of this parameter change in dependence on the changing cylindrical transducer diameter in the number of cases also differs significantly from linearity. This can be explained by calculation inaccuracy. It's possible that the reason for non-linearity is due to the fact that this amplitude depends no only on transducer diameter but also on some other factors. They may include integral influence of the whole scope of their dimensional characteristics (diameter, wall thickness and height) as well as such characteristics as the working band width and the "purity" of radial vibration itself. This "purity" is in complicated dependence on the extent of difference between radial vibration frequency and flexural vibration frequency (Table 2, 3). By taking this into account, the results of vibration amplitude calculation can obviously give only the idea on their corresponding tendency to change in selection of different transducer diameters (Figures 4, 5).

It is also clear from Figures 5 and 6 that resonance frequencies in radial vibration of the cylindrical transducers, having the same outer diameter, decrease appropriately with the increase of the cylinder inner diameter (diminishing the wall thickness) regardless of the material from which they are made. The influence of this transducer parameter on their vibration frequency increases with the diminishing the cylinder outer diameter. It is also noteworthy that when the wall thickness increases the working bands of the cylindrical transducers widen significantly (in two or three times) (see Tables 1 and 2). As it follows from the tables this widening is in total dependence on cylindrical transducer diameter.

The change in wall thickness of the cylindrical transducer also influences significantly the vibration resonance amplitudes which significantly increase with the diminishing wall thickness of the transducer (Figures 5, 6). This agrees with an increase of the system Q-factor.

It also follows from Tables 1, 2 and Figures 5, 6 that the resonance frequencies of the cylindrical transducers significantly depend on their material nature. Thus, the dimensions being the same to cause the transducers made of PZTB-3 have higher values of resonance frequencies. In general, such resonance frequency dependence on the cylinder outer diameter well agrees with the sound velocity in the corresponding ceramic material (Table 1).

Similarly, the vibration resonance amplitudes are significantly higher for PZTB-3 compared with PZT-19 at the same dimensions of the transducers. This change of vibration resonance amplitudes depending on the piezomaterial nature may be explained by the higher values of PZTB-3 piezoelectric constant and permeability.

Thus, the results of cylindrical transducer research allow statement the following conclusions:

The computer model based on the adapted software ACELAN was created and applied to the calculation of AFC of the cylindrical hydro-acoustic transducer.

For the first time, on the base of ACELAN there was obtained a complete fragment of general relations between vibration frequencies and amplitudes on the one hand and dimension characteristics and piezomaterial nature, on the other hand. It was shown that in general transducer the AFC are determined by their dimension characteristics and strongly depend on piezomaterial nature.

It was found that the vibration resonance frequency and amplitude of hydro-acoustic cylindrical transducers, if the other conditions are equal, are higher for transducers made of PZTB-3 than for those of PZT-19.

There was found that the calculated and experimental AFCs of the hydro-acoustic cylindrical transducers agree.

Based on the data obtained the choice of cylindrical transducers with optimal AFCs is possible. The resonance frequency decrease can be gained by using piezoceramics with higher sound velocity, by increasing the transducer diameter or diminishing their wall thickness. Their efficiency increase can be gained by diminishing their wall thickness and using the material with higher piezoelectric characteristics.

3. BIMORPH TRANSDUCERS

3.1. Introduction

Due to their simplicity and reliability, among the acoustic oscillators widespread have sound bimorph-type elements, working on bending, low-frequency vibration mode (Figure 7). Despite the fact that this type of sound elements is known long ago and is widely used in electroacoustic, to date there is no common understanding of "possibilities" this class of transducers, and sound elements based on them.

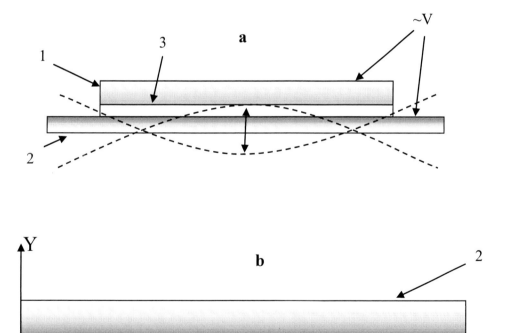

Figure 7. Schematic representation of the bimorph-type transducer with flexural vibrations under influence of an alternating voltage V (a) and its computer model (b), where there are the next designations: 1 - piezoelectric ceramics, 2 - steel membrane; 3 - adhesive layer.

This section presents the results of studies based on computer simulation of AFC of bimorphs, nature of materials and dimension characteristics of piezoelectric elements. The used materials of piezoelectric elements in the model experiments were: PZT-19, PZTB-3, PZT-5H, PZTST-2.

In contrast to the material basis studied in previous section the novel considered material PZTST-2 has rhombohedral phase. By judging the ratio of Ti/Zr in the composition of this material, one is practically on the border of solid solutions of various modifications.

3.2. Experimental Procedure

The aim of this study was investigation of bimorphs, consisting of a thin disc-shaped piezoelectric, glued with epoxy binder to a thin steel disc membrane. The specific objects were simulated into the air at a variable voltage V. Due to the simplicity in geometry the bimorphs are modeled by using two-dimensional axi-symmetric harmonic analysis.

The membrane had thickness of 100 microns and diameter of 32 mm, thickness of adhesive layer was assumed to be 10 microns, and the varied dimensional characteristics of the piezoelectric elements were:

- diameter (d, mm) in the range from 20 up to 30 mm with pitch of 2 mm
- thickness (h, m) in the range from 100 up to 220 microns with pitch of 20 microns.

The experiment consists in investigation of the frequency (f, Hz) and sound pressure level (SPL, dB) of bimorph resonance oscillation, measured in air at 10 cm (Figure 7, a). Animated vibration analysis also allowed us to judge the degree of distortion of the vibrations.

Based on the calculation results of the experiment for the transducers there have been built three-dimensional diagrams (d, h, Y), where d is the bimorph diameter, h is the piezoelement thickness, and Y is the property value (frequency or SPL-amplitude of oscillation resonance of the emitter). The obtained surfaces of the properties in the three-dimensional diagram, orthogonally projected onto the plane of the piezoelement dimension characteristics with the lines of identical property values (see Figures 8 – 11).

An important aspect of the work was to compare the results of computer simulation of the piezo-bimorph resonance characteristics with the results of the experimental study their similar dependences.

The piezo-bimorph manufacture includes a number of technological operations each of them may affect their resonance characteristics. Therefore, there was defined the goal to get the statistically most probable piezo-bimorph resonance frequency with the fixed dimension characteristics. With this aim it was made set of piezobimorphs in quantity of 30 pieces from the material PZTST-2, with a membrane diameter of 32 mm, a thickness of 100 microns, by using the standard technology, which included the following stages:

i.   Processing piezo-films by slip casting and rolling with size of thickness.
ii. Cutting discs and their preparation for firing and sintering.
iii. Sintering. These billets were heated initially at 50 °/h up to 700°C, and then at 100 °/h up to 1250°C and held at this temperature during 2 hours. As a result there have been obtained ceramic discs with diameter of 22.6 ± 0.3 mm and thickness of 115 ± 2 microns.

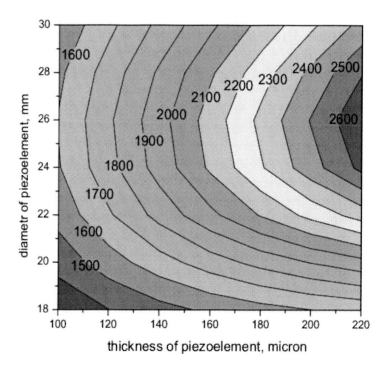

Figure 8. Dependence of the frequencies of bimorph oscillation resonace on the diameter and thickness of piezoelectric material PZT-19.

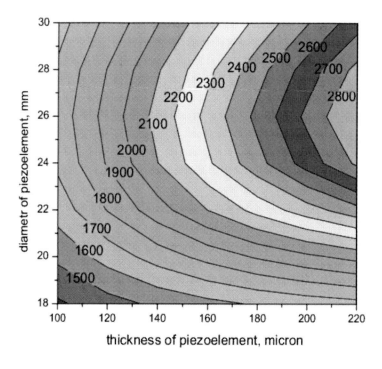

Figure 9. Dependence of the frequencies of bimorph oscillation resonace on the diameter and thickness of piezoelectric material PZTB-3.

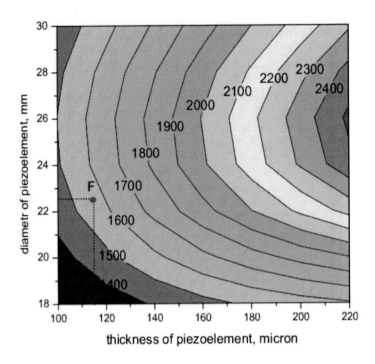

Figure 10. Dependence of the frequencies of bimorph oscillation resonace on the diameter and thickness of piezoelectric material PZT-5H. Point F corresponds to the value of experimentally observed frequency.

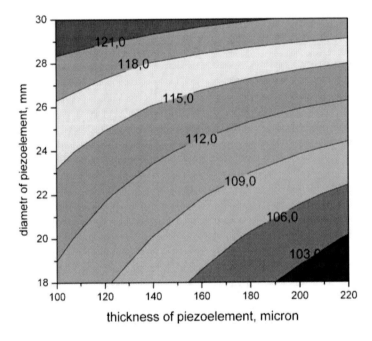

Figure 11. Dependence of the SPL of bimorph oscillation resonace on the changes in the diameter and thickness of piezoelectric material PZT-19.

iv. The electrodes were constructed on the ceramic disc by using silver brazing in accordance with the standard technique. The sample polarization was carried out in the air under *dc*-field of 2 kV/mm during 40 seconds in the pulsed mode.

v. Bimorphs were obtained by gluing the membrane to the prepared ceramic discs by using adhesive epoxy resin ED-20 in accordance with the standard technology. The adhesive film in these bimorphs was 10 – 15 microns.

vi. Measurements of the resonance frequencies of piezo-bimorphs were made on a special stand, which included sound generator, sound-level meter, and special acoustic chamber. The results of the measurements of the resonance frequencies are given in the form of appropriate graphs of statistical processing of the piezo-bimorph set in quantity of 30 pieces.

3.3. Discussion

The results of the figures show that for bimorphs with piezoelements from PZTST-2 with thickness of 115 ± 2 microns and diameter of 22.6 ± 0.3 mm the statistically most probable resonance frequency to be into range of 1550 ± 50 Hz.

In Figure 9, the experimental data on the bimorph resonance frequency (point F) are compared with the results of the computer simulation. As it is seen, between the calculated data and the experimental results there is a good coincidence. It is also clear that rather laborious experiment of the manufacturing the 30 bimorphs, allowed us to get only one point in the following diagram of computer calculation.

In Figures 8 – 10, which are typical diagrams for the result presentation of computer simulation, there are shown the changes in resonance frequencies in vibrations of piezo-bimorphs depending on the thickness and diameter of their piezoelements made of three different materials at fixed values other bimorph characteristics (the adhesive material and the membrane, the thickness of adhesive layer, the size of the membrane). As can be seen from the figures, the resonance frequency of bimorphs in increasing the diameter of the piezoelectric elements from 18 up to 26 mm increases, and with further increase of diameter begins to decrease. By regarding the effect of the piezoelectric element thickness on the resonance frequency we conclude that generally with an increase in the thickness frequency increases, and its impact on the frequency reaches maximum when the diameter of the piezoelectric element approaches 26 mm.

Figure 11 shows a typical pattern of variation of the SPL of piezo-bimorph resonant oscillations depending on the thickness and diameter of piezoelectric elements at fixed values other parameters. As it can be seen from the figure, the influence of thickness and diameter of the piezoelectric elements on the bimorph output is also to be complex, as in the case with the change of resonance frequencies. The maximum bimorph output, in the studied range of the sizes of the piezoelectric elements, was detected by minimum thickness and maximum diameter. The Figure 11 also shows that further decrease in the thickness of the piezoelectric element should lead to further increase the bimorph output.

Figures 8 and 9 show that, although in the transition from bimorph from piezoelectric material PZT-19 to bimorph from piezoelectric material PZTB-3 general character of the dependences of the resonance frequencies and SPL on the thickness and diameter of the piezoelectric element retains, however for the same dimension characteristics of piezo-bimorph components its resonance frequency increases on about 200 Hz, in the whole.

By summing up the results of computer simulation of the bimorphs with piezoelements from different materials, one may conclude that in identical characteristics of bimorph components, the increase of the resonance frequency is observed in the next transition: PZT-5H → PZT-19 → PZTB-3, and increase in the SPL of resonance oscillations occurs in the transition: PZT-19 → PZTB-3 → PZT-5H.

This behavior of the dependence of the resonance frequency is well explained by corresponding change in the velocity of sound in these piezoelectric materials, and the nature of the output change of piezo-bimorph is directly related with the piezomaterial effectiveness and corresponding change in piezoelectric modulus.

Thus, the results of the bimorph research allow us to make the following conclusions:

i.   For computer simulation of the piezo-bimorph resonance characteristics it has been developed the piezo-bimorph model based on the ANSYS software adapted for the purpose and method of processing the results.

ii.   For the first time, it has been obtained a complete picture for the dependence of the frequency and SPL of the piezo-bimorph oscillation resonances on the material nature and dimension characteristics of the used piezoelectric elements. As materials we used PZT-19, PZTB-3 and PZT-5H, and as varying dimension characteristics of piezoelectric elements was the bimorph diameter into range of 18 – 30 mm and thickness into range of 100 – 220 micrometers.

iii. It has been shown that the maximum SPL of the vibration resonances of the bimorph increased in the transition from the material PZT-19 to PZTB-3 and further to the material PZT-5H, and the frequency of oscillation resonances increased in the transition from PZT-5H to PZT-19, and further to PZTB-3.

iv.   There was made the comparison of the calculated and experimental results for the frequencies of piezo-bimorph oscillation resonances and found their good coincidence.

CONCLUSION

The results obtained and presented in the chapter show that ACELAN and ANSYS software allow computer modeling and calculation of resonance amplitudes as well as acoustic pressures that are developed by piezoelectric transducers. The numerically obtained transducer resonance characteristics well agree with their experimental data.

It is also shown that the computer modeling opens completely new opportunities for the research and optimization of piezoelectric transducer characteristics. It is noteworthy that, compared with their experimental research, computer modeling allows us to obtain the necessary results with lower expenses and time consumptions.

ACKNOWLEDGMENTS

This research was supported in part by the Russian Foundation for Basic Research (Grant No. 10-08-13300).

REFERENCES

[1] Dobrucki, A. B.; Pruchnickia, P. *Sensors and Actuators A: Physical.* 1997, *vol. 58, Issue 3,* 203-212.

[2] Fernandes, A.; Pouget, J. *Int. J. Solids and Struct.* 2003, *vol. 40,* 4331–4352.

[3] Li, X.-F.; Peng, X.-L.; Lee, K.Y. Europ. J. *Mechanics A: Solids.* 2010, *vol. 29,* 704-713.

[4] Wu, X.-H.; Chen, C.; Shen, Y.-P.; Tian X.-G. *Int. J. Solids Struct.* 2002, *vol. 39,* 5325–5344.

[5] Wang, Q.; Liew, K.M. *Int. J. Solids and Struct.* 2003, *vol. 40,* 6653–6667.

[6] Roncari, E.; Galassi, C.; Craciun, F.; et al. *J. Europ. Ceram. Soc.* 2001, *vol. 21,* 409-417.

[7] Takeshi, M.; Kurosawa, M. K.; Higuchi, T. *Sensors and Actuators.* 2000, *vol. 83,* 225–230.

[8] Kanda, T.; Makino, A.; Ono, T.; et al. *Sensors and Actuators A.* 2006, *vol. 127,* 131–138.

[9] Lu, P.; Lee, H. P.; Lu, C. *Int. J. Mech. Sci.* 2005, *vol. 47,* 437–458.

[10] Lu, P.; Lee, H. P.; Lu, C. *Composite Struct.* 2006, *vol. 72,* 352–363.

[11] Montealegre, R. W.; Buiochi, F.; Adamowski, J. C.; et al. *Ultrasonics.* 2009, *vol. 49,* 484–494.

[12] Jalili, H.; Goudarzi, H. *Sensors and Actuators A.* 2009, *vol. 149,* 266–276.

[13] Shen, M.H.; Chen, F.M.; Chen, S.N. *Int. J. Solids and Struct.* 2006, *vol. 43,* 2336–2350

[14] Lee, K. H.; Lim, S. P. *J. Sound Vibrat.* 2003, *vol. 259(2),* 427–443.

[15] Li, H.; Yang, F.; Du, H.; et al. *Acta Mechanica Solida Sinica.* 2009, *vol. 22, No. 5,* 499-509.

[16] Kim, J. O.; Lee, J. G.; *J. Sound and Vibrat.* 2007, *vol. 300,* 241–249.

[17] Wang, H.M. *Appl. Math. Modeling.* 2011, *vol. 35,* 1765–1781.

[18] Lin, S. *Sensors and Actuators A.* 2007, *vol. 134,* 505–512.

[19] Yang, Z.T.; Guo, S.H. *Ultrasonics.* 2008, *vol. 48,* 716–723.

[20] Kim, J. O.; Hwang, K. K.; Jeong, H. G. *J. Sound Vibrat.* 2004, *vol. 276,* 1135–1144.

[21] Lin, S. *Ultrasonics.* 2007, *vol. 46,* 51–59.

[22] Liu, S.; Lin, S. *Sensors and Actuators A.* 2009, *vol. 155,* 175–18.

[23] Alippi, A.; Albino, M.; Angelici, M.; et al. *Ultrasonics.* 2004, *vol. 43,* 1–3.

[24] Nasedkin, A.V. In: *Modern Problems of Continuum Mechanics,* Proc. VII Int. Conf., Rostov-on-Don, 22-24 Oct. 2001. Novaja Kniga: Rostov on Don, 2002, *vol. 1,* pp 182-188.

[25] Yi, S.; Ling, S. E.; Ying, M.; Hilton, H. H.; Vinson, J. R. *Int. J. Num. Meth. Eng.* 1999, *vol. 45,* 1531-1546.

[26] Bathe, K.-J.; Wilson, E. L. *Numerical Methods in Finite Element Analysis.* Prentice-Hall, Englewood Cliffs, New Jersey, 1976.

[27] Akopov, O. N.; Belokon', A. V.; Yeremeev, V. A.; Nadolin, K. A.; Nasedkin, A. V.; Skaliukh, A. S.; Soloviev, A. N. In: *Proc. Int. Sci.-Pract. Conf. "Fundamental*

Problems of Piezoelectric Instrument Construction" ("P'ezotekhnika-99"), Rostov-on-Don, Azov. 1999, *vol. 2*, pp 241-251.

[28] Belokon', A. V.; Nasedkin, A. V.; Nikitaev, A. V.; Petushkov, A. L.; Skaliukh, A. S.; Soloviev, A. N. In: *Proc. Int. Sci.-Pract. Conf. "Fundamental Problems of Piezoelectric Instrument Construction",* ("P'ezotekhnika"), Tver', 2002, p 171.

[29] Ramesh, R. ; Kara, H. ; Bowen, C. R. *Ultrasonics.* 2005, *vol. 43, Issue 3*, 173-181.

[30] Sekouri, E. M.; Hu, Y.-R.; Ngo, A. D. *Mechatronics.* 2004, *vol. 14, Issue 9*, 1007-1020

[31] Dong, X.-J.; Meng, G.; Peng, J.-Ch. *J. Sound and Vibrat.* 2006, *vol. 297, Issues 3-5*, 680-693.

[32] Iona, F.; Shirane, G. *Ferroelectric Crystals.* Pergamon Press: New York, 1962.

In: Ferroelectrics and Superconductors
Editor: Ivan A. Parinov

ISBN: 978-1-61324-518-7
©2012 Nova Science Publishers, Inc.

Chapter 9

SUPERCONDUCTING MAGNETIC ENERGY STORAGE COIL INTEGRATION TO AUTONOMOUS RAIL LOCOMOTIVE ELECTRIC POWER TRANSMISSION

V. N. Noskov[*,1] *and M. Yu. Pustovetov*[≠1,2]

[1]Rostov-on-Don State University of Transport Communications, Rostov-on-Don, Russia
[2]Department of Machine-Tool Systems Automation and Electric Drives,
Power Engineering and Machine-Building Institute,
Don State Technical University, Rostov-on-Don, Russia

ABSTRACT

We developed a scheme for superconducting magnetic energy storage (SMES) coil integration to autonomous rail locomotive electric power transmission. This scheme is based on DC-DC converter technology. This work describes regimes of the suggested scheme. The sufficient energy capacitance for SMES on board of diesel-electric shunt locomotive is discussed.

1. INTRODUCTION

It is no new but actual idea about integration of energy storage device to autonomous rail locomotive electric power transmission [1]. While the trip of the locomotive the motor traction power changes widely. The goal is recuperation to SMES braking power of traction motors. In SMES this energy can be stored for a time. And, while the power generated by traction generator is insufficient for traction force realization, we can use stored in SMES power for additional simultaneous with generator supply of traction motors. Such a way it is possible by SMES to minimize installed power of diesel-generator and provide its work with constant power at mostly high efficiency. For a time it is possible to supply traction motors

---

*E-mail: nvn_nis@sci.rgups.ru, 1, Rostovskogo Polka Narodnogo Opolchenia Square, 344038 Rostov-on-Don, Russia.
≠E-mail: mgsn2006@rambler.ru, 6, Studencheskaya Street, 344000 Rostov-on-Don, Russia.

from SMES only, for example for ecological aims while locomotive starting from station with turned off diesel.

It is a trend in last 10 – 15 years to suggest DC-DC converter topology for energy storage devices such as SMES or ultra-capacitors integration to different electric power transmissions [2, 3]. The developed scheme described below is based on Reference [4].

2. POWER STATIC CONVERTER SCHEME AND ITS MODES

The suggested scheme of power static converter is presented in Figure 1. This converter converts input DC-voltage V_{IN} to output DC-voltage V_{OUT} with other characteristics. There are shown L the inductance of SMES coil, and Z_{LOAD} the complexor of load resistance.

In Figure 1 the following symbolic is used: $VD5$, $VD6$ are the diodes; $VD1$ and $K1$, $VD2$ and $K2$, $VD3$ and $K3$, $VD4$ and $K4$ present the combinations of diode and key full-controlled semiconductor triodes such as GTO.

This converter can fulfill several modes, namely: (i) to charge SMES from generator, (ii) to discharge SMES to load, (iii) to store DC-current in SMES coil, and (iv) to recuperate energy from load to SMES. The converter can be used not only with DC-DC electric power transmission but with AC-DC (in presence of input rectifier) or with AC-AC (in presence of input rectifier and output autonomous inverter) transmissions.

In Tables 1 – 6 the states of keys are shown for different modes of converter. The state "1" means the key is "ON". The state "0" means the key is "OFF".

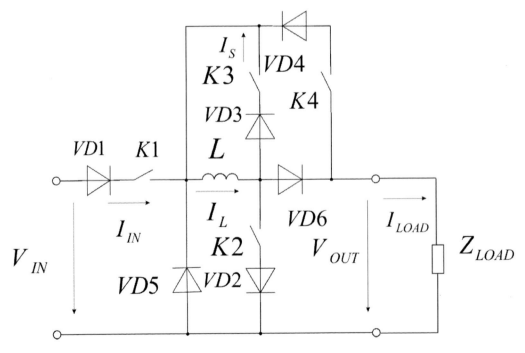

Figure 1. Power static converter with SMES coil electric scheme.

Table 1. States of keys in Mode 1

Key name	K1	K2	K3	K4
Key state	1	1	0	0

Mode 1: SMES is charged from voltage source V_{IN} (circuit $V_{IN}+ \; \rightarrow \; VD1 \rightarrow L \rightarrow VD2 \rightarrow V_{IN}-$)

The equation for circuit has form:

$$v_{IN}(t) = 2R_{VD} \cdot i_L(t) + L\frac{di_L(t)}{dt},$$ (1)

where $i_L(t)$ is the current through SMES coil, R_{VD} is the diode forward resistance when conducting.

Table 2. States of keys in Mode 2

Key name	K1	K2	K3	K4
Key state	0	0	0	0

Mode 2: SMES is discharged to load (circuit $L \rightarrow VD6 \rightarrow Z_{LOAD} \rightarrow VD5$)

The equation for circuit has form:

$$0 = 2R_{VD} \cdot i_L(t) + L\frac{di_L(t)}{dt} + v_{OUT}(t).$$ (2)

Table 3. States of keys in Mode 3

Key name	K1	K2	K3	K4
Key state	0	0	1	0

Mode 3: It is the mode of current storage in the SMES with partition discharge SMES to load (circuits $L \rightarrow VD3$ and $L \rightarrow VD6 \rightarrow Z_{LOAD} \rightarrow VD5$)

The equations for circuits are written as

$$\begin{cases} 0 = R_{VD} \cdot i_S(t) + L\frac{di_S(t)}{dt}; \\ 0 = 2R_{VD} \cdot i_{LOAD}(t) + L\frac{di_{LOAD}(t)}{dt} + v_{OUT}(t); \\ i_L(t) = i_{LOAD}(t) + i_S(t), \end{cases}$$ (3)

where $i_S(t)$ is the current stored in SMES-coil, $i_{LOAD}(t)$ is the current through the load.

Table 4. States of keys in Mode 4

Key name	K1	K2	K3	K4
Key state	0	1	0	1

Mode 4: SMES is charged from the load which to be in recuperating mode (voltage source V_{IN} is turned off, circuit $Z_{LOAD} \rightarrow VD4 \rightarrow L \rightarrow VD2$)

The equation for circuit has form:

$$v_{OUT}(t) = 2R_{VD} \cdot i_L(t) + L\frac{di_L(t)}{dt}. \tag{4}$$

Table 5. States of keys in Mode 5

Key name	K1	K2	K3	K4
Key state	1	1	0	1

Mode 5: SMES is charged from recuperating load and from voltage source V_{IN} at the same time (circuits $Z_{LOAD} \rightarrow VD4 \rightarrow L \rightarrow VD2$ and $U_{IN} + \rightarrow VD1 \rightarrow L \rightarrow VD2 \rightarrow U_{IN} -$)

The equations for circuits are written as

$$\begin{cases} v_{OUT}(t) = R_{VD} \cdot i_{LOAD}(t) + L\dfrac{di_{LOAD}(t)}{dt}; \\ v_{IN}(t) = 2R_{VD} \cdot i_{IN}(t) + L\dfrac{di_{IN}(t)}{dt}; \\ i_L(t) = i_{LOAD}(t) + i_{IN}(t), \end{cases} \tag{5}$$

where $i_{IN}(t)$ is the current from voltage source V_{IN}.

Table 6. States of keys in Mode 6

Key name	K1	K2	K3	K4
Key state	1	0	1	0

Mode 6: it is the mode of current storage in the SMES with the load supplying from voltage source V_{IN}, at the same time (circuits $L \rightarrow VD3$ and $U_{IN} + \rightarrow VD1 \rightarrow L \rightarrow VD6 \rightarrow Z_{LOAD} \rightarrow U_{IN} -$)

The equations for circuits are written as

$$\begin{cases} 0 = R_{VD} \cdot i_S(t) + L\dfrac{di_S(t)}{dt}; \\ v_{IN}(t) = 2R_{VD} \cdot i_{LOAD}(t) + L\dfrac{di_{LOAD}(t)}{dt} + v_{OUT}(t); \\ i_L(t) = i_{LOAD}(t) + i_S(t). \end{cases} \qquad (6)$$

We have to outline that in scheme in Figure 1 the absence of *K*4 is possible because of current through the *VD*4 will exist only in case of $V_{OUT} > V_{IN}$ in any event. The modes 1 – 6 alternation can be in arbitrary order.

3. ABOUT ENERGY CAPACITANCE FOR SMES ON BOARD OF DIESEL-ELECTRIC SHUNT LOCOMOTIVE

It is mostly likely to use the SMES on board of diesel-electric shunt locomotive because working modes are not even and include continuous work with small load and without load, short (10 – 20 s) duty cycles. Our analysis of five typical working modes of Soviet shunt locomotive supports that sufficient energy capacitance for SMES will be 120 MJ in presence of 400 kW on board electric generator and peak power of traction electric motors 650 kW. That result is in good accordance with data other specialists. For example, in [5] energy capacitance of super-capacitive energy storage device for GTW diesel-train calculated as 103 – 122 MJ in presence of 380 kW on board electric generator and peak power of traction electric motors equal to 620 kW. The calculated curve of energy stored in SMES on board of shunt locomotive in working mode "shunt by push" is shown in Figure 2.

In case of shunt locomotive of TEM-2 type we calculate *L* = 37.1 H.

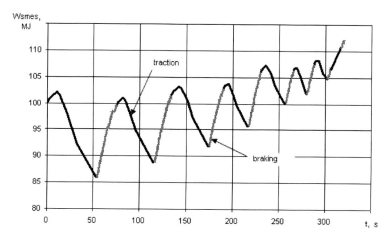

Figure 2. Calculated curve of energy stored in SMES on board of shunt locomotive in working mode "shunt by push".

4. CONVERTOR'S MODES COMPUTER MODELING

The working modes of the suggested convertor with SMES coil were computer modeled. Figure 3 presents the modeling results of currents by PSpice for some of modes when load imitates six traction electric motors of diesel-electric shunt locomotive TEM-2 with transmission of DC-DC type. It is assumed that the magnetic flux of motors is constant and parameters of motors are identical.

In this case we can write equation for the load branch as

$$v_{OUT}(t) = e_{OUT}(t) + R_{LOAD} \cdot i_{LOAD}(t) + L_{LOAD} \frac{di_{LOAD}}{dt} = Z_{LOAD} \cdot i_{LOAD}(t), \qquad (7)$$

where e_{OUT} is the electromotive force of traction motor armature, R_{LOAD} and L_{LOAD} are the equivalent resistance and equivalent inductance of six traction motor armatures connected in parallel sections.

For more clearance, Figure 4 presents results of current in SMES-coil modeling only where sections of SMES discharge and SMES charge from recuperating traction motors are alternates.

Then, Figure 5 presents results of voltage modeling developed at the same conditions that curves of currents in Figures 3 and 4.

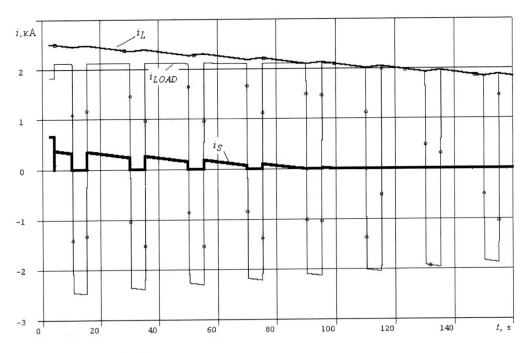

Figure 3. Modeling results of currents.

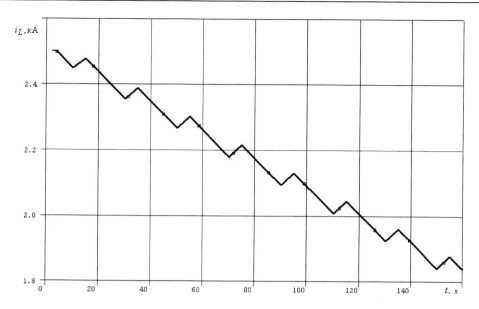

Figure 4. Modeling results of current in SMES-coil.

As initial conditions the scheme works at the static Mode 6. Traction motors work in a motor mode. Before $K1$ will be turned off the diodes $VD2$ and $VD4$ must be prepared. The key full-controlled semiconductor triodes $K2$ and $K4$ are controlled by signals v_{OUT} and i_{LOAD} such a way to provide braking current recuperation to SMES coil at high e_{OUT} in a generator mode of electric motors and avoid extra over-voltage on the load. The triode $K3$ is governed by signal i_{LOAD} to keep i_{LOAD} in a set corridor. After the triode $K1$ will be turned off it is possible to provide Modes 2, 3 and 4. In case of our modeling the triodes $K2$ and $K4$ operates simultaneously.

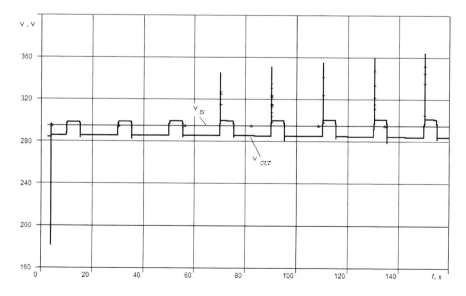

Figure 5. Modeling results of voltages.

Conclusion

The scheme of DC-voltage convertor with SMES-coil for integration of energy storage device to autonomous rail locomotive electric power transmission is suggested. Equations of main modes of this scheme are presented. Energy capacitance of SMES in case of shunt locomotive is calculated. The scheme ability to work is confirmed by results of computer modeling.

References

[1] Johnson, B.; Low, J.; Saw, G. Using a superconducting magnetic energy storage coil to improve efficiency of gas turbine powered high speed rail locomotive. *IEEE Trans. Appl. Supercond.* 2001, *vol. 11*, 1900-1903.

[2] United States Patent US2007/0108956A1 Int. Cl. G05F1/00. Assigner: Robert Louis Staigerwald. (US) Appl. No. 11/273984, Filed: Nov. 15, 2005. Publication date: May 17, 2007. *System and Method for Charging and Discharging a Superconducting Coil.*

[3] United States Patent 5,373,195 Int. Cl. H02M5/458. Assigner: General Electric Company (US) Appl. No. 993879, Filed: Dec. 3, 1992. Publication date: Dec.13, 1994. *Technique for Decoupling the Energy Storage System Voltage from the DC-link Voltage in AC Electric Drive Systems.*

[4] Russian Federation Patent 2370875 Int. Cl. H 02 M 3/135. Inventor: Klimenko, E.Yu.; Noskov, V. N.; Poltanov, A. E.; Polulyakh, E. P.; Pustovetov, M. Yu.; Rindin, V. N.; Flegontov, N. S. Assigner: JSC Russian Railways (RU) Priority date: Apr. 17, 2008. *DC-Voltage Converter.*

[5] Destraz, B.; Barrade, P.; Rufer, A. *Supercapacitive energy storage for diesel-electric locomotives.* SPEEDAM 2004 June 16th – 18th, Capri (Italy).

INDEX

A

absorption spectra, 131, 143, 152
acid, 4, 6
acoustics, 286
activation energy, 16, 22, 28, 29, 42, 45
actuators, 52, 110
additives, 118
adhesion, 202
adjustment, 52
ammonium, 4
amplitude, 15, 30, 93, 99, 101, 142, 143, 231, 237, 239, 241, 242, 283, 285, 288, 289, 292, 293, 294, 296
anisotropy, 2, 52, 53, 54, 64, 66, 67, 71, 72, 81, 85, 86, 87, 88, 258, 269
annealing, 12, 17, 18, 20, 24, 27, 33, 34, 42, 46, 47, 117, 118, 119, 133, 152, 153
antiferroelectrics, ix, 91
arithmetic, 166, 171
atmosphere, 106
atoms, 20, 33, 34, 160, 231, 232, 233, 234, 235, 236, 237, 238, 239, 240, 241, 242, 243, 244, 245, 246, 247, 248, 249, 250, 251, 252, 253, 254

B

backscattering, 231, 232, 233, 235, 236, 244, 248
barium, 92, 235
barriers, 237
base, ix, x, 1, 4, 5, 30, 36, 47, 105, 106, 108, 126, 166, 197, 220, 279, 286, 294
behaviors, 225
bending, 19, 284, 294
bias, 55, 91, 96, 113, 114, 116, 118
binary oxides, 119
bismuth, 124, 125, 135, 217

bounds, 63, 166, 167, 171, 172, 173, 176, 177, 181, 182, 183, 184, 185, 186, 187, 188, 189, 190, 191, 192, 193, 194, 195, 196, 204, 205, 213, 214, 217
brittleness, 4
bulk materials, 222

C

cables, 196
calcination temperature, 128
calcium, 4
casting, 296
ceramic, 47, 64, 67, 77, 87, 91, 96, 97, 99, 100, 105, 106, 107, 109, 110, 113, 116, 118, 125, 129, 130, 135, 136, 139, 144, 145, 146, 152, 153, 217, 258, 280, 285, 293, 296, 299
Ceramics, 7, 35, 96, 99, 105, 109, 116, 120, 228
chemical, 3, 6, 33, 127, 129, 143, 159, 160, 162, 285
chemical properties, 159
China, 165
circulation, 183
City, 165
clarity, 249
classes, 88
classification, 257, 258
clustering, 179
clusters, 119, 186, 206, 214, 216
coherence, 3
collisions, 250
color, 191, 245, 277, 278
commercial, 4
communication, 255
compatibility, 3, 19
competition, 85
compliance, 57, 166, 218, 258, 264
composites, ix, x, 1, 2, 3, 4, 30, 34, 36, 45, 47, 51, 52, 53, 64, 66, 67, 71, 76, 77, 78, 79, 80, 81, 83, 84, 85, 86, 87, 88, 165, 166, 167, 168, 169, 170, 173, 175, 176, 181, 182, 184, 185, 186, 191, 192,

194, 196, 197, 199, 202, 204, 205, 207, 208, 210, 211, 212, 214, 216, 217, 220, 221, 222, 224, 225, 257, 258, 259

composition, 4, 6, 42, 46, 71, 84, 87, 98, 102, 113, 114, 115, 116, 125, 128, 130, 135, 137, 141, 142, 158, 159, 160, 161, 249, 295

compounds, 124, 125, 126, 127, 128, 129, 146, 149, 152, 153, 156, 233

compression, 2, 4, 5, 6, 7, 10, 13, 14, 15, 19, 21, 25, 27, 28, 38, 39, 40, 42, 46, 217, 222

computation, 166, 199

computer, x, 96, 100, 106, 116, 232, 241, 249, 250, 254, 285, 287, 289, 292, 293, 295, 296, 299, 300, 301, 308, 311

computing, 212

conception, 2

Concise, 90

conductivity, ix, 105, 119, 138, 141, 165, 167, 168, 169, 170, 171, 172, 173, 174, 175, 176, 177, 178, 179, 181, 182, 183, 184, 186, 187, 188, 189, 190, 191, 192, 193, 194, 195, 196, 197, 198, 199, 200, 201, 202, 203, 205, 206, 207, 208, 210, 211, 212, 213, 214, 215, 216

configuration, 222, 277

conformity, 265, 280

connectivity, x, 17, 52, 53, 55, 77, 78, 85, 87, 168, 183, 257, 258, 259, 277, 278, 280

connectivity patterns, 77, 87

consent, 239

conservation, 197

constant load, 5

construction, 160

consumption, 126, 133

containers, 127

contamination, 133

contour, 245

convergence, 184, 284

cooling, 3, 96, 97, 98, 99, 100, 101, 103, 106, 107, 108, 109, 113, 114, 116, 118, 119, 133, 143, 144, 145, 149, 150, 151, 152

coordination, 160

copolymer, 53, 81

correlation, 19, 22, 63, 65, 98, 99, 167, 206, 214, 291

cracks, 4

creep, 7, 19

critical state, 7

critical value, 206

crystal growth, 95

crystal structure, x, 124, 231, 232, 235, 244, 248, 251

crystalline, 33, 119, 160, 249, 259

crystallites, 258

crystals, x, 33, 51, 52, 91, 92, 93, 94, 98, 99, 105, 108, 113, 125, 127, 222, 231, 232, 240, 244, 249, 252, 255

cubic system, 171

cycles, 116, 118, 308

D

Dagestan, 123

damping, 224, 225

data collection, 100, 116

data set, 289

decomposition, 2, 15, 23, 24, 25, 27, 31, 32, 33, 35, 43, 45, 46, 47, 135

defects, 2, 3, 23, 24, 25, 31, 32, 33, 35, 42, 43, 44, 45, 95, 104, 105, 119, 129, 133, 232

deformation, ix, 1, 2, 3, 4, 5, 6, 7, 8, 10, 11, 12, 13, 15, 16, 17, 18, 19, 20, 21, 22, 23, 24, 25, 26, 27, 28, 29, 30, 31, 32, 33, 34, 35, 36, 37, 38, 39, 40, 41, 42, 43, 44, 45, 46, 110, 126, 156, 215, 216, 217, 222, 223, 224, 259

degradation, 3, 138

depolarization, 101, 110, 119

depth, 232, 233, 237, 238, 239, 245, 246, 247, 250, 257, 258

derivatives, 197

derivatography, 119

destruction, 2

detection, 96

dielectric constant, 276, 279

dielectric permittivity, 54, 68, 72, 81, 100, 102, 110, 113, 114, 116, 264

differential equations, 167

diffraction, 5, 36, 43, 96, 98, 99, 104, 107, 108, 119, 130, 141, 145, 146, 152, 153, 156, 240

diffusion, x, 19, 20, 45, 92, 98, 102, 116, 128, 233, 234, 237, 239, 241, 248

diodes, 304, 310

discontinuity, 168, 196, 202

discs, 171, 172, 178, 186, 187, 188, 296, 299

dislocation, 3, 4, 19, 25, 31, 34, 35, 44, 280

disorder, 125, 126, 242

dispersion, 102, 103, 105, 107, 133, 177

dispersity, 187

displacement, 79, 80, 160, 218, 219, 222, 232, 236, 237, 239, 240, 249, 250, 251, 253, 258, 259, 264, 266, 284

disposition, 172, 203, 206, 213, 214

distortions, 92, 110, 240, 249, 279

distribution, 3, 27, 33, 34, 87, 92, 93, 94, 129, 168, 173, 178, 221, 232, 233, 235, 237, 239, 240, 241, 244, 245, 246

divergence, 250, 253, 260, 261

Index

283

domain structure, 51, 52, 53, 56, 67, 72, 87
doping, 30, 34, 36, 107
DTA curve, 128
dynamic pyroelectric measurements, ix, 91

E

EGB, 7
elastic deformation, 223
electric charge, 260, 261, 266
electric conductivity, 210
electric field, 55, 56, 78, 79, 80, 91, 93, 95, 96, 100,
   113, 114, 115, 116, 124, 259, 264
electrodes, 58, 92, 95, 96, 100, 106, 119, 276, 285,
   286, 299
electron, 6, 12, 130, 242
e-mail, 1, 51, 165, 257, 283, 303
emitters, x, 286, 287, 289, 291
energy, x, 2, 16, 22, 23, 28, 29, 30, 31, 33, 34, 35,
   36, 42, 45, 119, 126, 132, 133, 152, 184, 217,
   218, 222, 223, 224, 225, 234, 235, 236, 237, 239,
   240, 241, 244, 247, 249, 250, 266, 285, 303, 304,
   308, 311
energy consumption, 126, 133
energy expenditure, 225
engineering, 257, 259
environment, 127, 142
epoxy binder, 295
equality, 72, 85, 170, 171, 176, 197, 202, 216
equilibrium, 80, 124, 157, 186, 188, 189, 190, 191,
   196, 216, 217, 222, 239, 252, 262, 266
evidence, 95, 156
evolution, 46
excitation, 30
experimental condition, 232, 253
exposure, 116, 130, 144
external constraints, 223

F

FEM, 79, 86, 87, 258, 259, 273, 274, 279, 280, 284
ferrite, 124, 125, 135
ferroelectrics, ix, 91, 113, 166, 168, 216, 224
ferroelectrics–relaxors, ix
ferromagnetism, 125
FGM, 284
fibers, ix, 165, 167, 185
filament, 198, 202, 203, 205
films, 126, 296
financial support, 120
finite element method, 258, 274
fixation, 100, 279
fluid, 284

foams, 217, 222
force, 34, 36, 45, 133, 216, 222, 223, 284, 304, 309
formation, 1, 8, 13, 19, 20, 21, 25, 26, 31, 34, 36, 42,
   44, 45, 46, 92, 107, 119, 124, 126, 127, 140, 143,
   162, 165, 168, 187, 203, 206, 213, 217, 224, 258
formula, 15, 124, 125, 130, 131, 171, 177, 180, 181,
   184, 189, 200, 204, 206, 207, 214, 215, 220, 221,
   285
fractal structure, 206
fracture processes, 167
fragments, 21, 100, 141, 148
free energy, 217
freedom, 79, 80

G

gel, 4, 34, 36, 37, 38, 39, 40, 41, 42, 43, 44, 45, 46
geometry, 15, 22, 31, 202, 216, 283, 285, 295
Germany, 132
grades, 130
grain boundaries, 7, 20, 31
grain size, 22, 129
graphite, 130
gravity, 133
growth, 3, 4, 12, 13, 19, 20, 21, 26, 27, 31, 39, 40,
   43, 45, 95, 167

H

heat capacity, 217
heating rate, 5
height, 5, 8, 129, 287, 288, 293
helium, 130
heterogeneity, 197, 273
hexagonal lattice, 198, 199, 200, 203, 205
history, 125, 145
homogeneity, 138
House, 163, 255
housing, 133
hypothesis, 276
hysteresis, 7, 15, 22, 97, 99, 108, 110, 114, 115, 116,
   119
hysteresis loop, 7, 15, 22

I

ideal, ix, 33, 165, 166, 167, 168, 197, 199, 200, 202,
   203, 208, 211, 213, 216, 217, 273
identification, 24, 126
identity, 56
images, 161

impurities, 30, 36, 116, 127, 137, 146, 232, 243, 244, 248
incidence, 245
independent variable, 263
inequality, 191
inertia, 284
initial state, 55
initiation, 4, 20, 222
integration, x, 199, 233, 261, 303, 304, 311
interaction process, 237
interface, 3, 64, 197, 203, 211, 213, 216, 218
interphase, 45
ions, x, 125, 126, 137, 138, 140, 142, 231, 232, 233, 234, 235, 236, 237, 239, 240, 244, 249, 250, 253
iron, 138, 153
Italy, x, 51, 311

J

Japan, 48, 164

K

kinks, 105
knots, 171, 172

L

laminar, 15, 55, 72
Landau theory, 217
lattice parameters, 104, 141
lattices, 171, 190, 198, 199, 200, 204, 208, 210, 211, 214
laws, 224
lead, 25, 31, 34, 53, 63, 67, 85, 104, 105, 116, 125, 126, 135, 162, 167, 192, 202, 212, 216, 222, 223, 239, 284, 299
lifetime, 133
light, 43, 92, 244
linear dependence, 217
liquid phase, 128, 135
lithium, 107
low temperatures, 2, 34, 35, 125, 127, 130, 217
Luo, 89

M

magnetic field, 2, 7, 24, 28, 131, 143, 145
magnetic moment, 7
magnetic properties, ix, 124, 125, 126, 141
magnetic relaxation, 22
magnetization, 15, 124

magnitude, 54, 67, 102, 110, 133, 225
majority, 44
manufacturing, 55, 66, 129, 299
mass, 133, 233
materials, ix, x, 1, 3, 4, 5, 37, 38, 40, 42, 46, 47, 56, 76, 87, 88, 100, 105, 110, 113, 124, 125, 126, 127, 145, 162, 165, 166, 167, 169, 181, 182, 216, 217, 222, 223, 257, 258, 259, 264, 283, 284, 287, 295, 299, 300
matrix, ix, 1, 3, 35, 36, 37, 39, 42, 46, 56, 57, 77, 78, 79, 80, 81, 83, 85, 86, 87, 119, 129, 130, 138, 165, 166, 168, 169, 172, 173, 174, 175, 176, 177, 178, 183, 190, 192, 196, 197, 200, 202, 203, 206, 207, 211, 212, 213, 216, 218, 219, 221, 224, 225, 259, 270, 284
matrixes, 266
matter, 34
measurements, ix, 5, 7, 17, 22, 91, 98, 99, 130, 131, 152, 240, 299
mechanical properties, ix, 165, 167, 217
mechanical stress, 167, 258, 260, 279
media, x, 167, 171, 172, 173, 190, 257, 259, 260, 265, 266, 269, 274, 284
melting, 1, 2, 3, 4, 8, 23, 24, 25, 26, 31, 32, 35, 43, 46, 47, 127, 128, 135
melting temperature, 1, 2, 4, 8, 23, 26, 31, 47, 135
membranes, 283
MES, x, 303
meter, 66, 96, 106, 116, 131, 299
micrometer, 217
microscope, 12, 130
microscopy, 43
microstructure, ix, 1, 3, 4, 6, 7, 12, 13, 15, 20, 21, 22, 26, 31, 34, 36, 38, 42, 43, 46, 124, 129, 138, 141, 145, 146, 152, 165, 166, 168, 169, 170, 172, 174, 180, 181, 183, 186, 191, 220, 284
microstructure features, 180
microstructures, 172, 192, 220
migration, 21, 43
mixing, 33
models, 87, 166, 172, 175, 186, 216, 222, 258, 259, 274, 275, 278, 279, 280, 283
modifications, 295
modulus, 101, 107, 168, 175, 176, 177, 185, 186, 191, 192, 193, 194, 195, 196, 215, 216, 217, 218, 219, 220, 221, 222, 223, 224, 279, 300
mold, 129
mole, 4
momentum, 233, 234, 241
morphology, 24, 220
Moscow, 1, 49, 162, 164, 226, 229, 255, 281
Mössbauer data, 140
Mössbauer effect, 125

N

Nd, 124, 125, 128, 133, 135, 137, 138, 140, 232, 244, 245, 246, 248, 249, 252
neutral, 222
nitrates, 4
nitrogen, 217
nodes, 222, 223
non-polar, 156
nuclei, 231, 232

O

obstacles, 31
octane, 130
oil, x, 303, 311
operations, 129, 296
opportunities, 301
optimization, ix, 47, 124, 125, 126, 259, 283, 301
ordinary differential equations, 167
oscillation, 296, 297, 298, 300
oscillograph, 92
oxygen, 15, 34, 36, 105, 138, 156, 160, 162, 231, 232, 233, 234, 235, 236, 237, 238, 239, 240, 241, 242, 243, 249, 250, 251, 252, 253, 254

P

parallel, 6, 13, 19, 25, 27, 55, 56, 58, 66, 71, 72, 73, 78, 85, 86, 95, 161, 169, 240, 309
partition, 306
PCR, 107, 279
percolation, 167, 168, 170, 172, 173, 174, 181, 187, 206, 214, 215
periodicity, 124, 196, 197, 198, 199, 200, 201, 207, 208, 209, 211
permeability, 293
permittivity, 54, 68, 72, 81, 96, 100, 102, 105, 107, 110, 113, 114, 116, 130, 131, 138, 264, 275
perovskite ferroelectrics, ix, 91
phase boundaries, 92, 156
phase decomposition, 44
phase diagram, 91, 98, 99, 103, 104, 106, 108, 109, 110, 112, 113, 126, 156, 157
phase transformation, 23, 34, 105, 216
phase transitions, ix, 91, 96, 99, 100, 105, 119, 124, 125, 126, 127, 153, 156, 159, 160, 162, 166, 168, 216
physical characteristics, 287
physical features, 196
physical laws, 224
physical phenomena, 105

physical properties, 52, 87, 243
piezoelectric properties, 56, 72, 77, 87, 88, 113, 259, 277
piezoelectricity, 284
pitch, 296
plastic deformation, 2, 3, 4, 25, 36
plasticity, 19
plasticizer, 129
point defects, 2, 23, 24, 25, 31, 32, 35, 45, 104, 119
polar, 119, 125, 145, 156, 198
polarization, 55, 56, 60, 62, 63, 64, 65, 67, 68, 69, 71, 72, 73, 76, 77, 78, 79, 81, 83, 84, 86, 87, 88, 91, 92, 95, 96, 99, 101, 105, 116, 117, 118, 119, 258, 276, 279, 284, 285, 299
polymer, ix, 51, 53, 54, 55, 56, 57, 58, 64, 67, 68, 71, 72, 73, 77, 78, 79, 80, 81, 83, 84, 85, 86, 87, 258, 259
polymer composites, ix, 51, 53, 64, 67, 77, 80, 87
polymer matrix, 77, 78, 79, 81, 83, 85, 86, 87
polyurethane, 60, 62, 63, 65, 66, 76
polyvinyl alcohol, 129
polyvinylidene fluoride, 53, 68
porosity, x, 21, 258, 273, 274, 275, 276, 277, 278, 279, 280
precipitation, 23, 43
preparation, ix, 18, 22, 45, 105, 125, 127, 128, 129, 132, 296
preservation, 8, 24, 25, 175
principles, 184
probability, 184, 186, 190, 237
probe, 92, 93, 94
project, 47, 120
propagation, 284, 285
protons, 236, 240, 241, 244
prototype, 116
purity, 100, 106, 128, 130, 293
PVA, 129

R

radial distribution, 178
radiation, 5, 100, 107, 110, 119, 130, 232
radius, 6, 13, 19, 78, 79, 80, 137, 183, 198, 202, 205, 206, 208, 214, 285
reactions, 106, 126, 130, 231
reality, 23, 183
reasoning, 42
recommendations, 127
recovery, 23, 27, 31, 32, 33, 34, 35, 45, 101
recrystallization, ix, 20, 25, 124
reflexes, 156
relative size, 196
relaxation, 7, 22, 80, 93, 96, 105, 107, 143

reliability, 141, 294
renormalization, 178
researchers, 132
resistance, 217, 304, 305, 309
resolution, 130
response, 52, 65, 67, 72, 85, 86, 87, 91, 105, 113, 279
restoration, 222
restrictions, 173
rings, 47, 133
rods, 78, 81, 86, 269
room temperature, 5, 44, 51, 52, 54, 64, 67, 68, 81, 106, 107, 109, 116, 126, 137, 138, 139, 140, 141, 143, 146, 149, 153, 157, 217, 235
root, 173
rotations, 55, 72, 156
Russia, x, 1, 51, 88, 91, 120, 122, 123, 165, 231, 257, 280, 283, 303

S

safety, 129
scatter, 13, 217
scattering, 119, 232, 233, 234, 235, 236, 237, 239, 240, 244, 248, 249, 254
scope, 293
self-consistency, 176
SEM micrographs, 14, 27
semiconductor, 304, 310
sensitivity, 51, 66, 68, 84, 180, 291
sensors, 52, 257, 258
shape, 15, 22, 23, 27, 32, 38, 39, 40, 41, 46, 63, 78, 79, 80, 104, 151, 172, 176, 198, 208, 224
shear, 104, 175, 176, 177, 185, 186, 191, 192, 193, 215, 216, 217, 218, 220, 221, 222, 223, 224, 284
shortage, 166
signals, 310
signal-to-noise ratio, 68
signs, 81, 169
silver, 96, 106, 119, 299
simulation, 232, 241, 249, 250, 254, 295, 296, 299, 300
single crystals, ix, 51, 52, 98, 99, 108, 113, 125, 232, 240, 254
sintering, ix, 8, 27, 33, 40, 45, 46, 116, 124, 125, 127, 128, 129, 130, 133, 135, 137, 152, 153, 289, 296
skeleton, 206, 258, 277, 278, 280
software, x, 96, 100, 116, 131, 182, 285, 293, 300
sol-gel, 4, 34, 36, 37, 38, 39, 40, 41, 42, 43, 45, 46
solid phase, 130

solid solutions, ix, 51, 52, 54, 91, 105, 106, 107, 108, 109, 118, 119, 124, 125, 126, 128, 132, 135, 137, 138, 141, 156, 157, 295
solid state, 125, 126
solubility, 110, 130
solution, x, 4, 6, 25, 92, 100, 129, 133, 138, 140, 167, 172, 173, 174, 177, 196, 197, 198, 199, 201, 204, 205, 207, 208, 210, 211, 213, 215, 220, 223, 233, 259, 264, 267, 271, 273, 274, 284, 285
specialists, 77, 88, 308
spectroscopy, 152
spin, 138, 140
spindle, 133
square lattice, 78, 169, 171, 198, 200, 203
St. Petersburg, 120
stability, 15, 109, 125, 135, 137, 144, 145, 217, 222, 223, 224, 257, 258
stabilization, 92, 100, 118
state, 7, 8, 13, 25, 35, 42, 46, 55, 84, 92, 98, 100, 105, 110, 117, 118, 124, 125, 126, 127, 129, 138, 141, 142, 143, 152, 153, 156, 181, 206, 217, 219, 220, 222, 223, 232, 278, 289, 304, 305, 306, 307
states, 25, 53, 67, 104, 105, 125, 126, 127, 140, 158, 159, 181, 220, 260, 304
statistical processing, 299
steel, 295
stoichiometry, 33, 34, 135
storage, x, 303, 304, 306, 307, 308, 311
stratification, 149
stress, 19, 21, 33, 56, 80, 217, 218, 219, 220, 258, 259, 260, 261, 262, 264, 266
stress concentrators, 21, 33
stress fields, 33
stretching, 162
strong interaction, 210
strontium, 4
structure, x, 2, 3, 24, 34, 52, 53, 55, 56, 67, 72, 79, 80, 86, 87, 92, 100, 104, 105, 107, 110, 116, 119, 124, 125, 126, 127, 130, 137, 138, 140, 141, 153, 161, 165, 166, 167, 168, 170, 171, 176, 179, 182, 184, 192, 196, 203, 204, 205, 206, 213, 216, 217, 218, 222, 224, 231, 232, 235, 236, 240, 244, 248, 249, 258, 259, 273, 277, 278, 287
substitution, 36, 205, 214, 258
substrate, 129, 235
substrates, 106
succession, 109
superconducting magnetic energy storage (SMES), x, 303
superconducting materials, 3
superconductivity, 45
superconductor, 3
surface region, 246

Index

susceptibility, 7, 17, 30, 138, 140
suspensions, 183
symmetry, 54, 58, 67, 69, 71, 78, 88, 95, 104, 107, 116, 124, 126, 138, 162, 168, 169, 170, 173, 181, 221, 240
synthesis, ix, 42, 106, 116, 124, 125, 126, 127, 128, 130, 132, 162

T

Taiwan, x, 165, 225
tangles, 3, 34
target, 235
techniques, ix, 259, 260
technology, 96, 100, 110, 116, 124, 125, 129, 279, 296, 299, 303
TEM, 24, 308
temperature dependence, 23, 30, 32, 96, 98, 100, 106, 107, 108, 110, 113, 116, 119, 142, 143, 145, 149
tension, 222
test data, 176, 177, 187, 199, 210
texture, 1, 2, 3, 4, 5, 7, 8, 9, 10, 13, 15, 19, 20, 21, 22, 23, 24, 25, 26, 31, 32, 33, 34, 35, 36, 37, 38, 39, 42, 44, 45, 46
thermal activation, 20
thermal analysis, 127
thermal expansion, 3
thermal stability, 15, 135, 137
thermal treatment, 119
thin films, 126
tin, 225
titanate, 92, 284
titanium, 113
topology, 191, 304
torsion, ix, 1, 2, 4, 5, 6, 15, 26, 27, 36, 37, 46
training, 36
trajectory, 250
transducer, 87, 88, 257, 258, 283, 284, 285, 286, 288, 289, 290, 291, 293, 294, 295, 301
transformation, 34, 127, 159, 160, 217
transition elements, 127
transition temperature, 2, 38, 98, 108, 142, 143, 145
translation, 181, 192, 278
transmission, x, 285, 303, 304, 308, 311
transport, 30, 33, 168
treatment, 119, 127, 130, 133, 241
trial, 183
tungsten, 42, 46
twinning, 95, 96
twins, 19, 34
twist, 5, 7, 10, 12, 13, 15, 18, 19, 21, 46

U

unification, 222
uniform, 1, 13, 15, 27, 34, 39, 273
United Kingdom, 88
United States, 311

V

vacancies, 104, 240
vacuum, 96, 100, 116, 130, 275, 276
valence, 138, 141, 152, 153
vanadium, 225
vapor, 114
variables, 167, 184, 263, 273
variations, 92, 94, 95, 105, 110
VDF, 68, 81, 83, 84, 86
vector, 55, 56, 58, 60, 62, 63, 64, 65, 67, 68, 69, 71, 72, 77, 78, 79, 81, 83, 84, 86, 88, 116, 160, 191, 258, 259, 260, 261, 262, 266, 270
velocity, 284, 293, 294, 300
vibration, 239, 241, 284, 285, 286, 288, 289, 290, 291, 292, 293, 294, 296, 300
vinylidene fluoride, 53, 81

W

Washington, 164
water, 133, 286, 287, 291, 292, 293
wave propagation, 284, 285
wavelengths, 57
weak interaction, 152, 153
weight ratio, 127
wires, 196
witnesses, 130, 153

X

X-ray analysis, 116, 156, 157
X-ray diffraction, 5, 43, 96, 98, 99, 107, 108, 119, 130, 141, 145, 146, 152, 153
X-ray diffraction (XRD), 5
X-ray diffraction data, 96, 98, 99, 141, 146
XRD, 5, 8, 10, 25, 26, 36, 37

Y

yield, 79, 80, 231, 232, 233, 235, 236, 237, 238, 239, 240, 241, 243, 244, 246, 247, 248, 249, 250, 251, 252, 253, 254